A Practical Guide to Reliable Finite Element Modelling

A Practical Guide to Reliable Finite Element Modelling

Alan Morris
Emeritus Professor of Computational Structural Analysis,
Cranfield University, UK

With contributions from

Ahmed Rahman
QinetiQ, UK

John Wiley & Sons, Ltd

Telephone (+44) 1243 779777

Email (for orders and customer service enquiries): cs-books@wiley.co.uk
Visit our Home Page on www.wileyeurope.com or www.wiley.com

Other Wiley Editorial Offices

John Wiley & Sons Inc., 111 River Street, Hoboken, NJ 07030, USA

Jossey-Bass, 989 Market Street, San Francisco, CA 94103-1741, USA

Wiley-VCH Verlag GmbH, Boschstr. 12, D-69469 Weinheim, Germany

John Wiley & Sons Australia Ltd, 42 McDougall Street, Milton, Queensland 4064, Australia

John Wiley & Sons (Asia) Pte Ltd, 2 Clementi Loop #02-01, Jin Xing Distripark, Singapore 129809

John Wiley & Sons Canada Ltd, 6045 Freemont Blvd, Mississauga, ONT, L5R 4J3

Wiley also publishes its books in a variety of electronic formats. Some content that appears in print may not be available in electronic books.

Library of Congress Cataloging-in-Publication Data

Morris, Alan.
A practical guide to reliable finite element modelling / Alan Morris ; with contributions from Ahmed Rahman.
p. cm.
Includes bibliographical references and index.
ISBN 978-0-470-01832-3 (cloth)
1. Finite element method. 2. Error analysis (Mathematics) I. Title.
TA347.F5M675 2007
620.001′51825–dc22 2007044555

British Library Cataloguing in Publication Data

A catalogue record for this book is available from the British Library

ISBN-13 978-0470-01832-3

Typeset in 10.5/13pt Sabon by Thomson Digital, New Delhi.
Printed and bound in Great Britain by TJ International Ltd, Padstow, Cornwall
This book is printed on acid-free paper responsibly manufactured from sustainable forestry in which at least two trees are planted for each one used for paper production.

To Lilian,
for endless patience.

Contents

Preface XIII

1 Introduction **1**
- 1.1 Aim of the book 1
- 1.2 Finite element types – a brief overview 3
- 1.3 Finite element analysis and finite element representations 6
- 1.4 Multi-model analyses 8
- 1.5 Consistency, logic and error control 8
- 1.6 Chapter contents 11
  - 1.6.1 Chapter 2 Overview of static finite element analysis 11
  - 1.6.2 Chapter 3 Overview of dynamic analysis 12
  - 1.6.3 Chapter 4 What's energy got to do with it? 13
  - 1.6.4 Chapter 5 Preliminary review of errors and error control 14
  - 1.6.5 Chapter 6 Discretisation: elements and meshes or some ways to avoid generated error 14
  - 1.6.6 Chapter 7 Idealisation error types and sources 15
  - 1.6.7 Chapter 8 Error control 15
  - 1.6.8 Chapter 9 Error-controlled analyses 16
  - 1.6.9 Chapter 10 FEMEC walkthrough example 17
- References 17

2 Overview of Static Finite Element Analysis 19
2.1 Introduction 19
2.2 The direct method for static analyses 20
2.2.1 Element matrices 20
2.2.2 Assembled global stiffness matrix for static analyses 21
2.2.3 Global coordinates 26
2.2.4 Some typical elements 29
2.2.5 The analysis loop 29
2.3 Reducing the problem size 33
2.3.1 Symmetry 33
2.3.2 Condensation and superelements 40
2.3.3 Sub-structures 46
References 49

3 Overview of Dynamic Analysis 51
3.1 Introduction 51
3.2 Element mass matrix 51
3.2.1 Free undamped vibrations 53
3.3 Additional information that can be extracted to support a dynamic finite element analysis 59
3.3.1 Sturm property check 59
3.3.2 Rayleigh Quotient 61
3.4 Forced responses 62
3.4.1 Modal analysis 62
3.4.2 Direct integration 65
3.5 Damped forced responses 66
3.5.1 Modal analysis with damping 66
3.5.2 Modal damping ratio 68
3.5.3 Direct integration 69
3.6 Reducing the problem size 70
3.6.1 Symmetry 71
3.6.2 Reducing the number of variables 72
3.6.3 Sub-structure analysis (component mode synthesis) 76
References 78

4 What's Energy Got to Do with It? 79
4.1 Introduction 79
4.2 Strain energy 81

4.3 Potential energy 82
4.4 Simple bar 83
4.5 General case 83
4.6 Minimum potential energy 84
4.7 The principle of minimum potential energy applied to a simple finite element problem 86
4.8 Finite element formulation 88
4.9 Direct application to an axial bar element 90
4.10 Convergence in energy and convergence in stress 92
4.10.1 Single bar element model 94
4.10.2 2-bar element model 96
4.10.3 4-bar element model 98
4.10.4 8-bar element model 101
4.10.5 16-bar element model 102
4.11 Results interpretation 103
4.11.1 Potential energy convergence 103
4.11.2 Stress improvement 104
4.11.3 Displacement convergence 106
4.12 Kinetic energy 106
4.13 Final remarks 111
References 111

5 Preliminary Review of Errors and Error Control 113

5.1 Introduction 113
5.2 The finite element process 114
5.3 Error and uncertainty 117
5.4 Novelty, complexity and experience 120
5.4.1 Analysis novelty 120
5.4.2 Degree of complexity 123
5.4.3 Experience 125
5.5 Role of testing 128
5.6 Initial steps 129
5.6.1 Qualification process 129
5.6.2 Acceptable magnitude of error or uncertainty 132
5.7 Analysis Validation Plan (AVP) 135
5.8 Applied common sense 138
5.9 The process 140
References 141

6 **Discretisation: Elements and Meshes or Some Ways to Avoid Generated Error** **143**
6.1 Introduction 143
6.2 Element delivery 144
6.2.1 Two-dimensional elements 145
6.2.2 Three-dimensional elements 148
6.2.3 Why do this? 151
6.2.4 Optimal stress points and making the most of them 151
6.3 Mesh grading and mesh distortion 155
6.3.1 Mesh grading 155
6.3.2 Element distortions 158
6.3.3 Group 3 distortions (quadratic distortions) 161
6.3.4 Group 4 distortions (cubic and higher order distortions) 162
6.3.5 Distortions of other element types 162
6.3.6 General principles with respect to element distortions 163
6.4 The accuracy ladder 164
6.4.1 The stress ladder 164
6.4.2 The mesh ladder 165
6.4.3 Automatically moving up the accuracy ladder 165
References 174

7 **Idealisation Error Types and Sources** **175**
7.1 Design reduction and idealisation errors 175
7.2 Analysis features 178
7.3 The domain 180
7.3.1 The domain of analysis 180
7.3.2 Domain reduction 183
7.4 Levels of abstraction 184
7.5 Boundary conditions 192
7.6 Material properties 198
7.7 Loads and masses 200
7.7.1 Loads 200
7.7.2 Masses 201
References 201

8 **Error Control** **203**
8.1 Introduction 203

8.2 Approach and techniques 204
8.2.1 Approach 204
8.2.2 Techniques 207
8.3 Accumulation of errors and uncertainties 241
8.4 The role of testing 242
8.4.1 Analyst-requested tests 244
8.4.2 Validation test 245
References 247

9 Error-Controlled Analyses 249
9.1 Introduction 249
9.2 Is the finite element system fit for purpose? 251
9.2.1 What does it do? 251
9.2.2 How do we know it does it? 252
9.2.3 Is size important? 254
9.3 Quality Report 256
9.4 The error and uncertainty control method 259
9.4.1 Introduction 259
9.4.2 FEMEC 260
9.4.3 FEMEC implementation process 261

10 FEMEC Walkthrough Example 303
10.1 Introduction 303
10.2 FEMEC static analysis illustrative problem 304
10.2.1 The design requirement 304
10.2.2 Application of FEMEC via the Quality Report 306
10.3 A brief look at the dynamic FEMEC world 345
10.3.1 Modified design and qualification parameters 345
10.3.2 FEMEC approach 346
References 351

Index 353

Preface

This book is not a standard finite element text that can be used to provide all the information required to obtain a well-grounded understanding of all aspects of the Finite Element Method. There are already many excellent books on this topic and a number of these are referenced in Chapter 1, Introduction. The question that the book attempts to answer is 'How can an error-controlled finite element analysis be performed?' It is tempting to think that with the development of comprehensive finite element packages there is no need to worry about errors and uncertainties. Sadly this is not the case.

The Sleipner oilfield within the Norwegian sector of the North Sea is one of the major sources of oil and gas for Europe. The Sleipner A platform is a concrete gravity base structure consisting of 24 cells that rest on the sea bed at a depth of 82 m with a total base area of 16,000 m^2. Four of these cells are elongated so that they reach above the surface of the sea and support a deck that weighs 57,000 tons and drilling equipment weighing 40,000 tons. On 23 August 1991, while being prepared for deck mating through a controlled ballasting operation in the Gandsfforden outside Stravanger, the first Sleipner A platform sprang a leak and sank. The crash caused a seismic event of 3.0 on the Richter scale, left a pile of debris at a depth of 220 m and an economic loss of £700 million. The cause of the crash was traced to an inaccurate finite element analysis that underestimated the shear stress in the cells by 47%, so that certain concrete walls were not thick enough. After the accident, a more careful finite element analysis was performed on the original Sleipner A platform which predicted a structural failure at 62 m matching well with the actual failure depth of 64 m. Had an effective quality system been in place that allowed the analysis team to control the errors and uncertainties in the analysis, this failure could have been avoided. It may be thought that because the event took place some time ago the current situation would be much better, but, to the author's personal knowledge, other serious analysis failures have taken place recently. These have not been publicised as legal action was taken but were resolved at the courtroom door following an agreed compensation package.

In order to avoid such distressing consequences an analyst needs to have both sufficient basic knowledge of the Finite Element Method and a procedure for systematically performing a finite element analysis. This book aims to satisfy both these needs by providing essential background knowledge and information and a sequential application process. The book draws on two sources. One is information from lectures developed at Cranfield University and given to postgraduate aeronautics students and industrial short courses given both at Cranfield and in-house at international aerospace companies in the UK and elsewhere. The second source is the research output from a major UK government-funded project, within the Safety Critical Systems initiative, entitled SAFESA™, under contract DTI/EPSRC project 9034. Five organisations were involved in the project: Cranfield University, Lloyds Register, W.S. Atkins, Nuclear Electric (now British Nuclear Group) and Assessment Services (now Siemens). The author is particularly grateful to a number of colleagues involved in the SAFESA project who worked for these companies: Dr Mike Fox, Dr John Maguire, Dr Nigel Knowles and Professor Rade Vignjvec. Through creative and innovative thinking these engineers came forward with concepts and ideas that have significantly influenced the contents of this book.

Although Chapter 9 draws on the output of the SAFESA project, the method presented therein is distinctive. Nevertheless, the reader may wish to take advantage of the earlier work and this can be done, at one level, by consulting references [1] and [2] which present a synopsis of the main SAFESA results. A fuller description can be found in the *SAFESA Technical Manual* that was issued to the technical community, at the conclusion of the project, by the Minister then in charge of the Department of Industry and Science, the Rt Hon. Michael Heseltine MP (now Lord Heseltine). The SAFESA project team subsequently gave the NAFEMS organisation permission to reprint the *Manual* and copies can be obtained from that organisation through its offices in East Kilbride, Glasgow, UK.

The companion website for the book is http://wiley.com/go/morrisfem

REFERENCES

1. Morris, A.J. and Vignjvec, R., Consistent finite element analysis and error control. *Comput. Methods Appl. Mech. Eng.*, 1997. **140**: 87–108.
2. Morris, A.J., The qualification of safety critical structures by finite element analysis methods. In *Proc. Inst. Mech. Eng. G: J. Aerosp. Eng.*, 1996. **210**: 203–208.

1

Introduction

1.1 AIM OF THE BOOK

There are many excellent text books on finite element theory incorporating the development of specific types of finite elements and describing the associated solution processes. This book has a different purpose from these standard texts as it provides a practical guide for the reliable use of the Finite Element Method in supporting the design of complex structures. Within this broad framework it gives an introduction to the Finite Element Method and links it to the problems associated with creating an effective and relatively error-free finite element model for solving a real-world structural design problem. By error is meant the difference between the finite element analysis's predicted behaviour and response of a structure subjected to applied loads and that which occurs when the structure enters service where the in-service loads come into play.

In practical terms the book is intended to assist engineers and companies involved with finite element analysis on a regular basis to operate in a manner that:

1. Reduces the possibility that any type of error is introduced into a finite element analysis.
2. Ensures that analyses undertaken by an individual analyst or analysis team are performed to a consistent and reliable standard.
3. Provides documentary evidence of having adhered to a consistent error control process as a basis for a defence in legal proceedings

A Practical Guide to Reliable Finite Element Modelling A. Morris

should a structural failure occur after a finite element analysed product has entered service.

Clearly one of the key aspects of the book is the provision of a methodology that allows a finite element analysis of a structure to be undertaken in such a manner that potential differences between the values for specific behaviour parameters obtained from the analysis and the measured values from operational use are identified and controlled. This requires that the analyst is not only able to identify the sources of error that may give rise to such differences, but also able to provide bounds on their maximum likely value. The targeted parameters should be selected by a process that clearly and explicitly defines the qualification criteria that, when satisfied, allow the structure to be constructed and enter service in a manner that renders it fit for purpose. In essence, the process is attempting to generate a procedure that places analysis as the primary route for the qualification of a structure. This creates a new environment in which testing is analysis controlled and is employed to support the analyst, providing information for the bounding or control of potential errors. In this situation, testing is a subservient activity because the analyst defines specific requirements for test data to compensate for identified deficiencies in the finite element analysis. If a test is now used in the proof of a structure, it is there simply to validate the analysis which has become the actual validating machine.

In attempting to satisfy the requirements listed above the book offers a basis for constructing a logical approach to finite element analysis. This is ambitious and it is not claimed that it provides a complete and totally comprehensive method for satisfying this requirement. Rather it provides a door through which the reader is invited to step and after crossing the threshold develop the ideas presented herein into a more comprehensive and authoritative method that is personal to an individual analyst or analysis team. In the case of an inexperienced or new finite element analyst, it provides a starting point. For an experienced analyst or a company that regularly undertakes finite element analyses, it should be taken as an input into what should be a regular review of their finite element qualification process.

In order to keep the length and complexity of the book under control the problem domain is restricted to linear static and linear dynamic structural analyses. Nevertheless, the broad approach adopted in the chapters devoted specifically to error control and treatment has general applicability.

Finally, it is worth noting that this book is not intended as a broad introduction to the use of finite element analysis in engineering design; this is covered by Adams and Askenazi [1]. Nor does it focus on the development of internal error bounds and the use of this type of bounding process in h- and p-type adaptive meshing codes. However, the use of such codes is touched on as they provide one component in a total error and uncertainty control methodology. Details of error estimation techniques based on internal and self-referencing procedures are covered in the excellent book by Szabó and Babuška cited as reference [2] and, in more detail, by reference [3].

1.2 FINITE ELEMENT TYPES – A BRIEF OVERVIEW

The underlying principle of the Finite Element Method is that a physical structure is modelled as an assemblage of individual elements as outlined in Chapters 2 and 3 but more fully in books addressing the mathematical fundamentals such as references [4], [5], [6]. All finite element models employ polynomial approximations to at least one of the main fields employed in describing the physical phenomena that are the focus of the analysis. In this book, attention is restricted to the analysis of loaded structures responding in a manner that can be modelled using elasticity theory. For this class of modelling problems there are three basic element types: displacement elements, equilibrium elements and hybrid elements. All commercially available finite element packages and systems employ displacement finite elements, many employ some hybrid elements and a few have equilibrium elements. Chapters 2 and 3, in outlining some of the fundamentals of the method, use displacement elements. However, most of the arguments advanced in this book apply equally to all three types.

A schematic of a displacement finite element is shown in Figure 1.1. The displacement on the interior of the element is approximated using relatively low-order polynomials. These polynomials must have a form that ensures the displacements at the edge or edge surfaces of the element can link up with adjacent elements in such a way that certain components of the displacement field are continuous across adjacent element interfaces. In the case of plates and shells the polynomials must be able to ensure continuity of the appropriate rotation terms. The

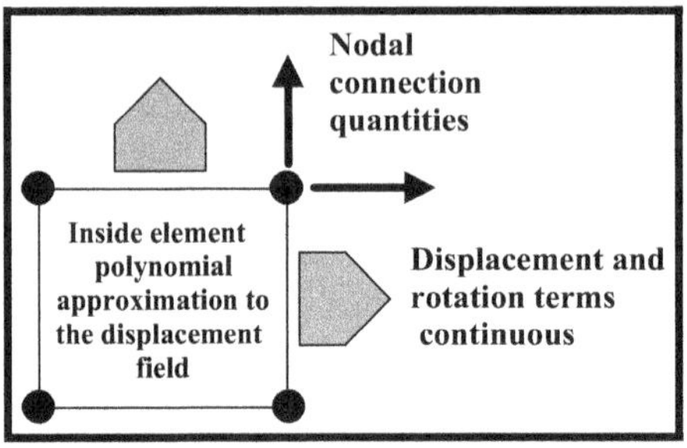

Figure 1.1 Schematic of a displacement element.

polynomials are then defined in terms of nodal values that can be specified at a vertex, as shown in Figure 1.1, or at specific points along element edges or surfaces. Adjacent elements are now connected to each other through these nodes and because of this the nodal displacements or rotations are called connection quantities. Loads are applied to the finite element model through these same nodes. It is worth noting that the displacement finite element formulation degenerates the structure under analysis into a set of points distributed through the space occupied by the structure and there is no longer any explicit representation of the actual structure nor any explicit representation of the physically distributed load system. As shown in Chapters 2 and 3, the resulting nodal model is then solved in terms of the initially unknown nodal connection quantities and terms such as element stresses are derived from this solution.

The formulation for an equilibrium finite element is similar, in principle, to that of the displacement element as shown in Figure 1.2. In the case of an equilibrium element, the interior stress field is approximated by polynomials and the connection from one element to the next is via side or surface forces that are distributed along the element edges and surfaces. As shown in Chapter 4, the displacement

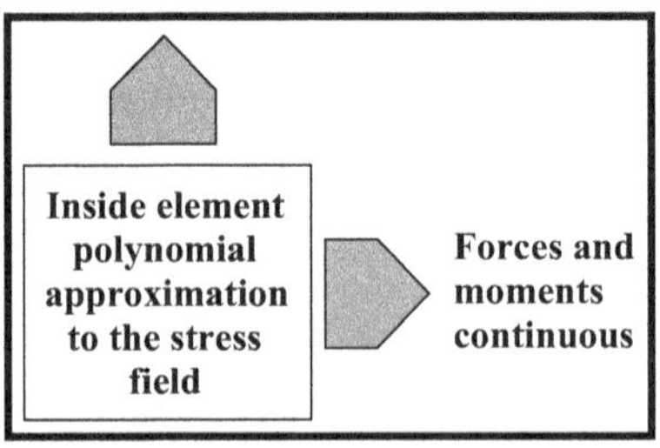

Figure 1.2 Schematic of an equilibrium element.

formulation gives rise to stress discontinuities across the inter-element interface boundaries. Equilibrium elements, on the other hand, generate continuous stresses as the solution crosses from one element to an adjacent element but inter-element compatibility is not preserved. In the early days of finite element analysis this stress continuity property made equilibrium elements popular with aircraft stressmen in allowing them to track the internal load paths. This property was also mistakenly thought to mean that equilibrium elements were more accurate than displacement elements. An early example of the use of equilibrium elements can be found in a publication by one of the pioneers of the Finite Element Method, Fraeijs de Veubeke, and his gifted assistant Guy Sander in reference [7]. A very good description of this type of element can be found in the book by Tong and Rossettos [8] which is, unfortunately, now out of print.

Reference [8] is also a good starting point for a description of the third type of finite element, known as the hybrid element, which also receives a brief description in reference [6]. This element is shown schematically in Figure 1.3 where it can be seen that there are two fields being deployed for the element. Inside the element it looks like an equilibrium element but there is also a line distribution of displacement along the edge of the element shown in Figure 1.3 or a surface distribution if a three-dimensional solid element is employed. This additional displacement field is approximated by either a one- or two-dimensional polynomial depending on the dimensionality of the element. This approximation is formed in terms of nodal displacement or rotation values which then form the connection quantities for attaching adjacent elements. The element

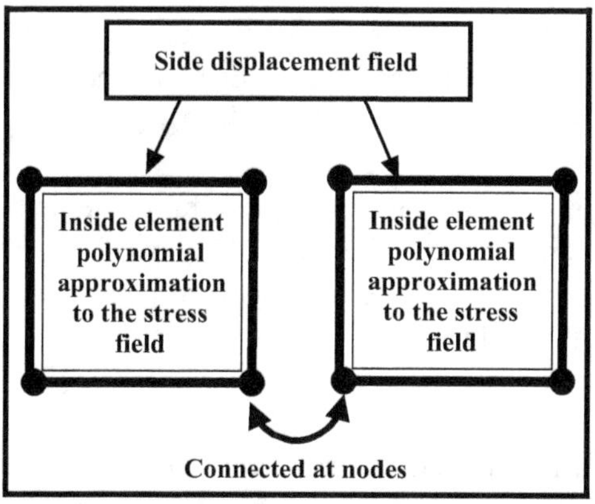

Figure 1.3 Schematic of a hybrid element.

appears to the outside world as a standard displacement element. The displacement fields are playing a subtle role as they act as Lagrange multipliers on the continuity condition that element stress equilibrium is maintained across element boundaries as with the equilibrium element. These elements find application in the development of finite elements for plate and shell analysis problems.

1.3 FINITE ELEMENT ANALYSIS AND FINITE ELEMENT REPRESENTATIONS

A finite element analysis is a numerical simulation of the behaviour of a real-world structure which is intended to provide information that can be used by a designer or design team to ensure a structural design is fit for purpose when it enters in-service operation. The process of setting up an analysis requires that such factors as the loads applied to the structure, the structural behaviour and responses, the boundary conditions, etc., are all represented by a set of mathematical functions or operations. This is an important concept to understand because the focus of the analysis is the real world, which is *not* a mathematical model.

A finite element analysis is, therefore, a process that takes an actual structure, subject to its constraints including attachments to other structures – including the Earth which is simply a very big structure. It then has to perform the following tasks:

1. Convert the real-world system into a mathematical description.
2. Turn this description into a form which allows a computer to be brought into the picture to solve this mathematical problem.
3. Take this output and turn it back into parameters that relate to the real-world structural behaviour.

In undertaking this series of operations the finite element analysis can be envisaged as passing through a series of 'worlds' or, more accurately, representations. Although these representations and their ramifications are covered in detail in later chapters, it is worth setting the scene in this introductory chapter.

The first of these is the *Real-World* representation which constitutes the object (structure) which is to be analysed and its environment. It is a representation because the structure itself often does not physically exist at the time of the analysis. Nevertheless this representation models the in-service structure and the way that it will be actually loaded, supported, etc., and the way that it responds to the loading and support environment. For the purposes of the present book it is assumed that there are no errors associated with this representation, even though this may not be the case in many instances.

The *Reduced Real-World* representation is the first level of abstraction and is a modified version of the real world in which uncertainties concerning the real structure and its environment are taken into account. The level of abstraction may be quite large such that the loads may not be the actual loads seen by the structure, nor the structural form of the real structure. Often this type of abstraction is imposed by the qualification requirements which attempt to account for uncertainties in, for example, the loads acting on the structure by imposing stereotype loads together with safety factors.

The third representation is the *Idealised World* which takes the structural world model and turns it into a form which can be analysed by the Finite Element Method. This is a very profound level of abstraction which converts the structural model with its welds, rivets, bounded joints, etc., into a smooth model in which each component, together with its boundary condition, loading situation, etc., can be mathematically defined. Thus the decisions concerning factors such as the linearity or otherwise of the structural behaviour are made at this stage. This is the most critical part of the whole finite element analysis process as, in a loose sense, the construction of an idealised world represents a transition from a world 'exterior' to the computer to an 'interior' world.

Once the idealisation process has been performed a number of closely related representations are constructed. First is the *Finite Element World* which maps onto the idealisation a set of specific finite elements which can adequately represent the mathematical behaviour defined in the idealised world. This also includes the selection of the element boundary conditions and the element loads. Second is the creation of a *Meshed World* in which the elements have a specific location, shape, etc. Finally in this sequence of 'interior' worlds is the *Solution World* where a procedure is employed to obtain a solution for the idealised structure represented by the idealised world.

As far as the analysis process is concerned, the final stage transforms the results expressed in the solution world and reinterprets them so that they provide results for the structural world problem. This is often called *De-Idealisation.* The results obtained from this process can be used to correlate with the structural tests performed when the structure has, eventually, been constructed. This book is endeavouring to create a methodology that ensures no significant differences appear when this correlation is performed.

1.4 MULTI-MODEL ANALYSES

There are many reasons why the evaluation of a structural design might require the use of several finite element models with different levels of fidelity. If the structure is completely new, it is necessary to get some feel for its basic characteristics if only, for no other reason, to undertake an initial sizing of the structural components and parameters. The starting point could be a low-fidelity analysis model with relatively few finite elements or low-order elements. This provides an initial 'view' of the structure, identifying major load paths, inertia characteristics, the mechanisms through which externally applied loads enter the structure, etc. Once the initial configuration has been established, higher fidelity models can come into play where either more or different elements – or, indeed, both – are employed. This first move up the fidelity ladder could be a transitional step where the analyst 'zooms in' on certain parts of the structure using local high-fidelity models, leaving the rest of the structure to be handled by the initial low-fidelity model. A further step could be the development of one or more high-fidelity models for the entire structure. In the case where the structural design is based on an existing design this same pathway may be required depending on the complexity of the structure as discussed in Chapter 5. The number and type of finite element models required depend on many factors with each analysis focused on clear objectives.

Irrespective of the number of analyses required, the levels of representation outlined in Section 1.3 are present in all of them. This situation provides ample opportunity for the introduction of errors so that the results from the final or intermediate analyses are significantly different from the measured performance of the in-service structure. Thus error control procedures are required for all analyses during the build-up process for the results.

1.5 CONSISTENCY, LOGIC AND ERROR CONTROL

The process of analysing a structure has to start with a clear definition of the real-world design problem and end with a finite element model that accurately reflects the behaviour of structure when in operational use and subject to the in-service loads. In seeking to control or eliminate the differences between the prediction from the finite element simulation and the actual operational performance it is necessary to employ a process-based approach. This process has to be both consistent and reliable if it is to inspire confidence. The case for such an approach can be argued, as done from a consistency viewpoint following the earlier work in reference [9] or from a reliability viewpoint as done in reference [10].

Several implications are associated with the concept of consistency. First, there is the requirement that the analysis process can be decomposed into a logical and coherent set of steps. These must give rise to a sequence of operations that provide a linked pathway whereby information controlling the analysis is passed from one step to its successor and backwards through any required feedback loops. Thus, the logical sequence which the method must follow is defined in a consistent manner. Second, consistency requires error control; if the errors cannot be controlled, the method cannot be consistent. Lack of error control would mean that an identical problem could be solved on two separate occasions and produce different results. Finally, and following on from the second point, a consistent method for performing finite element analyses should demonstrate repeatability with respect to results produced for a specific problem when run at different times by different analysts.

The second aspect associated with consistency identifies the need for the creation of a finite element error control and error treatment methodology. This is difficult, particularly when the complete analysis process is taken into account. Appealing to the Church–Turing theorem establishes that it is perfectly reasonable to ask that a computer program *perfectly* simulate the behaviour of a physically realisable system – such as a structure subject to loads. Thus, if the elastic behavioural response of the actual real-world structure, under static loading conditions, at a finite number of points is represented by a stiffness matrix K_R and a finite element model with n elements by K_n, the theorem implies that a measurable error e_n exists such that:

$$K_R = K_n + e_n$$

and that the error can be made vanishingly small for an appropriate choice of K_n. The corollary to the theorem is that a process must, in principle, exist where the difference between the response of a real-world structure and that of a finite element model of this structure can be controlled. This principle would imply that, while in practical terms it is impossible to reduce the error to zero, it should be possible to provide absolute bounds on e_n.

There are two problems associated with this principle. The first is the question of uncertainties. The implication in writing the real-world stiffness matrix K_R is that a unique real-world structure exists against which the finite element result is being compared. Unfortunately this is not the case since all the attributes of an actual structure are subject to variability: variability in material properties, fabrication, loading, etc. Such variability, discussed in Chapter 5, is termed uncertainty when applied to a new structure at the design stage and cannot be accurately assessed. Thus uncertainty constitutes a residual term in e_n which cannot be computed and, thus, this term cannot be driven to zero. As we shall see later, it is possible to consider ways in which uncertainties can be introduced into a formal error assessment process.

The second of these two problems concerns the difficulty of choosing an error control mechanism which avoids violating Gödel's theorems. The first of these theorems essentially states that if a formal theory, in which proofs are expressible by mathematical formulae, is proved to be consistent, it is not possible to prove completeness. The second asserts the impossibility of proving the consistency of such a theory by methods 'formalisable within the theory'. The essence of these theorems, in the present context, is that error bounds cannot be achieved by using finite element results to self-reference. This poses particular difficulties for methods proposed in this book which are attempting to create error control and error bounding methods for a new structure which, at the time the analysis is performed, does not exist. While the Church–Turing theorem tells us that a finite element model of the (as yet non-existent) structure under design must exist, Gödel's theorems imply that we have no adequate way of deciding how that model should be constructed using unaided finite element data. Thus, adequate error bounding procedures cannot be obtained using finite element information only. The questions raised in this section are clearly linked to questions of computability which are discussed by Belytchko and Mish in reference [11].

The arguments introduced in this section underpin the rationale for creating a consistent method to undertake finite element analyses of real-world structures. The process to achieve this is introduced in later chapters and represents one approach to the problem ensuring analyses are accurate and repeatable. In some ways it can be argued that what is advanced is simply common sense. However, common-sense solutions only become common after they have been explained. Other expositions of approaches for controlling error propagation in a finite element analysis are available. One is the SAFESA^{TM} method introduced in the Preface, another is in the excellent report, reference [12], from NASA examining technologies for use in the analysis of the Space Shuttle's external tank.

1.6 CHAPTER CONTENTS

Most technical books are not read as a novel where it is essential to start at the beginning and proceed chronologically to the final chapter; rather, the reader selects those parts relevant to the technical issues being addressed. To assist the reader in making a judgement as to where relevant information can be found we can now describe what will be found in the other chapters of this book.

1.6.1 Chapter 2 Overview of Static Finite Element Analysis

1.6.1.1 Aim

To go through the entire process followed by a computer in solving a structural analysis problem using the Finite Element Method for statically loaded problems exhibiting a linear response.

1.6.1.2 Outline

Chapter 2 covers the entire process starting with the derivation of the individual element stiffness matrix, coordinate transformation, assembly of elements into the global stiffness matrix, the application of boundary conditions, solution of the problem at the global level and then the evaluation of element properties (e.g. stress). It uses a set of spring elements as the initial demonstration example which focuses on a 1-D problem, then moves to considering an assemblage of bar elements for a 2-D demonstration.

Finally, the chapter discusses how the size of a finite element static analysis problem can be reduced. This discussion covers the following:

- condensation
- sub-structures
- symmetry and anti-symmetry.

With the availability of modern computing power, the need to exploit condensation and symmetry/anti-symmetry to reduce the amount of computing time and storage space is often thought to be unnecessary – this is a mistake, it is always worth saving computing effort! Furthermore, the use of substructure techniques is essential when an analysis is being undertaken by a number of separate analysis teams, particularly when these are non-collocated.

1.6.2 Chapter 3 Overview of Dynamic Analysis

1.6.2.1 Aim

To go through the process followed by a computer in solving a finite element analysis problem using the Finite Element Method for dynamically loading structures.

1.6.2.2 Outline

Chapter 3 picks up from Chapter 2 by showing that the introduction of dynamic loads requires the construction of a mass matrix that introduces inertia loads into the analysis employing a single spring element to demonstrate the process. The chapter then shows that the free vibration problem for this simple structure reduces to the solution of an eigenvalue problem. The use of simple checks that assist the analyst in establishing that a robust solution has been found are then highlighted.

Forced responses are then discussed, employing both a modal and direct integration. Both of these solution techniques are developed to include the effects of damping.

Finally, as with Chapter 2, the chapter discusses how the size of a finite element dynamic analysis problem can be reduced. This discussion again covers the following:

- condensation
- substructures
- symmetry and anti-symmetry.

If it is claimed that modern computing power negates the need to exploit condensation and symmetry/anti-symmetry to reduce the size of static analysis problems, this argument cannot be deployed in the case of very large-scale dynamic analysis problems. As with the static analysis case, substructuring is required in the case of multiple analysis teams.

1.6.3 Chapter 4 What's Energy Got to Do with It?

1.6.3.1 Aim

This chapter emphasises the fact that the Finite Element Method, for a statically loaded structure, is actually minimising the potential energy (PE) of the structural system and, for the dynamically loaded structure, the kinetic energy (KE). This demonstrates that the Finite Element Method is a convergent process – at least in an integrated sense.

1.6.3.2 Outline

Chapter 4 begins by defining potential energy and using a simple spring as a demonstration vehicle. It then develops the concept that minimising the PE does, indeed, lead to a correct solution for a statically loaded structure operating in the linear elastic domain. Using a combination of springs, it is shown that using this minimising principle leads to the standard finite element matrix formulation introduced in Chapter 2. The chapter demonstrates that the Principle of Minimum PE is the underlying basis used to create the matrices for the construction of a displacement finite element system focused on the solution of linear static analysis problems. It also introduces the concept of the consistent load vector that allows distributed loads to be accommodated by a set of displacement finite elements that have degenerated the structure to a set of discrete points, distributed across the structural domain.

The discussion of PE concludes with a simple illustration of the use of PE for the generation of appropriate matrices and load vectors for a simple set of statically loaded bar elements. This illustrates:

- that the method converges as the number of elements is increased to the correct potential energy as the number of elements is increased;
- that the consistent method loses loads which are applied at displacement boundaries;
- that there are jumps in stress across the junctions of common elements.

The chapter then addresses the finite element analysis of structures exhibiting dynamic responses. The fact that the solution of this type of analysis problem also requires the minimisation of a specific function, in this case a Lagrangian function that can be directly related to kinetic energy, is illustrated.

The chapter concludes by illustrating the kinetic energy convergence and its relationship to the dynamic response of a structure through the free-vibration analysis of the simple bar structures used in the PE demonstration.

1.6.4 Chapter 5 Preliminary Review of Errors and Error Control

1.6.4.1 Aim

To introduce the reader to some of the basic considerations relating the likely causes of error that can be encountered in a finite element analysis.

1.6.4.2 Outline

Chapter 5 opens with a brief discussion of error and uncertainty and their location in the total analysis process. The concepts of novelty, complexity and experience are introduced and linked to the possibility of the introduction of error and uncertainty within the finite element analysis process. The role of testing within an error control process is discussed. The discussion then moves on to consider the overall qualification process that brings in questions relating to acceptable levels of error and the need for an analysis validation plan.

1.6.5 Chapter 6 Discretisation: Elements and Meshes or Some Ways to Avoid Generated Error

1.6.5.1 Aim

This chapter endeavours to provide simple rules and guidance information to assist an analyst in the selection of appropriate elements and mesh layouts.

1.6.5.2 Outline

Chapter 6 opens with a discussion on using simple rules to work out what a particular element can deliver in terms of stress output which can then be linked to the required level of accuracy. It moves on to consider the use of optimal stress points and the associated concept of superconvergent elements. Meshing issues relating to element shape distortions and element grading are discussed together with some popular abuses. The chapter concludes by indicating how an analyst can attempt to measure and improve the internal level of accuracy.

1.6.6 Chapter 7 Idealisation Error Types and Sources

1.6.6.1 Aim

The chapter discusses the types of errors that can occur in the idealisation process for a finite element analysis of a structure subject to loads that give rise to linear responses.

1.6.6.2 Outline

Chapter 7 covers the range of error types and error sources that can occur in the major stages of a finite element analysis. Although the term error is used extensively in this chapter, many of the 'error' types and sources are due to the presence of uncertainties in the problem definition or model data. The chapter considers error sources due to the need to select a specific form for the structural performance and a domain of analysis. Error sources occurring in the definition of the mathematical model upon which the finite element model will be built are then treated. Finally, the need to control uncertainties in the selection of the boundary conditions and the load definition is explored.

The chapter explains the nature of the error sources, their potential influence on the results and points to ways that these can be identified so that they can be controlled and treated using the methods and approach discussed in Chapter 8.

1.6.7 Chapter 8 Error Control

1.6.7.1 Aim

This chapter introduces methods that can be used to treat the errors and error sources identified in Chapter 7 and other parts of this book.

1.6.7.2 Outline

Chapter 8 approaches the process of controlling error through a hierarchical methodology starting with simple control methods and ending with sophisticated numerical methods. It starts by considering simple engineering-based methods involving 'hand calculations' and engineering formulae – these are also used to create an initial 'view' of the design problem being confronted. The use of different mathematical models or levels of abstraction to control errors associated with an incorrect selection for the structural model is developed. This is followed by a detailed consideration of sensitivity methods used to bound the impact of any errors or uncertainties, generated during the idealisation process, on the finite element predicted behaviour of the in-service structure. This exploits both direct and indirect sensitivity methods that are derived from methods developed by the structural optimisation community. Both direct and indirect methods are demonstrated for static and dynamic analysis problems.

The chapter lays the foundation for creating a methodology that can be used as the basis of a quality control methodology.

1.6.8 Chapter 9 Error-Controlled Analyses

1.6.8.1 Aim

This is the key chapter of the book. Its aim is to provide a coherent and logical process that allows potential error and uncertainty sources latent within a finite element analysis to be identified and their magnitude assessed and bounded. On a first reading of the book, this chapter should be consulted early in the process as it shows the target application for much of the discussion and methods developed in the first eight chapters of the book.

1.6.8.2 Outline

Chapter 9 commences by addressing some basic questions relating to assessing the fitness for purpose of a finite element system. It then discusses the requirement for the construction of a 'quality report' which, ultimately, provides the basis for establishing that an analysis has been performed to an adequate standard.

After these initial sections the chapter describes a multi-level and multi-stage analysis procedure, entitled FEMEC, that starts with a review of a structure (real world or as designed) and moves through an interactive set of processes resulting in the production of analysis results that attempt to predict accurately the behaviour of the structure when in service. This incorporates the processes for identifying and defining error sources introduced in Chapter 7 and links these to the error control and treatment methods from Chapter 8. Each stage is formalised into a number of specific tasks with clear inputs and outputs. The methodology recognises that experience is a distinct advantage as complex problems often require 'lateral vision' when executing stages where judgement is needed in assessing the usefulness or otherwise of the results. In order to encourage the use of experience and lateral vision, the process starts with a preliminary error assessment and progresses through levels of deeper error assessment that often involve feedback loops to earlier stages in the assessment processes.

1.6.9 Chapter 10 FEMEC Walkthrough Example

1.6.9.1 Aim

The aim of this chapter is to provide an illustration of the FEMEC procedure detailed in Chapter 9.

1.6.9.2 Outline

Chapter 10 uses a simple static analysis problem involving a pressure-loaded reinforced plate to illustrate the stages in the FEMEC process. It sets out the design requirements together with the required level of accuracy for the analysis. The process is demonstrated by developing an illustrative full Quality Report for this analysis problem. The chapter concludes by showing a limited application of the methodology to a dynamically responding structure.

REFERENCES

1. Adams, V. and Askenazi, A., *Building Better Products with Finite Element Analysis*. 1999: Onward Press, ISBN 156690160X.

2. Szabó, B. and Babuška, I., *Finite Element Analysis.* 1991: John Wiley & Sons, ISBN 0471502731.
3. Babushka, I. and Strouboulis, T., *The Finite Element Method and Reliability.* 2001: Clarendon Press, ISBN 0198502761.
4. Chandrupatia, T.R. and Belegundu, A.D., *Introduction to Finite Elements in Engineering, 3rd edn.* 2002: Prentice Hall, ISBN 10013065919.
5. Zienkiewicz, O.C. and Taylor, R.L., *The Finite Element Method Set.* 2005: Butterworth–Heinemann, ISBN 100750664312.
6. Barthe, K.-J., *Finite Element Procedures.* 1995: Prentice Hall, ISBN 100133014584.
7. Fraeijs de Veubeke, B. and Sander, G., An equilibrium model for plate bending. *Int. J. Solids Struct.*, 1968. **4**: 447–468.
8. Tong, P. and Rossettos, J.N., *Finite Element Method: Basic Techniques and Implementation.* 1968: The MIT Press, ISBN 0262200325.
9. Morris, A.J. and Vignjvec, R., Consistent finite element analysis and error control. *Comput. Methods Appl. Mech. Eng.*, 1997. **140**: 87–108.
10. Babuška, I. and Oden, J.T., The reliability of computer predictions: can they be trusted? *Int. J. Numer. Anal. Modeling*, 2005. **1**(1): 1–18.
11. Belytchko, T. and Mish, K., Computability in non-linear solid mechanics. *Int. J. Numer. Methods Eng.*, 2001. **52**: 3–12.
12. Knight, N.F., Nemeth, M.P. and Hilburger, M.W., *Assessment Technologies for the Space Shuttle External Tank Thermal Protection System and Recommendations for Technology Improvement: Part 2: Structural Analysis Technologies and Modeling Practices.* 2004: NASA, TM-2004-213256.

2

Overview of Static Finite Element Analysis

2.1 INTRODUCTION

As explained in Chapter 1, although the existence of equilibrium and hybrid finite element formulations is recognised, this book focuses exclusively on displacement finite elements for reasons explained earlier. Chapter 4 takes a more fundamental approach to the creation of the Finite Element Method but, in this chapter, a very simple approach is taken to the creation of individual element and global matrices. The aim is to provide a background to the Finite Element Method and illustrate the processes used in all computer-based finite element systems. The chapter does not provide a comprehensive development of the Finite Element Method as there are many excellent text books that cover the field, such as those referenced in Chapter 1 and referenced here as [1], [2], [3] and [4]. In addition, a very comprehensive description can be found in the finite element handbook reference [5]. It simply provides the necessary background knowledge to allow the rest of the book to be read without the need to have an additional text book by the side of the reader when employing the methods discussed in the later chapters. In essence, the chapter answers the question 'How do computers perform a static finite element analysis?' In answering the question it follows what is known as the 'direct method' which exploits matrix analysis that directly mimics the way that the computer operates when addressing a finite element analysis.

2.2 THE DIRECT METHOD FOR STATIC ANALYSES

2.2.1 Element Matrices

The 'direct approach' derives element and global stiffness matrices for displacement finite elements for analyses involving structures subjected to static loads, employing the long-established matrix method of structural analysis. It was used by the early pioneers of the Finite Element Method and is a very effective illustration of the method as it follows the steps employed by finite element computer codes when solving a problem. For more information, the reader can consult one of the early classics in the field cited at reference [6]. A more detailed description that still exploits matrix theory but provides a great deal more information on the computational processes required to support a modern finite element system is given in reference [7].

We can begin very simply by taking a single spring subject to a set of loads as shown in Figure 2.1. The spring has a stiffness denoted by 'k' and is subject to a set of forces f_1 and f_2 that give rise to the end displacements u_1 and u_2 at the two nodes 1 and 2.[1]

Applying Hooke's Law, the relationship between the force f_1 and the displacements u_1 and u_2 is given by:

$$f_1 = ku_1 - ku_2 \tag{2.1}$$

Similarly:

$$f_2 = ku_2 - ku_1 \tag{2.2}$$

Combining these two simple equations into a matrix formulation gives:

$$\begin{Bmatrix} f_1 \\ f_2 \end{Bmatrix} = \begin{Bmatrix} k & -k \\ -k & k \end{Bmatrix} \begin{Bmatrix} u_1 \\ u_2 \end{Bmatrix} \tag{2.3}$$

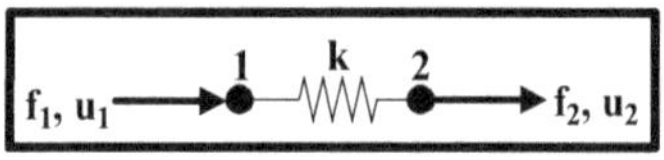

Figure 2.1 Single loaded spring.

[1] For an isolated spring in static equilibrium, such as the one shown here, the nodal forces f_1 and f_2 would be equal and opposite.

where $\begin{Bmatrix} f_1 \\ f_2 \end{Bmatrix}$ is the vector of element nodal forces, $\begin{Bmatrix} k & -k \\ -k & k \end{Bmatrix}$ is the element stiffness matrix and $\begin{Bmatrix} u_1 \\ u_2 \end{Bmatrix}$ is the vector of nodal displacements or nodal connection quantities.

Even though this is the simplest possible finite element, it illustrates some important facts relating to the Finite Element Method. First, the element stiffness matrix is singular and cannot be inverted so that the element displacements u_1 and u_2 cannot be derived from the applied forces f_1 and f_2. This means that, in the absence of inertia forces, the element is able to accommodate a rigid body motion without generating any interior element forces. It ensures that no stresses are generated in the element if the element nodes are given equal displacements, i.e. $u_1 = u_2$. Second, the element stiffness matrix is symmetric. Finally, it should be noted that forces and displacements act at nodes and are linked to each other through the stiffness matrix. In this way the displacement finite element formulation represents the behaviour of a structure as a set of nodal displacements and forces acting at a set of discrete points within the space of the structure. For this class of element everything – forces, displacements and loads – has to be accommodated at these discrete nodal points. Mapping everything to the element nodes is one of the key actions that the analyst or the finite element software has to undertake so that the analysis is represented by point terms.

This approach can be applied to create element matrices applicable to problems involving heat transfer, fluid flow, electrical networks, electromagnetic phenomena, etc. The Finite Element Method has a very wide field of application. It allows the analyst or analysis team to use a single package when confronting a design problem that requires a multidisciplinary analysis. One example would be the analysis of a structure subject to applied external forces and thermal loads. A modern analysis package handles this by employing a thermal analysis to create the thermal stress field, then supplementing it with a set of applied external loads to solve the complete problem.

2.2.2 Assembled Global Stiffness Matrix for Static Analyses

Although Section 2.2.1 has developed a stiffness matrix for a very simple finite element, all finite elements have the same matrix structure and properties. The next question is 'How does a computer create a matrix that represents a complete structure?'

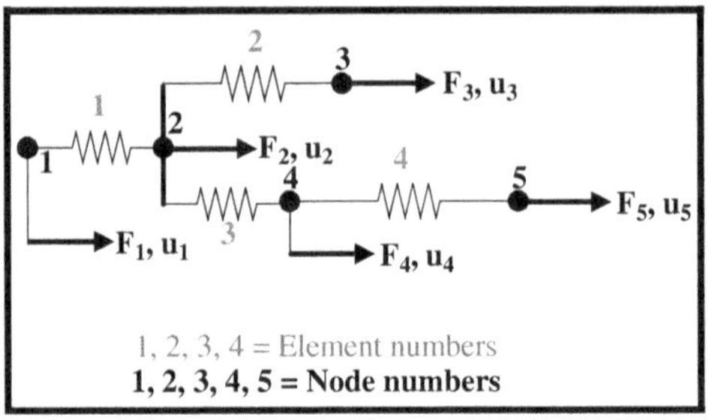

Figure 2.2 A four-spring global model.

Again we take a very simple representative structure to explain the principles underlying the construction of a global stiffness matrix to model the total behaviour of a complete structure. In order to develop the process, consider the simple spring problem shown in Figure 2.2.

It is now necessary to slightly generalise the nomenclature used in equation 2.3 so that:

$$\begin{Bmatrix} f_1 \\ f_2 \end{Bmatrix} = \begin{Bmatrix} k_{11} & k_{12} \\ k_{21} & k_{22} \end{Bmatrix} \begin{Bmatrix} u_1 \\ u_2 \end{Bmatrix}$$

where:

$$k_{11} = k_{22} = k$$

and:

$$k_{12} = k_{21} = -k$$

As we are dealing with a multi-element arrangement, a new nomenclature needs to be introduced, so when element 1 is being discussed, a superscript 1 is added to the element matrix, thus:

$$\begin{Bmatrix} f_1^{(1)} \\ f_2^{(1)} \end{Bmatrix} = \begin{Bmatrix} k_{11}^{(1)} & k_{12}^{(1)} \\ k_{21}^{(1)} & k_{22}^{(1)} \end{Bmatrix} \begin{Bmatrix} u_1 \\ u_2 \end{Bmatrix} \tag{2.4}$$

Similarly, for elements 2 to 4 appropriate superscripts are used to identify which element is making a contribution to the assembly process at each stage.

Taking equilibrium at each node in turn, starting with node 1 and using the matrix equation 2.4, then:

$$k_{11}^{(1)} u_1 + k_{12}^{(1)} u_2 = f_1^{(1)}$$

As there is no force applied to node 1 other than F_1, this term replaces $f_1^{(1)}$; note that the element displacements u_1 and u_2 have no superscript as these are global properties, thus:

$$k_{11}^{(1)}u_1 + k_{12}^{(1)}u_2 = F_1 \tag{2.5}$$

For node 2 there are three elements making a contribution and the equilibrium equation now has three contributions:

From element 1 at node 2: $k_{21}^{(1)}u_1 + k_{22}^{(1)}u_2 = f_2^{(1)}$

From element 2 at node 2: $k_{11}^{(2)}u_2 + k_{12}^{(2)}u_3 = f_1^{(2)}$

From element 3 at node 2: $k_{11}^{(3)}u_2 + k_{12}^{(3)}u_4 = f_1^{(3)}$

Equilibrium at node 2 requires that the element forces must combine to equilibrate the applied load F_2, hence:

$$F_2 = f_2^{(1)} + f_1^{(2)} + f_1^{(3)}$$

and therefore:

$$k_{21}^{(1)}u_1 + k_{22}^{(1)}u_2 + k_{11}^{(2)}u_2 + k_{12}^{(2)}u_4 + k_{11}^{(3)}u_2 + k_{12}^{(3)}u_3 = F_2$$

or:

$$k_{21}^{(1)}u_1 + (k_{22}^{(1)} + k_{11}^{(2)} + k_{11}^{(3)})u_2 + k_{12}^{(2)}u_3 + k_{12}^{(3)}u_4 = F_2 \tag{2.6}$$

Applying similar force-balance equilibrium at nodes 3, 4 and 5 gives three further equations.

For node 3:

$$k_{21}^{(3)}u_2 + k_{22}^{(3)}u_3 = F_3 \tag{2.7}$$

For node 4:

$$k_{21}^{(3)}u_2 + (k_{22}^{(3)} + k_{11}^{(4)})u_4 + k_{12}^{(4)}u_5 = F_4 \tag{2.8}$$

For node 5:

$$k_{21}^{(4)}u_4 + k_{22}^{(4)}u_5 = F_5 \tag{2.9}$$

Combining equations 2.5, 2.6, 2.7, 2.8 and 2.9 into matrix form gives:

$$\begin{bmatrix} k_{11}^{(1)} & k_{12}^{(1)} & 0 & 0 & 0 \\ k_{21}^{(1)} & k_{22}^{(1)} + k_{11}^{(2)} + k_{11}^{(3)} & k_{12}^{(2)} & k_{12}^{(3)} & 0 \\ 0 & k_{21}^{(2)} & k_{22}^{(2)} & 0 & 0 \\ 0 & k_{21}^{(3)} & 0 & k_{22}^{(3)} + k_{11}^{(4)} & k_{12}^{(4)} \\ 0 & 0 & 0 & k_{21}^{(4)} & k_{22}^{(4)} \end{bmatrix} \begin{Bmatrix} u_1 \\ u_2 \\ u_3 \\ u_4 \\ u_5 \end{Bmatrix} = \begin{Bmatrix} F_1 \\ F_2 \\ F_3 \\ F_4 \\ F_5 \end{Bmatrix} \tag{2.10}$$

This can be simply written as:

$$Ku = F$$

where K is the 5×5 matrix given in equation 2.10 and u is the vector of the displacements $u_1 \ldots u_5$ with F the vector of the applied loads $F_1 \ldots F_5$.

The matrix in 2.10 is the global stiffness matrix for the complete structure. As with the element stiffness matrix, it is singular because the global system has also to be able to accommodate rigid body movement. As with the element stiffness matrix this global stiffness matrix is symmetric about the leading diagonal; it is also banded with the non-zero terms clustering along the leading diagonal. The assembly process has torn apart the individual element stiffness matrices and then inserted the element components into the global stiffness matrix depending upon the element global node numbers. The next step is to apply the boundary conditions which specify the values for certain of the displacements thus rendering the stiffness matrix non-singular. Once the boundary conditions have been imposed and the applied loads inserted into the load vector, the resulting matrix equation can be solved to provide values for the nodal displacements. This global set of displacements can then be back-substituted to find the individual element displacements and forces.

In order to illustrate these steps, an example is used where the springs are assumed fixed at nodes 1 and 5 in Figure 2.2, hence $u_1 = u_5 = 0$, and the following loads are applied: $F_2 = 1000$ newtons, $F_3 = F_4 = 2000$ newtons, all the other nodal loads are zero. The individual spring stiffnesses are taken to be $k_1 = 1500$ newtons/metre, $k_2 = 2000$ newtons/metre, $k_3 = 2200$ newtons/metre and $k_4 = 1200$ newtons/metre.

Inserting element stiffness matrices into the global stiffness matrix 2.10 yields:

$$K = \begin{pmatrix} 1.5 \times 10^3 & -1.5 \times 10^3 & 0 & 0 & 0 \\ -1.5 \times 10^3 & 5.7 \times 10^3 & -2 \times 10^3 & -2.2 \times 10^3 & 0 \\ 0 & -2 \times 10^3 & 2 \times 10^3 & 0 & 0 \\ 0 & -2.2 \times 10^3 & 0 & 3.4 \times 10^3 & -1.2 \times 10^3 \\ 0 & 0 & 0 & -1.2 \times 10^3 & 1.2 \times 10^3 \end{pmatrix}$$

As this global stiffness matrix is singular, the boundary conditions must be incorporated into it in order to create a situation that allows the problem to be solved. Commercial systems undertake this task using a variety of procedures and these are discussed in the standard texts. In this instance, because the problem is very simple and is being used as an illustrative example, a simple procedure is employed. This is done by setting the terms u_1 and u_5 equal to zero in the matrix equation 2.10 and then placing a one at the appropriate diagonal position and zeros in the rows and columns associated with these two displacements. Thus the matrix K becomes the reduced matrix KR:

$$KR = \begin{Bmatrix} 1 & 0 & 0 & 0 & 0 \\ 0 & 5.7 \times 10^3 & -2 \times 10^3 & -2.2 \times 10^3 & 0 \\ 0 & -2 \times 10^3 & 2 \times 10^3 & 0 & 0 \\ 0 & -2.2 \times 10^3 & 0 & 3.4 \times 10^3 & 0 \\ 0 & 0 & 0 & 0 & 1 \end{Bmatrix}$$

The load vector also requires modification to give a new vector P with F_1 and F_2 set equal to zero. The problem is now solved by inverting the matrix KR and multiplying it with the modified load vector P to yield values of the nodal displacements, thus:

$$u = KR^{-1}P$$

to give a displacement vector in metres:

$$\begin{Bmatrix} u_1 \\ u_2 \\ u_3 \\ u_4 \\ u_5 \end{Bmatrix} = \begin{Bmatrix} 0 \\ 1.886 \\ 2.886 \\ 1.809 \\ 0 \end{Bmatrix}$$

These values of the nodal displacements can now be fed back into the individual elements to yield the internal forces. Thus, for element 1 using equation 2.4, these forces, in newtons, are given by:

$$\begin{Bmatrix} f_1^{(1)} \\ f_2^{(1)} \end{Bmatrix} = \begin{Bmatrix} k_{11}^{(1)} & k_{12}^{(1)} \\ k_{21}^{(1)} & k_{22}^{(1)} \end{Bmatrix} \begin{Bmatrix} 0 \\ 1.886 \end{Bmatrix} = \begin{Bmatrix} -2.829 \times 10^3 \\ 2.829 \times 10^3 \end{Bmatrix}$$

Applying this same process for the other three elements in turn gives the forces in these elements as:

$$\begin{Bmatrix} f_1^{(2)} \\ f_2^{(2)} \end{Bmatrix} = \begin{Bmatrix} -2 \times 10^3 \\ 2 \times 10^3 \end{Bmatrix} \quad \begin{Bmatrix} f_1^{(3)} \\ f_2^{(3)} \end{Bmatrix} = \begin{Bmatrix} 170.543 \\ -170.543 \end{Bmatrix}$$

$$\begin{Bmatrix} f_1^{(4)} \\ f_2^{(4)} \end{Bmatrix} = \begin{Bmatrix} 2.171 \times 10^3 \\ -2.171 \times 10^3 \end{Bmatrix}$$

No matter how complex a static structural analysis may be, the computer solves the problem following these simple steps. For a non-linear analysis, achieving equilibrium usually requires an inner looping process that ensures the element forces and the applied loads are brought into equilibrium.

Achieving equilibrium is a key property of the finite element solution process. Although error control processes are fully discussed in Chapter 8, it is worth drawing attention to the first error check that an analyst should perform when first examining the finite element solution output. The applied loads and the reactions calculated by the finite element system must be in agreement. In the case of this simple problem, the reactions are the calculated forces at the two restrained nodes 1 and 5 with forces of -2.829×10^3 newtons and -2.171×10^3 newtons, giving a summed value of 5.0×10^3 newtons. The question to be answered is: does this total reaction force balance the applied loads? The answer is clearly yes since $F_2 + F_3 + F_4$ also gives 5.0×10^3 newtons.

2.2.3 Global Coordinates

The description of the element creation and assembly process shown in Section 2.2.2 has used a one-dimensional coordinate system. The vast majority of finite element analyses relate to structures located in a global two- or three-dimensional coordinate system. Irrespective of the global coordinate system the element matrices, i.e. stiffness, mass, etc., are computed using a coordinate system local to the element as demonstrated with the simple spring element in Section 2.2.1. However, element matrices must be transformed so that they can be located within the structure's global coordinate system before the assembled global stiffness is constructed. In order to demonstrate the procedure for transforming matrices from one coordinate system to a second, we take the simple case of the spring shown in Figure 2.1 and transform the stiffness matrix from the local axis to a two-dimensional axis system involving two connection quantities or degrees of freedom at each node as shown in Figure 2.3. There has been a change in nomenclature as the local nodal forces and displacements previously denoted by f_1, u_1 and f_2, u_2 are now denoted by F_1, U_1 and F_2, U_2. The terms f_1, u_1 , f_2, u_2 now represent the forces and displacements at node 1 in the new x–y coordinate system and are therefore equivalent to F_1, U_1; similarly f_3, u_3, f_4, u_4 replace F_2 and U_2 and are equivalent to the force and displacement at node 2 measured in the element coordinate system.

In order to transform the 2 × 2 element stiffness matrix into a 4 × 4 global matrix we note that:

$$u_1 = U_1\cos\theta \quad u_2 = U_1\sin\theta$$
$$u_3 = U_2\cos\theta \quad u_4 = U_2\sin\theta$$

This allows us to write the transformation equations:

$$\begin{Bmatrix} u_1 \\ u_2 \\ u_3 \\ u_4 \end{Bmatrix} = \begin{Bmatrix} \cos\theta & 0 \\ \sin\theta & 0 \\ 0 & \cos\theta \\ 0 & \sin\theta \end{Bmatrix} \begin{Bmatrix} U_1 \\ U_2 \end{Bmatrix} \tag{2.11}$$

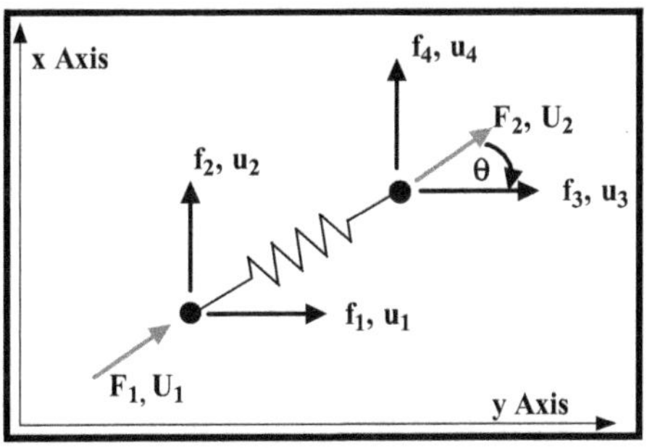

Figure 2.3 Spring transformed from a local one-dimensional axis system to a global two-dimensional axis system.

Similarly a transformation equation can be created that links the local and global nodal forces:

$$\begin{Bmatrix} f_1 \\ f_2 \\ f_3 \\ f_4 \end{Bmatrix} = \begin{Bmatrix} \cos\theta & 0 \\ \sin\theta & 0 \\ 0 & \cos\theta \\ 0 & \sin\theta \end{Bmatrix} \begin{Bmatrix} F_1 \\ F_2 \end{Bmatrix} \tag{2.12}$$

Recalling equation 2.3 and rewriting it in terms of the notation used in this section with k as the spring stiffness, we have:

$$\begin{Bmatrix} F_1 \\ F_2 \end{Bmatrix} = \begin{Bmatrix} k & -k \\ -k & k \end{Bmatrix} \begin{Bmatrix} U_1 \\ U_2 \end{Bmatrix}$$

Pre-multiplying this set by the transformation matrix gives:

$$\begin{Bmatrix} \cos\theta & 0 \\ \sin\theta & 0 \\ 0 & \cos\theta \\ 0 & \sin\theta \end{Bmatrix} \begin{Bmatrix} F_1 \\ F_2 \end{Bmatrix} = \begin{Bmatrix} \cos\theta & 0 \\ \sin\theta & 0 \\ 0 & \cos\theta \\ 0 & \sin\theta \end{Bmatrix} \begin{Bmatrix} k & -k \\ -k & k \end{Bmatrix} \begin{Bmatrix} U_1 \\ U_2 \end{Bmatrix}$$

and using equation 2.12:

$$\begin{Bmatrix} f_1 \\ f_2 \\ f_3 \\ f_4 \end{Bmatrix} = \begin{Bmatrix} \cos\theta & 0 \\ \sin\theta & 0 \\ 0 & \cos\theta \\ 0 & \sin\theta \end{Bmatrix} \begin{Bmatrix} k & -k \\ -k & k \end{Bmatrix} \begin{Bmatrix} U_1 \\ U_2 \end{Bmatrix} \tag{2.13}$$

Applying the transpose of equation 2.11 to matrix equation 2.13 yields:

$$\begin{Bmatrix} f_1 \\ f_2 \\ f_3 \\ f_4 \end{Bmatrix} = \begin{Bmatrix} \cos\theta & 0 \\ \sin\theta & 0 \\ 0 & \cos\theta \\ 0 & \sin\theta \end{Bmatrix} \begin{Bmatrix} k & -k \\ -k & k \end{Bmatrix} \begin{Bmatrix} \cos\theta & \sin\theta & 0 & 0 \\ 0 & 0 & \cos\theta & \sin\theta \end{Bmatrix} \begin{Bmatrix} u_1 \\ u_2 \\ u_3 \\ u_4 \end{Bmatrix} \tag{2.14}$$

Now writing:

$$f = \begin{Bmatrix} f_1 \\ f_2 \\ f_3 \\ f_4 \end{Bmatrix} \quad T^T = \begin{Bmatrix} \cos\theta & 0 \\ \sin\theta & 0 \\ 0 & \cos\theta \\ 0 & \sin\theta \end{Bmatrix} \quad k = \begin{Bmatrix} k & -k \\ -k & k \end{Bmatrix} \quad u = \begin{Bmatrix} u \\ u_2 \\ u_3 \\ u_4 \end{Bmatrix}$$

the matrix equation 2.14 can now be rewritten as:

$$f = T^T kTu$$

Or putting:

$$K = T^T kT \tag{2.15}$$

we have $f = Ku$ where K is the transformed stiffness matrix created by exploiting a transformation matrix that takes a 2×2 stiffness matrix and turns it into a 4×4 spring stiffness matrix within a two-dimensional x–y coordinate system. The matrix T is the smallest transformation matrix used in a finite element analysis system as it operates on the lowest possible order finite element. Clearly the size of the matrix will increase if we move to a three-dimensional space and if the number of nodes associated with an element increases as shown in Section 2.2.4.

The transformation process defined by equation 2.15 frequently turns up in the application of the Finite Element Method. In this case it is expanding a matrix in Section 2.3.2.1 that reduces the size of a matrix to condense out unwanted nodal connection quantities.

2.2.4 Some Typical Elements

Figure 2.4 lists some of the finite element types used in the finite element analysis of structures. It is not intended to be a comprehensive list but simply illustrates some of the popular elements used in industry. The number of nodes for each of the elements illustrated is the minimum as all of these elements can have more nodes as required by the analyst; this point is returned to in later sections. It is also worth remarking that the shell elements can have displacement degrees of freedom at certain nodes and rotation degrees of freedom at others. These elements are incorporated into the finite element process using the approach described in Section 2.2.2.

2.2.5 The Analysis Loop

The complete process described in Section 2.2.2 can be represented as a loop that starts by creating element stiffness matrices k_i and finishes by

Element type	Degree of freedom (connection quantities)	Illustration
Bar element	One degree of freedom at each node: displacement	
Torsion bar element	One degree of freedom at each node: rotation	
Membrane element	Two degrees of freedom at each node: displacements	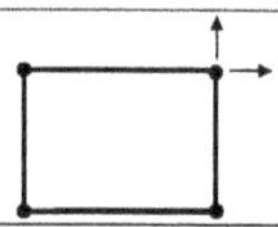
Three-dimensional solid element	Three degrees of freedom at each node: displacements	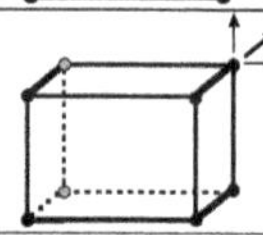
Two-dimensional beam bending element	Two degrees of freedom at each node: one displacement and one rotation	
Three-dimensional beam element	Six degrees of freedom at each node: three displacements and three rotations	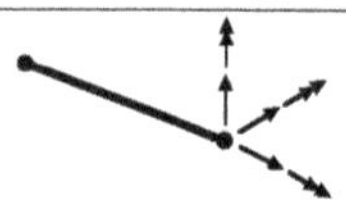
Flat plate element	Three degrees of freedom at each node: one displacement and two rotations	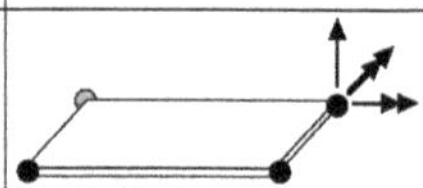
Shell element	Five degrees of freedom at each node:threedisplacements and two rotations	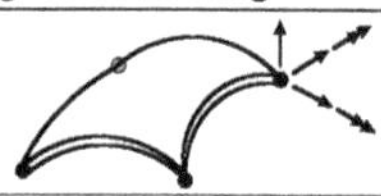

Figure 2.4 Some typical finite elements.

calculating element forces f_1^i and f_2^i (Figure 2.5). The result is accepted if the applied loads and the calculated reactions agree. In chapter 4 we note that some of the applied load may be lost if consistent loads are applied and in Chapters 8 and 9 a full range of checks are introduced to augment the simple load–reaction consistency check.

A spring represents the simplest structural finite element, yet most of the steps undertaken by mainstream finite element systems follow the steps outlined above. However, in most cases the analysis loop is extended as the process, for the static analysis case, concludes with the calculation of element strains and stresses. Thus the loop is augmented by additional steps that are undertaken at the element level as shown in Figure 2.6.

This direct approach employs a very simple method to create an element stiffness matrix. In reality, a more complex approach is used to create these matrices using shape functions that approximate the displacement field within an element. This is introduced in Chapter 4 where it is linked to the use of potential energy and also explains how a

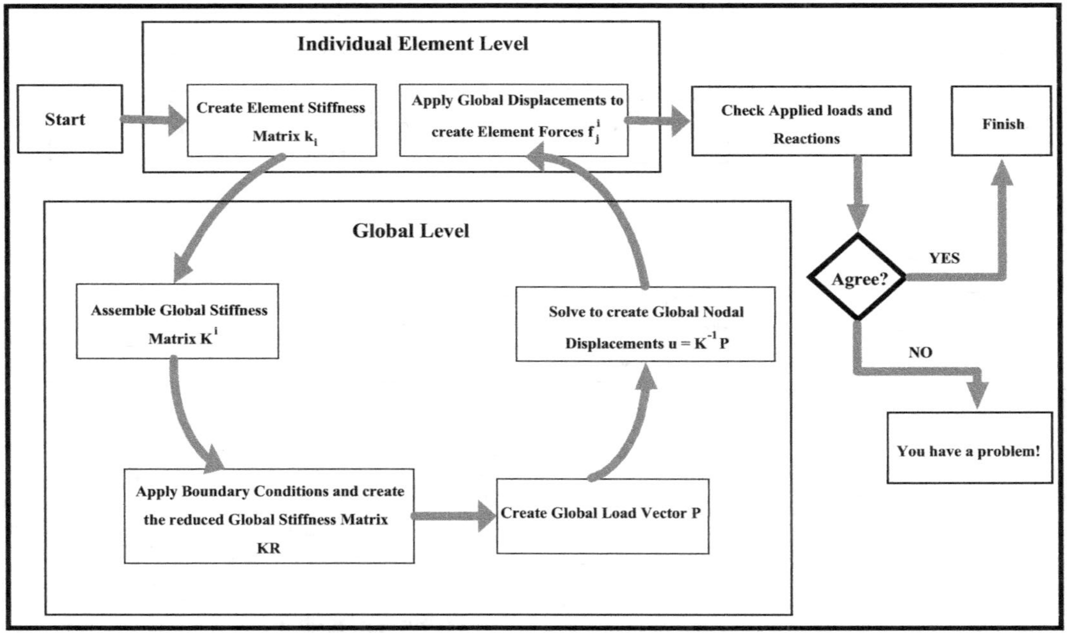

Figure 2.5 Analysis loop.

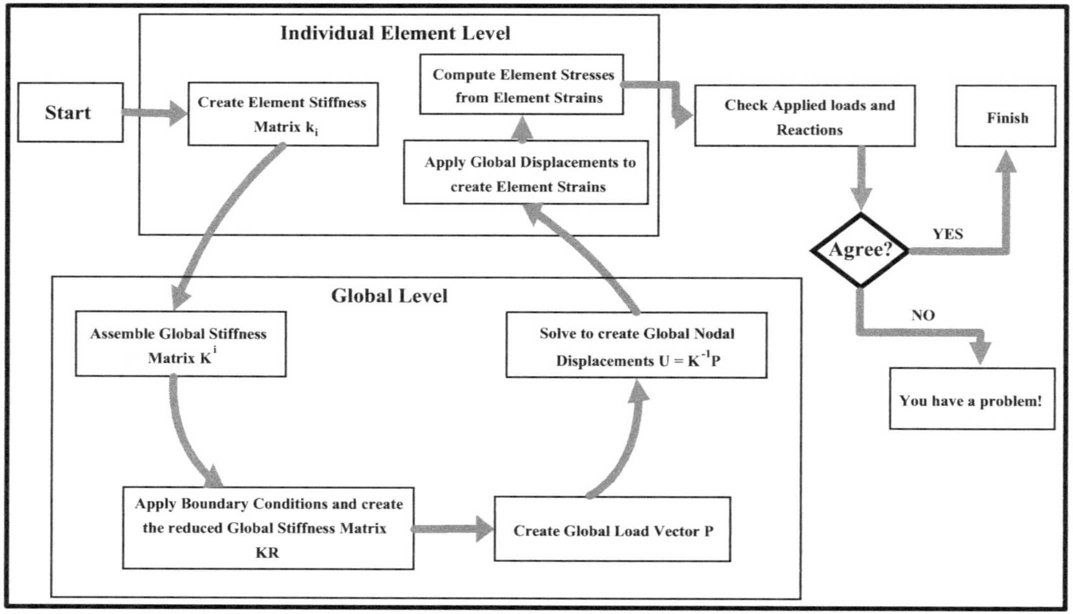

Figure 2.6 Augmented analysis loop.

load that is distributed over the surface of an element can be represented by equivalent nodal forces. The use of shape functions allows a complex element and a complex loading system to be represented by displacements and loads defined at element nodes so that, as above, a finite element analysis of a complex design problem is represented by nodal displacements and forces only.

2.3 REDUCING THE PROBLEM SIZE

In later chapters the question of the size of the analysis problem is discussed and it is indicated that reducing the size of a problem is normally advantageous. There are a number of ways that can be employed to achieve such a reduction. One method is to take advantage of any symmetry that the structure exhibits. Another is to slave some of the degrees of freedom to a master set of degrees of freedom and undertake the analysis using the reduced number of master degrees of freedom; this is known as condensation. Condensation was very popular in the past when computers were much less powerful than at present and in the early days of employing PCs to undertake a finite element analysis. Today condensation methods are still extensively used for large dynamic analysis problems (see Chapter 3, Section 3.3), for the construction of superelements,[2] and are the underpinning methodology for the application of sub-structuring techniques used when an analysis is performed by more than one team, particularly when the teams are physically separated.

2.3.1 Symmetry

Most analysis problems do not present themselves in a perfectly symmetric form, but some do come in a form where the structure is physically symmetric while the loads are not![3] In this situation the problem can be solved by exploiting structural symmetry and applying symmetric and anti-symmetric loads as illustrated in Figure 2.7.

---

[2] Superelements have a very useful role to play in situations where local analyses are combined with a global analysis.

[3] In certain situations the structure may deviate from physical symmetry by a sufficiently small amount so that the analyst may consider imposing symmetry in order to achieve the benefits of size reduction.

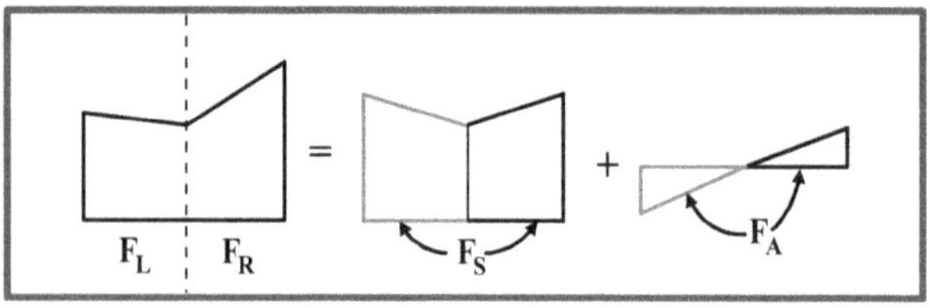

Figure 2.7 Symmetric and anti-symmetric loads.

Figure 2.7 represents, in a schematic form, the actual loads on the left- and right-hand sides of the structure by F_L and F_R respectively. The loads F_S and F_A represent the actual loads decomposed into symmetric and anti-symmetric components that can now be applied to the symmetric structure. Although these symmetric and anti-symmetric loads shown in Figure 2.7 are applied to the entire structure, in reality the analysis employs only half the structure, i.e. the part shown in black, while the part in grey is omitted. Along the line of structural symmetry, shown as a dashed line in the extreme left figure in Figure 2.7, symmetry conditions are applied when the symmetric loads are used and anti-symmetric conditions when the anti-symmetric loads are applied. Proceeding in this way has the advantage that half the structure is considered and, although the structure is now analysed twice, the solution time is still significantly reduced. Also, the smaller the analysis, the easier it is to control analysis errors! We now need to evaluate the new applied loads F_S and F_A.

We use the nomenclature that U_L and U_R represent the displacement fields on the left- and right-hand sides of the complete structure and U_S and U_A the displacement fields of the symmetric and anti-symmetric loaded structures respectively. On the assumption that U_L is defined in coordinates which are reflected about the plane of symmetry:

$$\begin{Bmatrix} U_R \\ U_L \end{Bmatrix} = \begin{Bmatrix} I & I \\ I & -I \end{Bmatrix} \begin{Bmatrix} U_S \\ U_A \end{Bmatrix} \tag{2.16}$$

or:

$$\begin{Bmatrix} U_S \\ U_A \end{Bmatrix} = \frac{1}{2} \begin{Bmatrix} I & I \\ I & -I \end{Bmatrix} \begin{Bmatrix} U_R \\ U_L \end{Bmatrix} \tag{2.17}$$

where I is the appropriate identity matrix.

To obtain F_S and F_A from F_L and F_R we use equivalent work, thus:

$$2F_S U_S + 2F_A U_A = F_R U_R + F_L U_L$$

Using the matrix equations 2.11:

$$2F_S U_S + 2F_A U_A = F_R(U_S + U_A) + F_L(U_S - U_A)$$

Comparing coefficients gives:

$$F_S = \frac{1}{2}(F_R + F_L)$$

and:

$$F_L = \frac{1}{2}(F_R - F_L)$$

or:

$$\begin{Bmatrix} F_S \\ F_A \end{Bmatrix} = \frac{1}{2}\begin{Bmatrix} I & I \\ I & -I \end{Bmatrix}\begin{Bmatrix} F_R \\ F_L \end{Bmatrix} \tag{2.18}$$

The matrix equation 2.18 defines the loads to be applied to half of the symmetric structure that is subjected to symmetric and anti-symmetric boundary conditions at the line of symmetry.

To illustrate the method, consider the simple symmetric structure shown in Figure 2.8 subjected to two loads of value 50,000 and 100,000 newtons. All the bars in this structure are identical and have a cross-sectional area of 0.1 m^2 and modulus of elasticity 200×10^9 N/m^2.

The symmetry line goes through nodes 4, 5 and 6 and gives rise to the anti-symmetric and symmetric problems shown in Figure 2.9. Note that the symmetric constraints along the symmetry line restrain the structure from moving in the x-direction at nodes 4 and 5 while the anti-symmetric constraints restrain the structure from moving in the y-direction at nodes 4 and 5. The two load cases corresponding to those defined in matrix equation 2.13 are a vertical load of 75 kN at node 7 for the symmetric case and 25 kN at node 7 for the anti-symmetric case. The two elements along the symmetry line, elements 9 and 10, have cross-sectional areas of 0.05 m^2, i.e. half the cross-sectional area of the actual members in the structure.

Performing the analyses gives the displacement fields shown in Table 2.1 for the symmetric and anti-symmetric loading cases using the numbering sequence of Figure 2.9.

The displacement field for the complete structure is found by making use of matrix equation 2.16 to allow symmetric and anti-symmetric fields

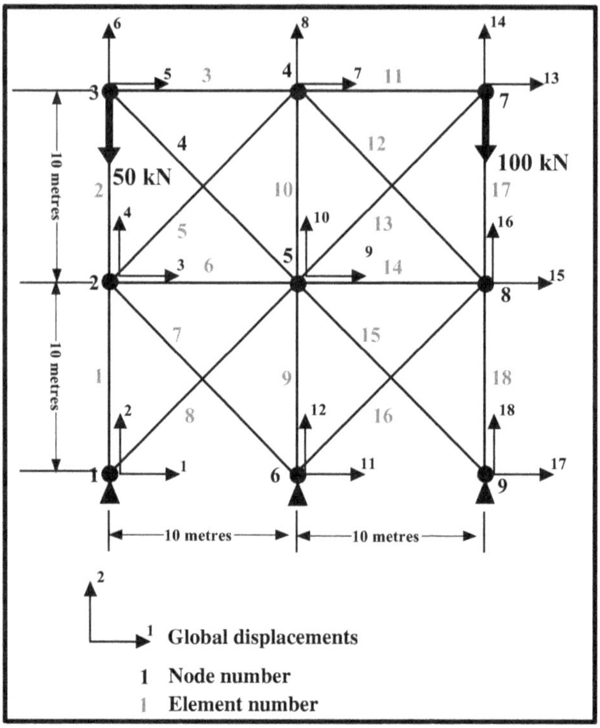

Figure 2.8 Multi-frame symmetric bar structure.

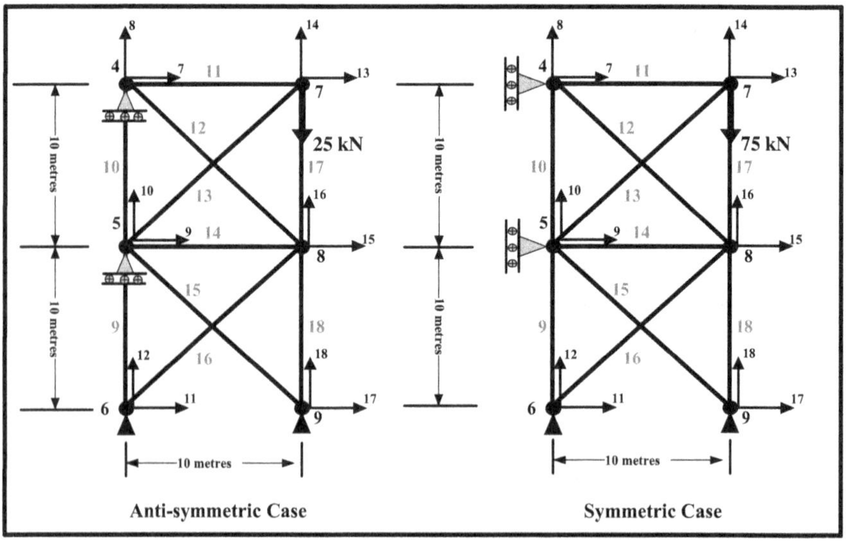

Figure 2.9 Anti-symmetric and symmetric analysis problems.

Table 2.1 Symmetric and anti-symmetric displacements.

Degree of freedom	Symmetric displacement U_S (metres)	Anti-symmetric displacement U_A (metres)
7	0.0	2.153×10^{-5}
8	-1.772×10^{-5}	0.0
9	0.0	4.147×10^{-6}
10	-1.334×10^{-5}	0.0
11	0.0	0.0
12	0.0	0.0
13	9.197×10^{-6}	2.28×10^{-5}
14	-4.855×10^{-5}	-2.226×10^{-5}
15	3.671×10^{-6}	6.887×10^{-6}
16	-2.025×10^{-5}	-1.103×10^{-5}
17	0.0	0.0
18	0.0	0.0

given above to be summed and differenced appropriately. This gives the displacement solution set shown in Table 2.2, using the numbering sequence of Figure 2.8.

Following the analysis loop shown in Figure 2.6, the solution can be completed by calculating the element forces and stresses from the displacements in Table 2.2 or from Table 2.1 where they will then need to be appropriately combined.

Table 2.2 Displacement field for complete structure.

Degree of freedom	Complete structure nodal displacements (metres)
1	0.0
2	0.0
3	3.217×10^{-6}
4	-9.216×10^{-6}
5	1.36×10^{-5}
6	-2.628×10^{-5}
7	2.153×10^{-5}
8	-1.772×10^{-5}
9	4.147×10^{-6}
10	-1.334×10^{-5}
11	0.0
12	0.0
13	3.2×10^{-5}
14	-7.081×10^{-5}
15	1.056×10^{-5}
16	-3.126×10^{-5}
17	0.0
18	0.0

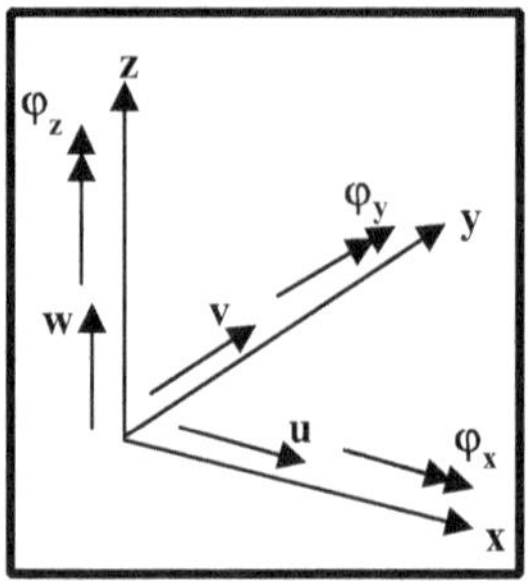

Figure 2.10 General three-dimensional display of the six rigid body degrees of freedom.

Although the use of symmetric and anti-symmetric loads is relatively straightforward, errors will occur in a complex analysis problem if mistakes are made in applying the appropriate displacement boundary conditions. In order to aid an engineer in undertaking an analysis we now define symmetry and anti-symmetry boundary conditions for a general three-dimensional body with six degrees of freedom as shown in Figure 2.10. If the x–y plane is selected as the plane of symmetry for a given problem as shown in Figure 2.11, then the restrained or fixed degrees of freedom are u, φ_y and φ_z.

Following this process a table can be set up that defines the fixed and free degrees of freedom for the three planes of symmetry x–y, y–z and z–x as shown in Table 2.3. Using this table it is possible to write down the conditions from the anti-symmetric case as these are simply the opposite of the symmetric conditions as displayed in Table 2.4. Most analysts find it easier to 'see' either symmetric or anti-symmetric configurations, so the easiest way to establish the displacement constraints for a given plane of symmetry/anti-symmetry is to write down

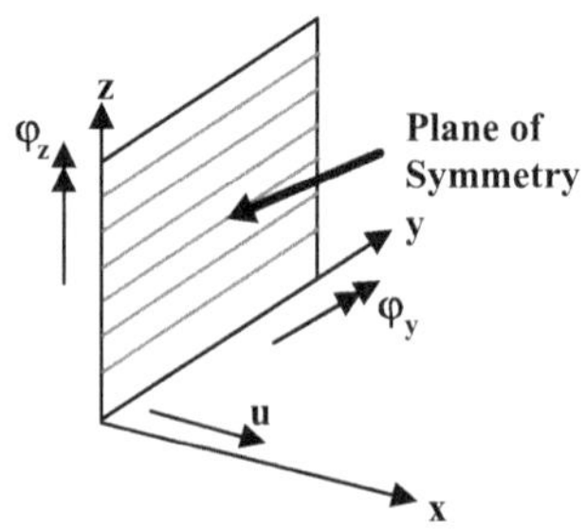

Figure 2.11 Taking the x–y plane as a plane of symmetry.

Table 2.3 Constraints on the displacement/rotation degrees of freedom: symmetry condition.

Plane of symmetry	u Displacement	v Displacement	w Displacement	ϕ_x Rotation	ϕ_y Rotation	ϕ_x Rotation
x–y	Free	Free	Fixed	Fixed	Fixed	Free
y–z	Fixed	Free	Free	Free	Fixed	Fixed
z–x	Free	Fixed	Free	Fixed	Free	Fixed

Table 2.4 Constraints on the displacement/rotation degrees of freedom: anti-symmetry condition.

Plane of symmetry	u Displacement	v Displacement	w Displacement	ϕ_x Rotation	ϕ_y Rotation	ϕ_x Rotation
x–y	Fixed	Fixed	Free	Free	Free	Fixed
y–z	Free	Fixed	Fixed	Fixed	Free	Free
z–x	Fixed	Free	Fixed	Free	Fixed	Free

the conditions associated with the case with which the analyst is most comfortable and then take the opposite values if required. In many cases the structural plane of symmetry will not be located in a direction parallel with one of the axes of the main coordinate systems adopted for the analysis. The most convenient way to handle this is to use the local axis facility available in commercial finite element software systems to set up a local coordinate system that lines up with the structural plane of symmetry.

This simple type of structural symmetry situation is only one of a range of structural symmetry conditions that include multiple or repeated symmetry as found in multi-span bridges and radial symmetry in cyclically repeated structures as, for example, in rotating machinery components. Although more complex, these are also handled in a similar manner to the case illustrated in this section by taking an individual segment with appropriate boundary conditions and loads resolved into suitable components.

Structures with components involving axially symmetric solids, such as thick shells of revolution, are very common in practice but the approach adopted for the analysis of this type of structure is different from that discussed above. This structural form requires special axially symmetric solid elements and the analysis problem is similar to plane-stress problems. In the case where an axially symmetric solid is subjected to asymmetric loads, Fourier series are employed to allow a set of component analyses to be superimposed.

Although the primary use of symmetry conditions is to reduce the size of the problem, there are certain situations where an analysis problem could not be solved without exploiting symmetry conditions. For example, the windows of some passenger-carrying aircraft are held in place by clips that do not allow the window to pick up stresses from the fuselage as the aircraft climbs to altitude where cabin pressurisation would impart hoop stresses in the fuselage structure. In this situation the boundary condition for the window suppresses vertical movement and in-plane rotation at the window's edge. This is insufficient to suppress rigid body movement in the finite element model of the window, but this can be counteracted by exploiting the double in-plane structural symmetry which aircraft windows exhibit.

Finally, a word of caution: if an analyst is dealing with a non-linear problem, then it is possible for a symmetric problem to degenerate into a non-symmetric problem as the non-linear analysis progresses. Similarly, it is possible for a symmetric structure to generate non-symmetric buckling modes. However, in this latter case the use of symmetry and anti-symmetry is usually sufficient to capture all the buckling modes.

2.3.2 Condensation and Superelements

The second type of size reduction is achieved when nodes on the interior of a finite element region or an element are slaved to a set of exterior elements as illustrated in Figure 2.12, where the nodes on the interior region disappear and only those on the exterior are retained; this process is called condensation and the resulting element is termed a superelement.

2.3.2.1 The condensation process

The process of condensation does not ignore the influence of the interior nodal forces and displacements but incorporates them through the creation of modified global or element stiffness matrices and load vectors. For the sake of generality the element layout shown in Figure 2.12 is assumed to involve a group of elements assembled to represent the behaviour of a structure subjected to static loads.

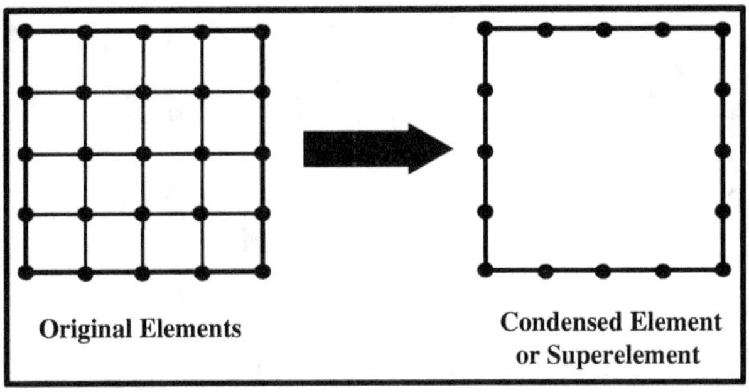

Figure 2.12 The condensation process following the arrow from left to right.

Taking K as the assembled global stiffness matrix, F as the assembled load vector and u as the nodal displacements, then the standard finite element matrix equation for the original element layout is:

$$Ku = F \tag{2.19}$$

The nodal displacements are now divided into those located on the boundary of the element domain denoted by u_1 and those on the interior nodes of the domain by u_2; similarly, the loads are divided into boundary and interior loads F_1 and F_2. In order to match this division, the stiffness matrix is partitioned so that equation 2.19 becomes:

$$\begin{Bmatrix} K_{11} & K_{12} \\ K_{12}^T & K_{22} \end{Bmatrix} \begin{Bmatrix} u_1 \\ u_2 \end{Bmatrix} = \begin{Bmatrix} F_1 \\ F_2 \end{Bmatrix} \tag{2.20}$$

Advantage has been taken in 2.20 of symmetry in the partitioned stiffness matrix. The first step is to solve for the interior displacements using the lower half of the partitioned stiffness matrix to yield:

$$u_2 = K_{22}^{-1}\{F_2 - K_{12}^T u_1\} \tag{2.21}$$

This can now be used in the upper half of the partitioned system in 2.20 to give:

$$K_{11}u_1 + K_{12}\{K_{22}^{-1}F_2 - K_{22}^{-1}K_{12}^T u_1\} = F_1 \tag{2.22}$$

Regrouping:

$$\{K_{11} - K_{12}K_{22}^{-1}K_{12}^T\}u_1 = F_1 - K_{12}K_{22}^{-1}F_2 \tag{2.23}$$

If we now put:

$$K_c = \{K_{11} - K_{12}K_{22}^{-1}K_{12}^T\} \tag{2.24}$$

and:

$$F_c = F_1 - K_{12}K_{22}^{-1}F_2 \tag{2.25}$$

this gives rise to a new matrix equation:

$$K_c u_1 = F_c \tag{2.26}$$

and the stiffness matrix K_c and the load vector F_c are now termed the condensed stiffness matrix and condensed load vector for the nodes associated with the displacements u_1. The process has transformed the original system, shown on the left side of Figure 2.13, to that on the right, thus forming a new superelement with no apparent interior nodes.

Because this system has symmetric stiffness matrices, the condensation process can be written as a simple transformation process with:

$$K_c = T^T K T \tag{2.27}$$

and:

$$F_c = T^T F \tag{2.28}$$

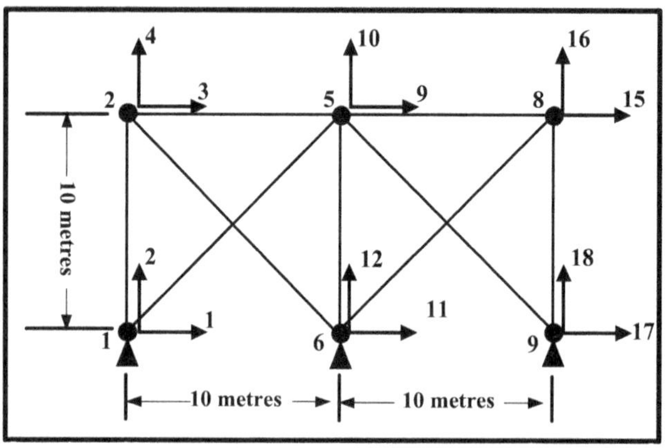

Figure 2.13 Condensed model.

where the transformation matrix T used in equations 2.27 and 2.28 is given by:

$$T = \left\{ \begin{matrix} I \\ -K_{22}^{-1}K_{12}^{T} \end{matrix} \right\} \tag{2.29}$$

As an illustration, consider the loaded frame structure used to demonstrate the use of symmetry in Section 2.3 and shown in Figure 2.8. The loads, dimensions and material properties of the structure are those used in the Section 2.3 demonstration problem. The target is to condense the problem down to one where the basic finite element layout is that shown in Figure 2.14, i.e. the nodes 3, 4 and 7 are condensed out leaving a problem with 12 degrees of freedom from the original 18 degrees of freedom. Thus the vector of master degrees of freedom u_1 relates to the degrees of freedom $1, 2, 3, 4, 9, 10, 11, 12, 15, 16, 17, 18$ and u_2 relates to the remaining condensed degrees of freedom, namely 5, 6, 7, 8, 13, 14.

The stiffness matrix for the entire structure K has now to be partitioned into the sub-matrices K_{11}, K_{12} and K_{22} required by equation 2.20. The locations that the components of these matrices take up in the global stiffness matrix is shown in Figure 2.14 where terms 11 represent the terms that are found in the matrix K_{11}, the terms 12 are the

$$\begin{bmatrix}
11 & 11 & 11 & 11 & 12 & 12 & 12 & 12 & 11 & 11 & 11 & 11 & 12 & 12 & 11 & 11 & 11 & 11 \\
11 & 11 & 11 & 11 & 12 & 12 & 12 & 12 & 11 & 11 & 11 & 11 & 12 & 12 & 11 & 11 & 11 & 11 \\
11 & 11 & 11 & 11 & 12 & 12 & 12 & 12 & 11 & 11 & 11 & 11 & 12 & 12 & 11 & 11 & 11 & 11 \\
11 & 11 & 11 & 11 & 12 & 12 & 12 & 12 & 11 & 11 & 11 & 11 & 12 & 12 & 11 & 11 & 11 & 11 \\
21 & 21 & 21 & 21 & 22 & 22 & 22 & 22 & 21 & 21 & 21 & 21 & 22 & 22 & 21 & 21 & 21 & 21 \\
21 & 21 & 21 & 21 & 22 & 22 & 22 & 22 & 21 & 21 & 21 & 21 & 22 & 22 & 21 & 21 & 21 & 21 \\
21 & 21 & 21 & 21 & 22 & 22 & 22 & 22 & 21 & 21 & 21 & 21 & 22 & 22 & 21 & 21 & 21 & 21 \\
21 & 21 & 21 & 21 & 22 & 22 & 22 & 22 & 21 & 21 & 21 & 21 & 22 & 22 & 21 & 21 & 21 & 21 \\
11 & 11 & 11 & 11 & 12 & 12 & 12 & 12 & 11 & 11 & 11 & 11 & 12 & 12 & 11 & 11 & 11 & 11 \\
11 & 11 & 11 & 11 & 12 & 12 & 12 & 12 & 11 & 11 & 11 & 11 & 12 & 12 & 11 & 11 & 11 & 11 \\
11 & 11 & 11 & 11 & 12 & 12 & 12 & 12 & 11 & 11 & 11 & 11 & 12 & 12 & 11 & 11 & 11 & 11 \\
11 & 11 & 11 & 11 & 12 & 12 & 12 & 12 & 11 & 11 & 11 & 11 & 12 & 12 & 11 & 11 & 11 & 11 \\
21 & 21 & 21 & 21 & 22 & 22 & 22 & 22 & 21 & 21 & 21 & 21 & 22 & 22 & 21 & 21 & 21 & 21 \\
21 & 21 & 21 & 21 & 2 & 22 & 22 & 22 & 21 & 21 & 21 & 21 & 22 & 22 & 21 & 21 & 21 & 21 \\
11 & 11 & 11 & 11 & 12 & 12 & 12 & 12 & 11 & 11 & 11 & 11 & 12 & 12 & 11 & 11 & 11 & 11 \\
11 & 11 & 11 & 11 & 12 & 12 & 12 & 12 & 11 & 11 & 11 & 11 & 12 & 12 & 11 & 11 & 11 & 11 \\
11 & 11 & 11 & 11 & 12 & 12 & 12 & 12 & 11 & 11 & 11 & 11 & 12 & 12 & 11 & 11 & 11 & 11 \\
11 & 11 & 11 & 11 & 12 & 12 & 12 & 12 & 11 & 11 & 11 & 11 & 12 & 12 & 11 & 11 & 11 & 11
\end{bmatrix}$$

Figure 2.14 Decomposed global matrix.

components of matrix K_{12}, the terms 21 are components of the matrix K_{12}^T and the terms 22 are the positions of the terms that make up the matrix K_{22}.

The transformation matrix T from equation 2.29 is given by:

$$T = \begin{Bmatrix} 1 & 0 & 0 & 0 & 0 & 0 & 0 & 0 & 0 & 0 & 0 & 0 \\ 0 & 1 & 0 & 0 & 0 & 0 & 0 & 0 & 0 & 0 & 0 & 0 \\ 0 & 0 & 1 & 0 & 0 & 0 & 0 & 0 & 0 & 0 & 0 & 0 \\ 0 & 0 & 0 & 1 & 0 & 0 & 0 & 0 & 0 & 0 & 0 & 0 \\ 0 & 0 & 0.25 & 0.604 & 0.5 & -0.207 & 0 & 0 & 0.25 & -0.396 & 0 & 0 \\ 0 & 0 & 0.065 & 0.896 & -0.131 & 0.207 & 0 & 0 & 0.065 & -0.104 & 0 & 0 \\ 0 & 0 & 0.315 & 0.5 & 0.396 & 0 & 0 & 0 & 0.315 & -0.5 & 0 & 0 \\ 0 & 0 & 0.207 & 0.207 & 0 & 0.586 & 0 & 0 & -0.207 & 0.207 & 0 & 0 \\ 0 & 0 & 0 & 0 & 1 & 0 & 0 & 0 & 0 & 0 & 0 & 0 \\ 0 & 0 & 0 & 0 & 0 & 1 & 0 & 0 & 0 & 0 & 0 & 0 \\ 0 & 0 & 0 & 0 & 0 & 0 & 1 & 0 & 0 & 0 & 0 & 0 \\ 0 & 0 & 0 & 0 & 0 & 0 & 0 & 1 & 0 & 0 & 0 & 0 \\ 0 & 0 & 0.25 & 0.396 & 0.5 & 0.207 & 0 & 0 & 0.25 & -0.604 & 0 & 0 \\ 0 & 0 & -0.065 & -0.104 & 0.131 & 0.207 & 0 & 0 & -0.065 & 0.896 & 0 & 0 \\ 0 & 0 & 0 & 0 & 0 & 0 & 0 & 0 & 1 & 0 & 0 & 0 \\ 0 & 0 & 0 & 0 & 0 & 0 & 0 & 0 & 0 & 1 & 0 & 0 \\ 0 & 0 & 0 & 0 & 0 & 0 & 0 & 0 & 0 & 0 & 1 & 0 \\ 0 & 0 & 0 & 0 & 0 & 0 & 0 & 0 & 0 & 0 & 0 & 1 \end{Bmatrix}$$

Using this transformation matrix and applying the reduction process given by equations 2.27 and 2.28, the reduced stiffness matrix K_c and load vector F_c can be created. When the boundary conditions are applied, these two terms reduce to:

$$K_c = \begin{Bmatrix} 3.045\times10^9 & -5\times10^8 & -2.261\times10^9 & -4.142\times10^8 & -7.65\times10^7 & 2.071\times10^8 \\ -5\times10^8 & 3.121\times10^9 & 0 & -8.284\times10^8 & -2.071\times10^8 & 4.142\times10^8 \\ -2.261\times10^9 & 0 & 5.937\times10^9 & 0 & -2.261\times10^9 & 0 \\ -4.142\times10^8 & -8.284\times10^8 & 0 & 5.071\times10^9 & 4.142\times10^8 & -8.284\times10^8 \\ -7.65\times10^7 & -2.071\times10^8 & -2.261\times10^9 & 4.142\times10^8 & 3.045\times10^9 & 5\times10^8 \\ 2.071\times10^8 & 4.142\times10^8 & 0 & -8.284\times10^8 & 5\times10^8 & 3.121\times10^9 \end{Bmatrix}$$

$$F_c = \begin{Bmatrix} 3.265\times10^3 \\ -3.447\times10^4 \\ -6.53\times10^3 \\ -3.107\times10^4 \\ 3.265\times10^3 \\ -8.446\times10^4 \end{Bmatrix}$$

These operate on the degrees of freedom numbered 3, 4, 9, 10, 15, 16 and by using K_c and F_c the reduced problem can be solved to create the displacement vector u_1 which, after restoring the fixed values on the degrees of freedom 1, 2, 11, 12, 17, 18, allows equation 2.21 to be used to create the displacement vector u_2. These two vectors are now given by:

$$u_1 = \begin{Bmatrix} 0 \\ 0 \\ 3.217 \times 10^{-6} \\ -9.216 \times 10^{-6} \\ 4.147 \times 10^{-6} \\ -1.334 \times 10^{-5} \\ 0 \\ 0 \\ 1.056 \times 10^{-5} \\ -3.3126 \times 10^{-5} \\ 0 \\ 0 \end{Bmatrix} \qquad u_2 = \begin{Bmatrix} 1.36 \times 10^{-5} \\ -2.629 \times 10^{-5} \\ 2.153 \times 10^{-5} \\ -1.772 \times 10^{-5} \\ 3.2 \times 10^{-5} \\ -7.081 \times 10^{-5} \end{Bmatrix}$$

These can be combined to given the complete set of nodal displacements as shown in Table 2.5 which is identical to results shown in Table 2.2

Table 2.5 Final displacement vector.

Degree of freedom	Nodal displacements (metres)
1	0.0
2	0.0
3	3.217×10^{-6}
4	-9.216×10^{-6}
5	1.36×10^{-5}
6	-2.628×10^{-5}
7	2.153×10^{-5}
8	-1.772×10^{-5}
9	4.147×10^{-6}
10	-1.334×10^{-5}
11	0.0
12	0.0
13	3.2×10^{-5}
14	-7.081×10^{-5}
15	1.056×10^{-5}
16	-3.126×10^{-5}
17	0.0
18	0.0

where the structural problem is solved using the complete structure and load vector with no reduction.

2.3.2.2 Temptations

Engineers often want to have a quick assessment of the influence of the loads on the structural deformations, stress levels, etc. In these circumstances it is tempting to reduce the problem size artificially by using an ad-hoc approach. A first thought might be to remove loads that are relatively insignificant when compared to the remaining set of dominant loads. This is equivalent to setting the loads $F_2 = 0$ in equations 2.21 and 2.22 so that the displacement fields have the following errors:

$$u_2(\text{error}) = K_{22}^{-1}F_2$$
$$u_1(\text{error}) = -\{K_{11} - K_{12}K_{22}^{-1}K_{12}^T\}^{-1}K_{12}K_{22}^{-1}F_2$$

This shows that omitting what are considered insignificant loads can have a significant effect on the solution depending on the terms in the stiffness matrix.

The second temptation is to set those nodal displacements, thought likely to be small in the solution field, equal to zero. This now means that loads associated with these fixed displacements, F_2 in the above analysis, are now the reactions at fixed nodes. Manipulating equation 2.21, these reaction terms are found to be given by:

$$F_2 = R = K_{12}K_{11}^{-1}F_1$$

If an analyst has gone down this path, then the reactions R will need to be examined in order to ensure that these are relatively small.

In the case of both of these 'temptations', the message is clear: resist temptation unless there are very good reasons for adopting one of these ad-hoc reduction methods and an error assessment is performed.

2.3.3 Sub-structures

A large aircraft such as the A380 is designed and manufactured by a large number of companies working in harmony. This process of using a multi-company consortium to create a complex component is not unique to the aircraft industry and is employed in many industrial sectors. It has the advantage that appropriate expertise can be drawn into the design and manufacturing process and it spreads the financial risk. Each consortium member has design responsibility and, in the case of the structural components, responsibility for creating an appropriate finite element model for its own component. Although these models are independently created, they do have to be combined, in some way, into a single representation of the entire structure. This is achieved by considering each of the models as a sub-structure or component model of the total finite element model.

Figure 2.15 gives an outline illustration of an aircraft wing in which five companies are involved and each has its own finite element analysis model. These individual models are termed sub-structures so that the model created by company 1 is sub-structure 1, that by company 2, sub-structure 2, and so forth. Eventually these sub-structures need to be linked so that the wing, as a whole, is correctly analysed. This is done through a process of condensation that allows one sub-structure to be inserted into the stiffness matrix and load vector of a second. Thus sub-structure 1 can be condensed and inserted into the matrix and load vectors of sub-structure 2; this combined stiffness and load vector can, in its turn, be condensed and inserted into the stiffness matrix and load vector for sub-structure 3. This is a sequential process that continues until all the sub-structural models have been combined in a 'master' model; in the present case the 'master' model would be with (possibly) company 5. Although for clarity of explanation it is convenient to describe this process as a linked sequence implemented chronologically, in practice, the assemblage of the 'master' model is often done in a more direct manner.

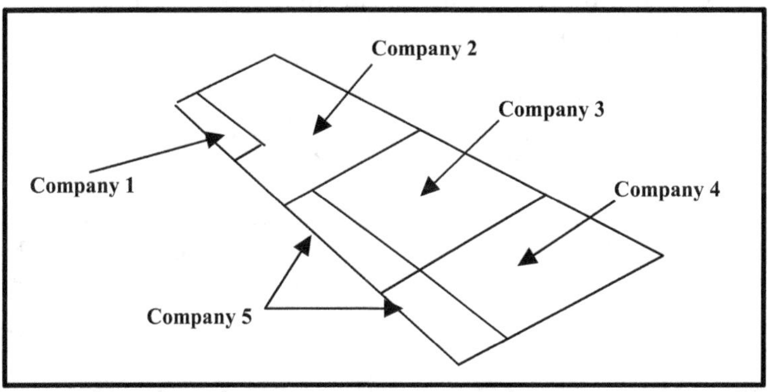

Figure 2.15 Multi-company wing analysis.

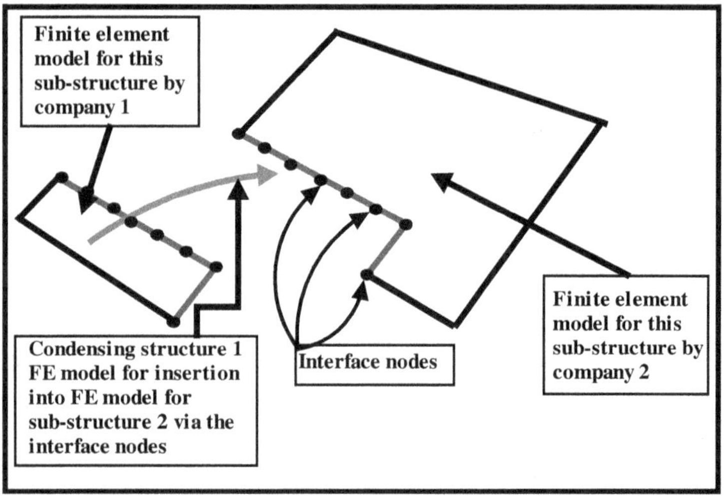

Figure 2.16 Sub-structure interfaces.

In order to see a little more clearly what is taking place in this process, consider the two sub-structures, sub-structure 1 and sub-structure 2, associated with the aileron and the outer wing structure. The finite element models are connected at the interface line indicated in Figure 2.16. Although the elements used in the two structural models may be different, the interface nodes are common to both structures and the degrees of freedom are the same at each corresponding interface node.

In order to develop the sub-structure process, we denote the displacements at the interface nodes as u_c for both sub-structure 1 and 2: u_1^1 for the remaining displacements in sub-structure 1 and u_1^2 for the corresponding displacements in sub-structure 2. Partitioning the stiffness matrix and load vector for sub-structure 1:

$$\begin{Bmatrix} K_{11}^1 & K_{12}^1 \\ (K_{12}^1)^T & K_{22}^1 \end{Bmatrix} \begin{Bmatrix} u_1^1 \\ u_c \end{Bmatrix} = \begin{Bmatrix} F_1^1 \\ F_2^1 \end{Bmatrix}$$

Condensing these equations with respect to the nodes associated with the interface displacements u_c gives an equation equivalent to 2.26:

$$K_c^1 u_c = F_c^1 \tag{2.30}$$

This process has condensed the stiffness matrix and load vector for sub-structure 1 into a stiffness matrix and load vector for an element defined in terms of the nodal displacements u_c.

Turning to sub-structure 2 and partitioning the stiffness matrix and load vector so that terms associated with the interface displacements can be identified gives:

$$\begin{Bmatrix} K_{11}^2 & K_{12}^2 \\ (K_{12}^2)^T & K_{22}^2 \end{Bmatrix} \begin{Bmatrix} u_1^2 \\ u_c \end{Bmatrix} = \begin{Bmatrix} F_1^2 \\ F_2^2 \end{Bmatrix} \tag{2.31}$$

Equation 2.30 can be added to 2.31 in the normal manner to give:

$$\begin{Bmatrix} K_{11}^2 & K_{12}^2 \\ (K_{12}^2)^T & K_{22}^2 + K_c^1 \end{Bmatrix} \begin{Bmatrix} u_1^2 \\ u_c \end{Bmatrix} = \begin{Bmatrix} F_1^2 \\ F_2^2 + F_2^2 \end{Bmatrix} \tag{2.32}$$

This matrix equation has combined the two models for sub-structure 1 and sub-structure 2 into a single matrix equation defined in terms of displacements and loads at the nodes of sub-structure 2 only. If it were required, this process could be repeated at the interface nodes between sub-structure 2 and sub-structure 3 to produce a set of equations equivalent to 2.31 but involving the nodes within sub-structure 3 only. As indicated above, this process can be repeated until a final boundary is encountered that allows the solution process to be commenced. This involves solving the final matrix equation to generate the displacements for the final structure and then back-substituting using equation 2.21, sub-structure by sub-structure, to generate the displacements associated with each independent structure. The process yields the full set of global displacements and the remaining tasks required to achieve a complete set of analysis results by picking up the stage in the 'analysis loop' of Figure 2.6 at the point where the global displacements are applied to the element nodes.

In order to keep the explanation simple, the discussion of the use of sub-structuring has omitted cases where some of the nodes within those elements to be condensed are also constrained. However, incorporating constraints within the condensation process is not a major task and the overall condensation process described above does not significantly change.

REFERENCES

1. Szabó, B. and Babuška, I., *Finite Element Analysis*. 1991: John Wiley & Sons, ISBN 0471502731.
2. Chandrupatia, T.R. and Belegundu, A.D., *Introduction to Finite Elements in Engineering, 3rd edn.* 2002: Prentice Hall, ISBN 10013065919.

3. Zienkiewicz, O.C. and Taylor, R.L., *The Finite Element Method Set*. 2005: Butterworth–Heinemann, ISBN 100750664312.
4. Barthe, K.-J., *Finite Element Procedures*. 1995: Prentice Hall, ISBN 100133014584.
5. Kardestuncer, H., *Finite Element Handbook*. 1987: McGraw-Hill, ISBN 007033305.
6. Pzemieniecki, J.S., *Theory of Matrix Structural Analysis*. 1985: Dover, ISBN 0486649482.
7. Dhatt, G. and Touzot, G., *The Finite Element Method Displayed*. 1984: John Wiley & Sons, ISBN 0471901105.

3

Overview of Dynamic Analysis

3.1 INTRODUCTION

This chapter continues the discussion opened in Chapter 2 and is concerned with how computers solve structural dynamics problems employing the Finite Element Method. As with Chapter 2 the displacement Finite Element Method only is addressed. The aim is to provide the reader with adequate information relating to the solution of structures modelled by finite elements exhibiting dynamic behaviour. More comprehensive treatments of this subject matter can be found in the standard texts on the Finite Element Method, some of which are referenced in Chapters 1 and 2. A book specifically devoted to finite element vibration analysis that is an excellent and comprehensive coverage of this subject is listed as reference [1] and this text amplifies the topics introduced in this overview. An alternative approach to this analysis area which also has a finite element component can be found in the book by Geradin [2]. Once again, the comprehensive handbook on the Finite Element Method can be consulted in reference [3].

3.2 ELEMENT MASS MATRIX

In Section 2.2.1, the fundamentals of the application of the Finite Element Method to the solution of static analysis problems were introduced via the 'direct approach' and used simple spring elements.

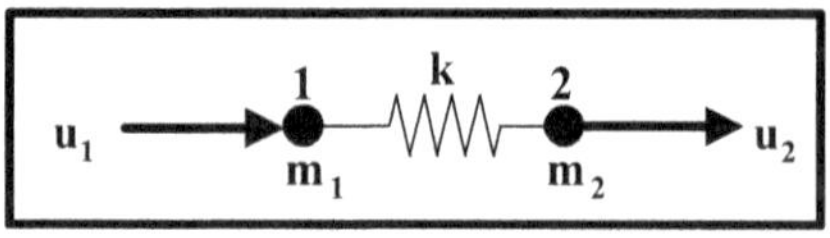

Figure 3.1 Inertia loaded spring.

This same approach is now used to discuss how finite elements can be employed in the solution of structural problems in the presence of inertia forces. Consider the simple spring–mass system shown in Figure 3.1 where, for simplicity, there are no externally applied forces, only inertia forces that are generated through the motion of the two masses m_1 and m_2 at nodes 1 and 2 and the system is undamped. The spring stiffness is given by k and the nodes have displacements u_1 and u_2 as shown in Figure 3.1 with associated nodal velocities and accelerations given by $\dot{u}_1, \dot{u}_2, \ddot{u}_1, \ddot{u}_2$ respectively. Because the problem now has velocities and accelerations, these terms and the displacements are functions of time and should, correctly, be written as $u_1(t)$, $u_2(t)$, $\dot{u}_1(t), \dot{u}_2(t), \ddot{u}_1(t), \ddot{u}_2(t)$. Unless required for clarity the bracketed time will be omitted from the equations developed herein.

In this dynamic situation inertia forces replace the applied forces used in equations 2.1 and 2.2 in Chapter 2 so that these become:

$$m_1\ddot{u}_1 + ku_1 - ku_2 = 0 \tag{3.1}$$

and:

$$m_2\ddot{u}_2 + ku_2 - ku_1 = 0 \tag{3.2}$$

Combining these simple equations into a matrix formulation gives:

$$\begin{Bmatrix} m_1 & 0 \\ 0 & m_2 \end{Bmatrix} \begin{Bmatrix} \ddot{u}_1 \\ \ddot{u}_2 \end{Bmatrix} + \begin{Bmatrix} k & -k \\ -k & k \end{Bmatrix} \begin{Bmatrix} u_1 \\ u_2 \end{Bmatrix} = \begin{Bmatrix} 0 \\ 0 \end{Bmatrix} \tag{3.3}$$

The term $\begin{Bmatrix} m_1 & 0 \\ 0 & m_2 \end{Bmatrix}$ is the mass matrix for this simple element, the element stiffness matrix and nodal displacement vector remain unchanged as defined in Section 2.2.1 in Chapter 2 and $\begin{Bmatrix} \ddot{u}_1 \\ \ddot{u}_2 \end{Bmatrix}$ is the vector of nodal accelerations. When the mass matrix is multiplied by the vector of nodal accelerations, it gives a set of inertia forces that are generated when the nodes of this simple spring–mass combination are moved from a position of rest.

In matrix equation 3.3 the stiffness matrix is singular as explained in Section 2.2.1; however, the mass matrix is non-singular so that this equation can represent a physically realisable dynamic structural response. This feature is exploited in the next section where the free vibration of a structure is considered.

If this simple spring is subjected to forcing terms then the nodal forces come back into play but, unlike in Chapter 2, these are now time-varying functions and equation 3.3 becomes:

$$\begin{Bmatrix} m_1 & 0 \\ 0 & m_2 \end{Bmatrix} \begin{Bmatrix} \ddot{u}_1(t) \\ \ddot{u}_2(t) \end{Bmatrix} + \begin{Bmatrix} k & -k \\ -k & k \end{Bmatrix} \begin{Bmatrix} u_1(t) \\ u_2(t) \end{Bmatrix} = \begin{Bmatrix} f_1(t) \\ f_2(t) \end{Bmatrix}$$

or, using a more general form for this term, we have:

$$M\ddot{u}(t) + Ku(t) = F(t) \tag{3.4}$$

Finally, the structure will normally have some form of damping associated with its dynamic movement. This may be due to the presence of an internal damping action as the material undergoes internal friction. It can also be due to the fact that the structure is moving in an external medium such as air, which also dampens the structural response. The general equation 3.4 is modified and becomes:

$$M\ddot{u}(t) + C\dot{u}(t) + Ku(t) = F(t) \tag{3.5}$$

The new matrix denoted by C is the damping matrix and changes its nature depending upon the type and nature of the damping action. Deciding on the terms that go into this matrix is one of the areas that require experience and one of the major sources of error when undertaking an analysis where damping plays a significant role.

In the normal analysis situation where many elements are employed, both the mass and the stiffness matrices are assembled into global matrices following the procedure illustrated in Section 2.2.2 in Chapter 2 where the stiffness matrix only is considered. This process is now illustrated for the special case of free vibration analysis in the next section.

3.2.1 Free Undamped Vibrations

In order to find the natural frequencies of a vibrating structure, two assumptions are made. First, that there are no external forcing terms or damping and, second, that the nodes move in a manner that can be represented by harmonic motion. If we stay with the single spring–mass system, the second assumption implies that:

$$\begin{Bmatrix} u_1(t) \\ u_2(t) \end{Bmatrix} = \begin{Bmatrix} u_1 e^{i\omega t} \\ u_2 e^{i\omega t} \end{Bmatrix} \tag{3.6}$$

Putting this into equation 3.3 yields:

$$\left\{ \begin{Bmatrix} k & -k \\ -k & k \end{Bmatrix} - \omega^2 \begin{Bmatrix} m_1 & 0 \\ 0 & m_2 \end{Bmatrix} \right\} \begin{Bmatrix} u_1 \\ u_2 \end{Bmatrix} = \begin{Bmatrix} 0 \\ 0 \end{Bmatrix} \tag{3.7}$$

where $\begin{Bmatrix} u_1 \\ u_2 \end{Bmatrix}$ is a vector of displacement amplitudes and ω a natural frequency. This is a classical eigenvalue problem with the natural frequencies being the eigenvalues and the mode shapes the eigenvectors. There are as many natural frequencies and modal vectors as there are degrees of freedom; for this simple single spring there are two. One of these represents the rigid body movement where both nodes move without any relative movement while the second is the only remaining mode where one node does move relative to the other.

We can now move to a more general case with a large number of elements giving rise to a problem with 'n' degrees of freedom. As in the static case the global stiffness matrix K is banded and symmetric but, in addition, there is an assembled global mass matrix M made up from the individual element mass matrices in an identical manner to that used for the construction of the matrix K. If we now represent the global vector of displacement amplitudes by the term U, where $U^T = \{u_1, u_2, \ldots, u_n\}$, a more generalised form for equation 3.7 can now be written:

$$\{K - \omega^2 M\} U = 0 \tag{3.8}$$

The solution of this problem will result in the generation of a set of natural frequencies ω_p with $p = 1 \ldots n$. Corresponding to each natural frequency there is a modal vector U_p which also has to be calculated. As implied above, the terms in the modal vector are *not absolute but relative values*; that is, we can only compute how the nodes move relatively to one another and there is no outside datum against which to calibrate values.

There are a wide range of methods to solve the eigenvalue problem defined by equation 3.8 and these are iterative but normally solved for the frequencies and mode shapes simultaneously. Because the methods

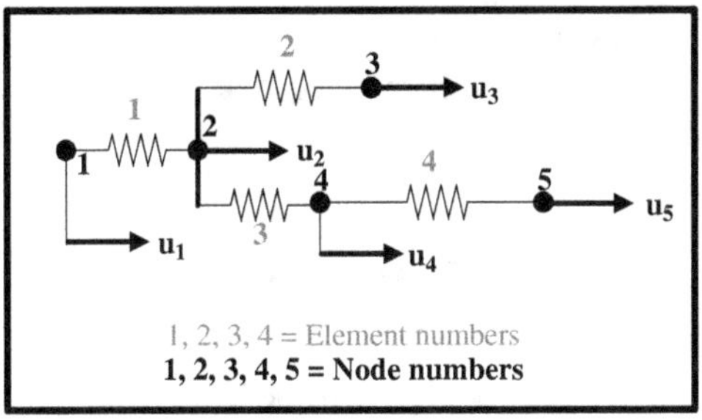

Figure 3.2 Spring–mass analysis problem.

are iterative, they can consume a considerable amount of computing time that *rapidly increases with increasing number of degrees of freedom*. If both a static and dynamic structural analysis is required, it is good practice, therefore, to use separate and different models for these analyses as the static case will normally require more elements than the dynamic case; we return to this point in Chapter 8.

In order to illustrate the solution and interpretation process let us consider again the spring problem used in Section 2.2.2 in Chapter 2, but remove the applied external nodal forces and place concentrated masses at these nodes as shown in Figure 3.2.

The spring element stiffnesses from Section 2.2.2 are retained and the masses at the five nodes are $m_1 = m_2 = m_3 = m_4 = m_5 =$ 100 kilograms. Using these values the global stiffness matrix K and mass matrix M in matrix equation 3.8 are:

$$K = \begin{Bmatrix} 1.5\times10^3 & -1.5\times10^3 & 0 & 0 & 0 \\ -1.5\times10^3 & 5.7\times10^3 & -2\times10^3 & -2.2\times10^3 & 0 \\ 0 & -2\times10^3 & 2\times10^3 & 0 & 0 \\ 0 & -2.2\times10^3 & 0 & 3.4\times10^3 & -1.2\times10^3 \\ 0 & 0 & 0 & -1.2\times10^3 & 1.2\times10^3 \end{Bmatrix}$$

$$M = \begin{Bmatrix} 100 & 0 & 0 & 0 & 0 \\ 0 & 300 & 0 & 0 & 0 \\ 0 & 0 & 100 & 0 & 0 \\ 0 & 0 & 0 & 200 & 0 \\ 0 & 0 & 0 & 0 & 100 \end{Bmatrix}$$

Placing these matrices into matrix equation 3.8 and solving the eigenproblem gives a set of five natural frequencies, as there are five nodes each with a single degree of freedom:

$$\begin{aligned}
\omega_1 &= 0\,\text{radians/second}\\
\omega_2 &= 2.793\,\text{radians/second}\\
\omega_3 &= 4.089\,\text{radians/second}\\
\omega_4 &= 4.744\,\text{radians/second}\\
\omega_5 &= 5.998\,\text{radians/second}
\end{aligned}$$

and the associated normalised modes (the normalisation process is explained in the next paragraph) are:

$$U_1=\begin{Bmatrix}0.035\\0.035\\0.035\\0.035\\0.035\end{Bmatrix} U_2=\begin{Bmatrix}-0.038\\-0.018\\-0.03\\0.025\\0.073\end{Bmatrix} U_3=\begin{Bmatrix}-0.079\\0.009\\0.056\\0.006\\-0.016\end{Bmatrix} U_4=\begin{Bmatrix}0.012\\-0.006\\0.047\\-0.048\\0.055\end{Bmatrix} U_5=\begin{Bmatrix}-0.029\\0.04\\-0.051\\-0.028\\0.014\end{Bmatrix}$$

The first frequency is simply a rigid body mode where the nodes move without relative movement to each other; the remaining four are vibration modes representing the first through to the fourth natural frequency. This analysis has taken place on the assumption that the structure is not restrained and this has given rise to the rigid body motion. If the structure had been located in such a manner that some of the nodes were fixed, then the stiffness and mass matrices would be reduced by the imposition of restraints on the appropriate nodal displacements. In such situations some or all of the rigid body modes would be suppressed.

There are a number of checks that can now be made to ensure that everything has worked well. Clearly the maximum strain and kinetic energies of the system must be equal for each of the frequencies, as a structure undergoing free vibration is simply exchanging kinetic energy for strain energy and vice versa. Thus:

$$0.5U_i^TKU_i = 0.5\omega_i^2U_i^TMU_i \qquad \text{for } i = 1\ldots5 \tag{3.9}$$

Taking the case of U_5, then:

$$\begin{aligned}
\text{the strain energy (SE)} &= 0.5U_5^TKU_5 = 17.987\,\text{N}\cdot\text{m}\\
\text{and the kinetic energy (KE)} &= 0.5\omega_5^2U_5^TMU_5 = 17.987\,\text{N}\cdot\text{m}
\end{aligned}$$

showing that these two quantities agree as required.

The second requirement is that the mode shapes should be mutually orthogonal, which can be demonstrated by showing that:

$$\begin{aligned} U_i^T M U_j &= 0 \qquad \text{if } i \neq j \\ &\text{and} \\ U_i^T M U_j &\neq 0 \qquad \text{if } i = j \end{aligned} \tag{3.10}$$

In fact, we can normalise the mode shapes such that $U_i^T M U_i = 1$, which is the usual practice in the analysis of vibrating structures, and, further, the vectors U_1 through to U_5 derived above have been normalised in this manner. Normalised mode shapes also have the useful property that:

$$\begin{aligned} U_i^T K U_j &= 0 \qquad \text{if } i \neq j \\ &\text{and} \\ U_i^T K U_j &= \omega_j^2 \quad \text{if } i = j \end{aligned}$$

It can be readily verified that the results from the above four-spring vibration analysis do satisfy these simple rules.

Having obtained a set of mode shapes it is now possible to calculate the forces in the springs for each of the modes separately by taking the components from the total modal vector for an individual element and multiplying these into the element stiffness matrix. As an example, the stiffness matrix for element 1 is:

$$k_1 = \begin{Bmatrix} 1.5 \times 10^3 & -1.5 \times 10^3 \\ -1.5 \times 10^3 & 1.5 \times 10^3 \end{Bmatrix}$$

To get the force in this spring associated with the first natural frequency we take the terms from U_2, i.e. $u_1 = -0.038$ and $u_2 = -0.018$, and take note of equation 2.31 to form the element modal vector:

$$\tilde{u}(t) = \begin{Bmatrix} u_1 \\ u_2 \end{Bmatrix} e^{i\omega t}$$

Then the bar forces are:

$$f(t) = k_1 \tilde{u}(t) = \begin{Bmatrix} -30 \\ 30 \end{Bmatrix} e^{i\omega t} \text{ newtons}$$

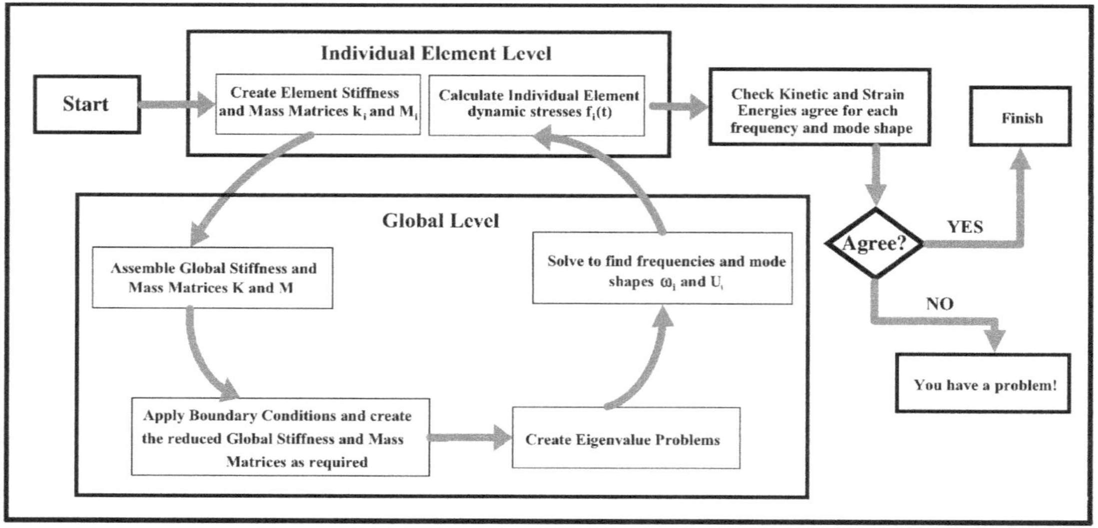

Figure 3.3 Free vibration analysis loop.

For more complex structural elements, bars, beams, etc., these forces would be replaced by structural dynamic strains and stresses.

The analysis loop associated with this type of free vibration analysis is shown in Figure 3.3.

3.3 ADDITIONAL INFORMATION THAT CAN BE EXTRACTED TO SUPPORT A DYNAMIC FINITE ELEMENT ANALYSIS

There is a great deal of information that lies hidden within the results of a finite element analysis of a freely vibrating structure. Within the limited context of this chapter two are noted as having value in the process of ensuring that the results are of high quality. The first of these is the Sturm Sequence property that allows us to interrogate the results for missing vibration modes. The other is the Rayleigh Quotient, which has a number of interesting properties and roles to play in supporting the solution of vibration problems but, in this section, it is deployed in a fairly limited manner.

3.3.1 Sturm Property Check

Although it is beyond the scope of this book to explain how finite element codes solve free vibration problems, it is worth recalling the fact that these methods are iterative. That is, they have a starting point and then employ an algorithm that stage by stage works its way to a solution that provides all the required frequencies and modes. A lot of effort has gone into the development of these solution methods and, by and large, they are very robust. However, for complex problems with a large number of degrees of freedom and, particularly, where there are sets of frequencies that are close, it is possible that the solution set of frequencies and modes can be incomplete with some of the frequencies and associated modes missing. It is, therefore, necessary to perform a check in order to see if an important frequency or set of frequencies is missing within the frequency range regarded by the designer as critical. For example, in the case of an analysis undertaken to support the design of a helicopter, a solution set that does not capture all the natural frequencies close to the blade-passing frequency or close to the human head-shake frequency could have serious consequences and the designer would target these as being critical ranges.

Most finite element systems have a facility that allows missing frequencies to be identified using a property noticed by the French mathematician Jacques Sturm in 1829. This states that the number of eigenvalues, i.e. the squares of the natural frequencies, less than a real number μ is given by the number of negative terms in a diagonal matrix D where:

$$K - \mu M = LDL^T$$

and where L is a lower triangular matrix with ones along the diagonal. A natural extension of this property is that the number of eigenvalues between two numbers μ_i and μ_j is given by the difference in the number of negative terms in D_i and D_j obtained from constructing:

$$K - \mu_i M = L_i D_i L_i^T$$

and:

$$K - \mu_j M = L_j D_j L_j^T$$

As an example, let us consider the simple spring problem shown in Figure 3.2 and test for the number of eigenvalues that lie between the number $\mu_i = 24$ and $\mu_j = 6$. For these two numbers, the D matrices are:

$$D_i = \begin{Bmatrix} -900 & 0 & 0 & 0 & 0 \\ 0 & 1 \times 10^3 & 0 & 0 & 0 \\ 0 & 0 & -4.4 \times 10^3 & 0 & 0 \\ 0 & 0 & 0 & -1.84 \times 10^3 & 0 \\ 0 & 0 & 0 & 0 & -417.391 \end{Bmatrix}$$

and:

$$D_j = \begin{Bmatrix} 900 & 0 & 0 & 0 & 0 \\ 0 & 1.4 \times 10^3 & 0 & 0 & 0 \\ 0 & 0 & -1.457 \times 10^3 & 0 & 0 \\ 0 & 0 & 0 & 5.522 \times 10^3 & 0 \\ 0 & 0 & 0 & 0 & 339.205 \end{Bmatrix}$$

From these results we know that below $\omega = 4.9$ radians/second there are four natural frequencies and below $\omega = 2.45$ there is one natural frequency (the rigid body mode), thus, between these two values, there are three actual natural frequencies. A glance at Section 3.2.1 reveals that there are, indeed, three natural frequencies between these two values.

3.3.2 Rayleigh Quotient

The Rayleigh Quotient, first noted by John Strutt, the 3rd Lord Rayleigh, is a number $R(v)$ obtained from:

$$R(v) = \frac{v^{\mathrm{T}} K v}{v^{\mathrm{T}} M v}$$

where v is an arbitrary vector and $R(v)$ has the property that it lies between the maximum and minimum of the squared natural frequencies of the structure. Thus if λ_1 is the lowest squared natural frequency of the structure and λ_n is the maximum, then:

$$\lambda_1 \leq R(v) \leq \lambda_n$$

If the vector v represents an actual mode shape for one of the natural frequencies, then $\sqrt{R(v)}$ gives the value of the natural frequency. Taking the spring problem used in Section 3.2.1 above, if the mode shape for the highest natural frequency is selected, i.e. $v^T = \{-0.029\ 0.04\ -0.051\ -0.028\ 0.014\}$, then $\sqrt{R(v)} = 5.998$ radians/second or ω_5. If we now change this vector to $u^T = \{-0.03\ 0.05\ -0.06\ -0.04\ 0.02\}$, then $\sqrt{R(u)} = 5.988$ radians/second, which is less than the value of the maximum natural frequency of this simple spring model as predicted. The fact that the Rayleigh Quotient has not changed much by varying the mode shape or eigenvector is not coincidental. If u is taken as an approximation to v to within a small error ε, then we have:

$$u = v + \varepsilon z$$

Then:

$$R(u) = (\omega_5)^2 + O(\varepsilon^2)$$

In other words, errors injected into the mode shapes have a relatively small impact on the natural frequency when this is estimated using the Rayleigh Quotient.

This quotient has a number of practical uses in supporting a finite element analysis of a complex structure:

1. If finite element results are being compared to test results, the Rayleigh Quotient can be computed using the test mode shapes and the finite element mass and stiffness matrices to assist in the correlation of test and finite element results.
2. If the Sturm property check indicates that the finite element solution procedure has missed certain modes within a specific range, this quotient can be used to explore the region within which the modes are known to exist in an effort to locate them.
3. In principle, the Rayleigh Quotient can provide upper and lower values on the natural frequencies of structure; however, this may not be practical in the case of large-scale structures.

3.4 FORCED RESPONSES

We now turn to dynamic problems involving forcing terms without damping so that the analysis problem is defined by equation 3.4 taken at the global level. Broadly, there are two approaches available to solve this type of structural analysis problem. One assumes that the responses are harmonic and employs the normalised mode shapes to act as basic shape functions for the structural response. This is known as modal analysis. The second makes no assumptions on the nature of the responses and directly solves the dynamic equation 3.5 by direct integration using finite differences.

3.4.1 Modal Analysis

In order to solve a forced vibration structural analysis problem for a given structure subject to specified external loads using modal analysis, it is first necessary to solve the free vibrations analysis problem for the same structure as described in Section 3.3. The solution of the free vibrations problem creates a set of normalised mode shapes defined above as U_i where $i = 1 \ldots n$ and the problem has 'n' degrees of freedom. Under the influence of the forcing term $F(t)$ the structure is assumed to respond with a time-dependent nodal displacement vector $U(t)$ as illustrated in equation 3.4. The approach assumes that the structural response displacement vector $U(t)$ can be replicated by using the mode shapes as trial functions, thus:

$$U(t) = \sum_{i=1}^{n} U_i Z_i(t) \tag{3.11}$$

where the terms Z_i are weighting functions that are currently of unknown value but will be given values during the modal solution process. If we now create a matrix of modal vectors and a vector of weighting functions, thus:

$$\phi = \{ U_1 \quad U_2 \quad \ldots \quad U_n \}$$

$$Z(t) = \begin{Bmatrix} Z_1(t) \\ Z_2(t) \\ \vdots \\ Z_n(t) \end{Bmatrix}$$

then we can write equation 3.11 as:

$$U(t) = \phi Z(t) \tag{3.12}$$

Substituting equation 3.12 into 3.4 gives:

$$M\phi\ddot{Z}(t) + K\phi Z(t) = F(t)$$

This can now be pre-multiplied by the matrix ϕ^T to give:

$$\phi^T M\phi\ddot{Z}(t) + \phi^T K\phi Z(t) = \phi^T F(t) \tag{3.13}$$

Because the normalised modes have been used, then:

$$\phi^T M\phi = I$$

and

$$\phi^T K\phi = [\omega^2]$$

where:

$$[\omega^2] = \begin{Bmatrix} \omega_1^2 & 0 & . & 0 \\ 0 & \omega_2^2 & . & 0 \\ . & . & . & . \\ 0 & . & . & \omega_n^2 \end{Bmatrix}$$

with ω_i the free vibration natural frequencies of the structure. Equation 3.13 now becomes:

$$I\ddot{Z} + [\omega^2]Z = \phi^T F \tag{3.14}$$

where, for convenience, the bracketed time (t) has been omitted. Equations 3.14 have a very advantageous property in that they are de-coupled and can be written in the form of n separate scalar equations:

$$\ddot{Z}_i + \omega_i^2 Z_i = \phi_i^T F \tag{3.15}$$

for $i = 1 \ldots n$. This can be further reduced by using:

$$P_i = \phi_i^T F \tag{3.16}$$

so that equations 3.15 become:

$$\ddot{Z}_i + \omega_i^2 Z_i = P_i \tag{3.17}$$

For known modal loading functions $P_i(t)$ these equations can be solved, as explained in standard vibration theory books such as references [4] and [5], by direct integration to yield the weighting terms Z_i and the resulting dynamic displacement field for the structure obtained from equation 3.12. The dynamic strains and stresses at the element level follow using the same approach as for the free vibration problem.

This method is very effective for solving forced vibration analysis problems but, as already noted, does increase the effort required as a full free vibration analysis has to be performed before the forced vibration process is brought into play. Nevertheless, it allows an analyst to examine the role and influence of the various free vibration modes on the forced response. If the response is not what the design requires, the analyst can look at how the modes are participating in the response and decide which frequencies to modify in order to achieve the desired structural response. The approach also gives, through the use of the modal loading functions, an indication of how the applied loads generate modal responses.

Modal methods have a very wide range of application outside of the structural field and are used, for example, by theoretical physicists in quantum mechanics. As a result, the methods underpinning modal analysis are very robust.

3.4.2 Direct Integration

Modal analysis breaks down when the forcing terms cannot be considered able to be represented by some form of harmonic approximation. This might occur in situations where the terms in the forcing function are highly discontinuous, in which case the analyst would need to employ a direct integration method. One method for undertaking a direct integration approach would be to use a central difference finite difference method where the 'n' nodal velocities and accelerations in the global model are approximated, respectively, in terms of the nodal displacements by:

$$\dot{u}_i(t) = \frac{u_i(t+\Delta t) - u_i(t-\Delta t)}{2\Delta t}$$
$$\ddot{u}_i(t) = \frac{u_i(t+\Delta t) - 2u_i(t) + u_i(t-\Delta t)}{(\Delta t)^2}$$

for $i = 1 \ldots n$. Inserting these terms into equation 3.4 gives:

$$\frac{M\{u(t+\Delta t) - 2u(t) + u(t-\Delta t)\}}{(\Delta t)^2} + Ku(t) = F(t)$$

and this can be manipulated to give the update formula:

$$u(t+\Delta t) = (\Delta t)^2 M^{-1}F(t) - \left\{(\Delta t)^2 M^{-1}K - 2\right\}u(t) - u(t-\Delta t) \quad (3.18)$$

This predicts what the vector of global nodal displacements will be at the end of the next time step providing the current and immediate past values of these displacements are known together with the current values of the nodal forcing terms contained in $F(t)$. Although not included in this form for the direct integration method, it would be possible to include time variations in the terms within the global mass and stiffness matrix if such changes are taking place.

During the middle phase of the direct method all the terms in equation 3.18 are known. However, at the start, when $t = 0$ the term $u(t - \Delta t)$ is not known because it does not exist. To get round this we note that the second term in our finite difference definitions for $t = 0$ yields:

$$u(0-\Delta t) = u(0) - (\Delta t)\dot{u}(0) + \frac{(\Delta t)^2\ddot{u}(0)}{2}$$

and from equation 3.4 we have:

$$\ddot{u}(0) = M^{-1}F(0) - M^{-1}Ku(0)$$

This requires that, in order to start the direct integration method, the analyst requires the initial displacement and velocity fields to be known.

This method seems, at first sight, to be very attractive as it is simple to understand and fairly simple to implement. But it does have drawbacks compared to the modal approach. It is opaque in that it is not easy to see how to change the structure if the responses do not satisfy the design requirements. Also a non-linear forcing function can give rise to a variety of numerical instabilities in the solution process that can cause the solution method to fail. However, things get worse when damping terms are introduced and these are now considered in the next section.

3.5 DAMPED FORCED RESPONSES

The finite element analysis of structures subject to forcing terms in the presence of damping is complex and this section simply touches on some of the key issues required for a background understanding of the methods used to find solutions in this area. It is not an area that an inexperienced analyst should enter without an experienced expert on hand to provide guidance and support – particularly when things start to go wrong.

The basic set of defining equations for this class of problem is those found in equation 3.5 and, following the approach in Section 3.4, the matrices and vectors are again taken at the global level. We also follow Section 3.4 in addressing the solution of this class of problem using both a modal and direct approach.

3.5.1 Modal Analysis with Damping

Damping is modelled using a variety of functions in commercial finite element systems but in this section we are content to employ one approach that does have a fairly wide range of applicability. This assumes that the terms in the damping matrix C can be represented by the expression:

$$C = \alpha M + \beta K \tag{3.19}$$

If $\alpha = 0$ then the higher vibration modes are heavily damped and the physical situation corresponds to that encountered when, for example, a structure is subject to viscous damping from an external fluid. If $\beta = 0$, then the higher modes are lightly damped and this can be interpreted as a situation in which the damping is due to internal material damping. Inserting equation 3.19 into 3.5 yields:

$$M\ddot{u}(t) + (\alpha M + \beta K)\dot{u}(t) + ku(t) = F(t) \tag{3.20}$$

Using equations 3.12, 3.16, together with normalised modes and again, for convenience, omitting the bracketed '*t*', then 3.20 can be reduced to a scalar form similar to equations 3.17:

$$\ddot{Z}_i + \left(\alpha + \beta\omega_i^2\right)\dot{Z}_i + \omega_i^2 Z_i = P_i \tag{3.21}$$

for $i = 1 \ldots n$. The term $\left(\alpha + \beta\omega_i^2\right)$ is a modal damping constant for the ith normal mode and we now introduce the modal damping ratio ζ_i defined by:

$$\zeta_i = \frac{\alpha + \beta\omega_i^2}{2\omega_i} \tag{3.22}$$

so equation 3.21 becomes:

$$\ddot{Z}_i + 2\zeta_i\omega_i\dot{Z}_i + \omega_i^2 = P_i \tag{3.23}$$

The solution to the analysis problem is now a matter of solving '*n*' equations defined by 3.23. As with the undamped forced response case, the prerequisite is that the free vibration problem is solved in order to generate the required mode shapes. However, solving these equations is not straightforward and depends on the size of the modal damping terms and if for $i = 1 \ldots n$:

- $\zeta_i < 1$, the system is underdamped and this means that if disturbed from rest by an impact force it would exhibit some harmonic motion before returning to rest.
- $\zeta_i = 1$, the system is critically damped and this means that if disturbed from rest by an impact force it would simply return to rest without exhibiting harmonic motion.

- $\zeta_i > 1$, the system is overdamped and if disturbed by an impact force would not return to its rest position.

It is possible that certain modes are underdamped while others are critically or overdamped, but in the case of the majority of structures it is likely that the system will be entirely underdamped. The reader is referred to references [1, 2] for an explanation of how the computer sets about solving these equations. It should be noted that for the underdamped structure the forcing functions, i.e. the applied loads and the responses, are not coincidental in time giving rise to a *phase angle* between the forcing and response terms.

As in the case of the undamped forced response case, this method allows the analyst to gain some understanding of how the various modes participate in the response of the structure to the applied loads. This insight can be helpful when the responses are out of the required design response range and structural changes are required to bring the design within the design envelope.

3.5.2 Modal Damping Ratio

In order to employ a modal analysis for a damped structure, the modal damping ratios need to be available. In order to proceed, the values for α and β are required to generate equation 3.22 in a form that allows values for ζ_i to be generated for any value of 'i'. However, it is not possible to solve this problem in a way that allows more than two modal damping ratios to be accurate.

In order to demonstrate the nature of the problem, consider a structure that has the first three natural frequencies given, in radians/second, by $\omega_1 = 2$, $\omega_2 = 5$ and $\omega_3 = 10$ and that these are known to have 2%, 4% and 6% critical damping, i.e. $\zeta_1 = 0.02$, $\zeta_2 = 0.04$ and $\zeta_3 = 0.06$. Using the first two of these frequencies to calculate α and β gives values to these two terms of 0.019 and 0.015 respectively. Thus, using equation 3.22 the generating expression for the modal damping ratios is:

$$\zeta_i = \frac{(0.019 + 0.015\omega_i^2)}{2\omega_i}$$

If this expression is now used to compute the modal damping ratio for the third natural frequency by making $i = 3$, this predicts that $\zeta_3 = 0.076$, which is in error by 27%.

If the first and third natural frequencies are used to calculate α and β, these two terms are now 0.058 and 0.005 respectively so that the modal damping expression is:

$$\zeta_i = \frac{(0.058 + 0.005\omega_i^2)}{2\omega_i}$$

Setting $i = 2$ in this expression predicts $\zeta_2 = 0.019$, an error of 52%.

Whilst this example has been created for demonstration purposes, it does clearly show that the expression in 3.22 cannot give an error-free representation of all the modal damping ratios that are likely to be encountered in a real analysis. In addition, the analyst has the problem of getting realistic values for damping factors. It is easy to understand that accurately analysing the behaviour of a major structure subjected to forcing terms where damping is a significant factor is extremely difficult and likely to be error prone.

3.5.3 Direct Integration

In order to undertake a direct integration of equation 3.5 it is tempting to use simply the same approach as that undertaken in the case of the undamped force response case in Section 3.4.2. That is, to employ a central finite differencing scheme and directly apply it to equation 3.5 and manipulate the resulting update formula to allow $u(t + \Delta t)$ to be generated from known values of $u(t)$ and $u(t - \Delta t)$ together with a known loading history $F(t)$. Unfortunately, this approach creates an updating procedure that is not unconditionally stable and can accumulate errors as the iteration process progresses down the time-line. A number of updating procedures have been put forward to circumvent this difficulty and these are similar to each other so that illustrating one is sufficient for our purpose and the one selected is that due to Newmark which is also very popular with developers of general purpose finite element systems.

Newmark's Method attempts to stabilise the update process by bringing into the updating formulae information from what is likely to happen at the next iteration. This is a common method in the case of computational fluid dynamics and in the fluids field is known as up-winding. The difference formulae used by Newmark are:

$$\begin{aligned} u(t + \Delta t) &= u(t) + (\Delta t)\dot{u}(t) + (\Delta t)^2\{(0.5 - \beta)\ddot{u}(t) + \beta\ddot{u}(t + \Delta t)\} \\ \dot{u}(t + \Delta t) &= \ddot{u}(t) + (\Delta t)\{(1 - \gamma)\ddot{u}(t) + \gamma\ddot{u}(t + \Delta t)\} \end{aligned}$$

Inserting these difference terms into 3.5 gives, after some manipulation, the following update formula:

$$
\begin{aligned}
&\left\{K + \frac{\gamma}{\beta(\Delta t)}C + \frac{1}{\beta(\Delta t)^2}M\right\}u(t+\Delta t) = F(t+\Delta t) \\
&\quad + C\left\{\frac{\gamma}{\beta(\Delta t)}u(t) + \left(\frac{\gamma}{\beta} - 1\right)\dot{u}(t) + (\Delta t)\left(\frac{\gamma}{2\beta} - 1\right)\ddot{u}(t)\right\} \\
&\quad + M\left\{\frac{1}{\beta(\Delta t)^2}u(t) + \frac{1}{\beta(\Delta t)}\dot{u}(t) + \left(\frac{1}{2\beta} - 1\right)\dot{u}(t)\right\}
\end{aligned}
$$

where the terms β and γ must be selected by the user. The process is stable if $\gamma > 0.5$ and $\beta > (2\gamma + 1)/16$ but slow! Essentially the term γ is an artificial damping term and if $\gamma > 0.5$ it gives rise to positive damping and if $\gamma < 0.5$ it gives rise to negative damping. A reasonable choice for the analyst is to set $\gamma = 0.5$ and $\beta = 0.25$.

As with the modal approach, direct integration methods for complex damped forced response analysis problems require expertise and should not be attempted by the unaccompanied novice analyst. It should also be noted that the use of direct integration has the same lack of clarity with respect to the modal approach that was commented upon in the section on undamped forced responses.

3.6 REDUCING THE PROBLEM SIZE

Procedures for reducing the size of a finite element model are given consideration in Chapter 2 for the case where the design problem can be simulated by a linear static analysis. Despite the power of modern computers there are excellent reasons for employing model reduction techniques in the case of large-scale or complex linear analyses. For structures exhibiting dynamic behaviour, the case is stronger because the computing effort required to solve dynamics problems is significantly larger than required for a linear static analysis. This difference in computing requirement is highlighted when a linear finite element model with a very large number of degrees of freedom that has been developed for linear analysis is then used for the solution of dynamic cases. Using a static finite element model for dynamic analysis is not 'best practice' as explained elsewhere in this book, but it is a common procedure. It is based on the understandable argument, 'I've spent a lot of time and effort developing high-quality CAD and finite element models for my static analysis, so why throw it away to start again and introduce new modelling errors?' If this policy of employing a single finite element model for both static and dynamic analysis is pursued, then reduction methods will be required when the dynamic analysis is initiated.

As in the case of linear static analysis, one way of reducing the problem size is to exploit symmetry, the other is to follow formal reduction methods that, essentially, exploit similar procedures to those used in the static case for the reduction of variables.

3.6.1 Symmetry

In Section 2.3.1 the use of symmetry in the analysis of statically loaded structures was described. The boundary conditions for both symmetric and anti-symmetric cases, delineated there, also apply in the case of a vibrating structure. As with the static case, if the structure under consideration is symmetrical and the loads in a forced response are entirely symmetrical or anti-symmetrical, then the analysis can be performed using symmetric or anti-symmetric modes only. If the forces are not symmetrical or anti-symmetrical but the structure is symmetrical, then the analysis proceeds by combining a symmetric and anti-symmetric analysis with the dynamic loads applied to these two cases constructed following the same principles as those used to define equation 2.13. The two sets of results are then combined as with the static case to provide a complete solution.

Multiple or repeated symmetry and cyclically repeated symmetry are represented by connected sets of identical cells consisting of a single or groups of finite elements depending on the complexity of the problem. Within each cell there are interior and boundary degrees of freedom and each cell is coupled to neighbouring cells by interface degrees of freedom and forces as with the simple symmetry condition. When this class of vibration problem is solved, it is found that the structure allows propagation within certain frequency bands. These are called propagation bands. Waves outside these bands are strongly attenuated.

As with the static case, structures with components involving axially symmetric solids, such as thick shells of revolution, are very common in practice but the approach adopted for the analysis of this type of structure is different from that discussed above. This structural form requires the same type of special axially symmetric solid elements as employed in the static case with the addition of a solid element mass matrix. In the case where the axially symmetric solid is subjected to asymmetric loads, Fourier series can be employed to allow a set of component analyses to be superimposed.

3.6.2 Reducing the Number of Variables

Reducing the number of variables can be achieved by following the principles laid down in Chapter 2, Section 2.3.2.1, where a finite element stiffness matrix is reduced by a condensation process. The primary difference between the static and dynamic analysis cases is that inertia terms are brought into play in the dynamic case and these have to be included in any reduction process. As in the static case, a number of the degrees of freedom from the total analysis problem are given dominant status and these are traditionally termed master degrees of freedom u_m and the remaining degrees of freedom u_s are slaved to the masters. If the free vibration problem is considered, then the stiffness and mass matrices and the mode vector for the complete free vibration problem can be partitioned in the following manner:

$$K = \begin{Bmatrix} K_{mm} & K_{ms} \\ K_{ms}^T & K_{ss} \end{Bmatrix} \qquad M = \begin{Bmatrix} M_{mm} & M_{ms} \\ M_{ms}^T & M_{ss} \end{Bmatrix} \qquad U = \begin{Bmatrix} u_m \\ u_s \end{Bmatrix}$$

And inserting these forms into equation 3.8 gives:

$$\left\{ \begin{Bmatrix} K_{mm} & K_{ms} \\ K_{ms}^T & K_{ss} \end{Bmatrix} - \omega^2 \begin{Bmatrix} M_{mm} & M_{ms} \\ M_{ms}^T & M_{ss} \end{Bmatrix} \right\} \begin{Bmatrix} u_m \\ u_s \end{Bmatrix} = \begin{Bmatrix} 0 \\ 0 \end{Bmatrix} \tag{3.24}$$

In order to obtain a relationship between the master and slave degrees of freedom, the problem is considered to be a static analysis and the master degrees of freedom are assumed to be subject to fictitious loads, thus:

$$\begin{Bmatrix} K_{mm} & K_{ms} \\ K_{ms}^T & K_{ss} \end{Bmatrix} \begin{Bmatrix} u_m \\ u_s \end{Bmatrix} = \begin{Bmatrix} F \\ 0 \end{Bmatrix} \tag{3.25}$$

Using the lower half of the matrix equation 3.25 to relate the masters to the slaves gives:

$$u_s = -K_{ss}^{-1} K_{ms}^T u_m \tag{3.26}$$

And the transformation matrix linking the master and slave degrees of freedom to the masters alone can now be written as:

$$\begin{Bmatrix} u_m \\ u_s \end{Bmatrix} = Tu_m = \begin{Bmatrix} I \\ -K_{ss}^{-1}K_{ms}^T \end{Bmatrix} u_m \tag{3.27}$$

The transformation matrix T in equation 3.27 is identical to that derived for static condensation displayed as equation 2.29. Substituting equation 3.27 into 3.8 yields:

$$\{K - \omega^2 M\} Tu_m = 0$$

If this equation is now pre-multiplied by the transformation matrix the problem is reduced from one having $m + s$ degrees of freedom to one possessing only m degrees of freedom, thus:

$$T^T\{K - \omega^2 M\} Tu_m = T\{0\}$$

or

$$\{K_r - \omega^2 M_r\} u_m = 0 \tag{3.28}$$

where:

$$K_r = T^T KT \quad M_r = T^T MT$$

The reduced problem defined by equation 3.28 can be solved to provide the eigenvalues and eigenvectors and, therefore, the natural frequencies and mode shapes for the structure for the master degrees of freedom. The complete solution requires that terms missing from the mode shapes due to non-appearance of the slave degrees of freedom in equation 3.28 are recovered. This cannot be done by using equation 3.27 as the influence of structural inertia is missing from this equation. Taking u_m^n as the known master degrees of freedom associated with nth natural frequency ω_n and u_s^n as the currently unknown values for the slave degrees of freedom for the same natural frequency and substituting these into equation 3.24 gives:

$$\left\{\begin{Bmatrix} K_{mm} & K_{ms} \\ K_{ms}^T & K_{ss} \end{Bmatrix} - \omega_n^2 \begin{Bmatrix} M_{mm} & M_{ms} \\ M_{ms}^T & M_{ss} \end{Bmatrix}\right\} \begin{Bmatrix} u_m^n \\ u_s^n \end{Bmatrix} = \begin{Bmatrix} 0 \\ 0 \end{Bmatrix} \tag{3.29}$$

Taking the bottom partitions from equation 3.29 to recover the slave degrees of freedom yields:

$$u_s^n = -\{K_{ss} - \omega_n^2 M_{ss}\}^{-1}\{K_{ms}^T - \omega_n^2 M_{ms}^T\}u_m^n \tag{3.30}$$

If this reduction process is to be employed, the first decision is to select the masters and slaves; it is recommended in reference [1] that the slave degrees of freedom should be selected from degrees of freedom located in regions of high stiffness. Most commercial systems have automatic methods for selecting the slave degrees of freedom that are based on choosing those degrees of freedom that give the largest values for the ratio K_{ii}/M_{ii}, see reference [1].

In order to illustrate this reduction technique, consider the spring–mass system shown in Figure 3.2 that has the following global stiffness and mass matrices:

$$K = \begin{Bmatrix} 1.5\times10^3 & -1.5\times10^3 & 0 & 0 & 0 \\ -1.5\times10^3 & 5.7\times10^3 & -2\times10^3 & -2.2\times10^3 & 0 \\ 0 & -2\times10^3 & 2\times10^3 & 0 & 0 \\ 0 & -2.2\times10^3 & 0 & 3.4\times10^3 & -1.2\times10^3 \\ 0 & 0 & 0 & -1.2\times10^3 & 1.2\times10^3 \end{Bmatrix}$$

$$M = \begin{Bmatrix} 100 & 0 & 0 & 0 & 0 \\ 0 & 300 & 0 & 0 & 0 \\ 0 & 0 & 100 & 0 & 0 \\ 0 & 0 & 0 & 200 & 0 \\ 0 & 0 & 0 & 0 & 100 \end{Bmatrix}$$

The problem has five nodes and the reduction process is focused on using two nodes only, thereby reducing the problem from one involving 5×5 matrices to another involving 2×2 matrices. This requires identifying and working with three slave degrees of freedom and to aid in their selection we note that:

$$\frac{K_{11}}{M_{11}} = 15 \qquad \frac{K_{22}}{M_{22}} = 19 \qquad \frac{K_{33}}{M_{33}} = 20 \qquad \frac{K_{44}}{M_{44}} = 17 \qquad \frac{K_{55}}{M_{55}} = 12$$

and, therefore, select nodes 2, 3 and 4 as the slaves. In order to construct the transformation matrix 3.24 the terms K_{ss} and K_{ms} are required and these are given by:

$$K_{ss} = \begin{Bmatrix} 5.7 \times 10^3 & -2 \times 10^3 & -2.2 \times 10^3 \\ -2 \times 10^3 & 2 \times 10^3 & 0 \\ -2.2 \times 10^3 & 0 & 3.4 \times 10^3 \end{Bmatrix}$$

$$K_{ms} = \begin{Bmatrix} -1.5 \times 10^3 & 0 & 0 \\ 0 & 0 & -1.2 \times 10^3 \end{Bmatrix}$$

Inserting these into equation 3.27 yields the required transformation matrix:

$$T = \begin{Bmatrix} 1 & 0 \\ 0.659 & 0.341 \\ 0.659 & 0.341 \\ 0.426 & 0.574 \\ 0 & 1 \end{Bmatrix}$$

Using this matrix the reduced mass matrix Mr and stiffness matrix Kr can be constructed and inserted into equation 3.28 which gives two natural frequencies, $\omega_1 = 0$ radians/second and $\omega_2 = 2.965$ radians/second. The first natural frequency is the rigid body mode while the second is in close agreement with the original first natural frequency of 2.793. There are two points to note. First, the method has computed the two lowest frequencies, which is a general property; if the analysis had used three master degrees of freedom, then the three lowest natural frequencies would have been computed. Second, the non-rigid body frequency is slightly larger than that computed by considering the complete system in Section 3.1.1. The reason why such a difference can occur is explained in Chapter 4.

Solving the eigenvalue problem defined by equation 3.28 creates values for the master degrees of freedom but the full set requires the use of equation 3.30. Performing these tasks and normalising the resulting mode shape with respect to the full mass matrix, the mode shape U is given by:

$$U = \begin{Bmatrix} 0.051 \\ 0.008 \\ 0.015 \\ -0.037 \\ -0.065 \end{Bmatrix}$$

Comparing this vector with the equivalent vector in the solution of the problem in Section 3.1.1 and ignoring the sign reversal, it is clear that the two vectors are similar but certainly not identical. However, if the normalised mode shape is used to evaluate the first natural frequency using the expression $\omega^2 = U^T K U$, an improved estimate is obtained with $\omega = 2.876$ radians/second. This reduction method offers the advantage of reducing the required computing time but does create some problems with respect to accuracy, particularly with respect to mode shapes.

3.6.3 Sub-structure Analysis (Component Mode Synthesis)

In Chapter 2, Section 2.3.3, the concept of using sub-structures either to reduce the size of a vibration analysis or to permit several teams to work simultaneously on a design distributed across several companies was introduced. Figures 2.15 and 2.16 display a structure divided into sub-structures and indicate that each sub-structure has its own set of finite elements on the interior of the domain with internal nodes together with interface nodes where sub-structures are connected. The approach adopted in the dynamics case copies that used in the static case where the nodes on the interior of each sub-structure are condensed onto the boundary.

The process starts with the construction of the individual finite element models for each sub-structure which should include all the required structural constraints. The second stage is concerned with reducing the number of degrees of freedom for each sub-structure. This is achieved by assuming that the motion of a sub-structure is due to two effects. The first is the motion of the sub-structure with the interface boundary nodes fixed. This is represented by a linear combination of the constrained normal modes of the structure (with the interface boundary fixed). The second is the deformation induced by the motion of the boundaries.

In order to reduce the problem size, the degrees of freedom for an entire sub-structure '*s*' is denoted by u_s and the degrees of freedom associated with the internal nodes and boundary nodes are denoted by u_I and u_B respectively. At this point it is worth noting that the logic of the dynamic sub-structure process follows that of the reduction process given in Section 3.6.2 where, in the present case, the internal nodes '*I*' are equivalent to the slave degrees of freedom '*s*' of Section 3.6.2 and the '*B*' degrees of freedom are equivalent to the masters '*m*' so that each sub-structure stiffness and mass matrix is partitioned together with the nodal vector in the same form as that in Section 3.6.2, thus:

$$K_s = \begin{Bmatrix} K_{BB} & K_{BI} \\ K_{BI}^T & K_{II} \end{Bmatrix} \quad M_s = \begin{Bmatrix} M_{BB} & M_{BI} \\ M_{BI}^T & M_{II} \end{Bmatrix} \quad U = \begin{Bmatrix} u_B \\ u_I \end{Bmatrix}$$

Then the following transformation is applied:

$$u_S = \begin{Bmatrix} u_I \\ u_B \end{Bmatrix} = \begin{Bmatrix} \alpha_n & \alpha_c \\ 0 & I \end{Bmatrix} \begin{Bmatrix} q_n \\ u_B \end{Bmatrix} = T \begin{Bmatrix} q_n \\ u_B \end{Bmatrix} \tag{3.31}$$

where the columns of the matrix α_n are a set of selected natural modes of the sub-structure, normally from the lowest end of the frequency spectrum, with the interface boundaries fixed and q_n are generalised coordinates obtained by solving the eigenvalue equation given by:

$$\{K_{II} - \omega^2 M_{II}\} u_I = 0 \tag{3.32}$$

The matrix α_c is constructed following the logic used in Chapter 2, Section 2.3.2.1, to create the transformation matrix for the static condensation process. Thus:

$$\alpha_c = -K_{II}^{-1} K_{IB}$$

which provides a relationship between the boundary and internal degrees of freedom of the form:

$$u_I = \alpha_c u_B$$

We can now form a set of reduced sub-structure mass and stiffness matrices that are operated on by the boundary degrees of freedom only and are equivalent to the reduced matrices M_r and K_r in Section 3.5.2, thus:

$$K_{sr} = T^T K_s T = \begin{Bmatrix} \bar{K}_{NN} & 0 \\ 0 & \bar{K}_{BB} \end{Bmatrix} \quad M_{sr} = T^T M_s T = \begin{Bmatrix} \bar{M}_{NN} & \bar{M}_{NB} \\ \bar{M}_{NB}^T & \bar{M}_{BB} \end{Bmatrix} \tag{3.33}$$

where the sub-matrices in equation 3.33 are given by:

$$\bar{K}_{NN} = \alpha_n^T K_{II} \alpha_n \tag{3.34}$$

$$\bar{K}_{BB} = K_{BB} + \alpha_c^T K_{IB} \tag{3.35}$$

$$\bar{M}_{NN} = \alpha_n^T M_{II} \alpha_n \tag{3.36}$$

$$\bar{M}_{BB} = M_{BB} + \alpha_c^T M_{IB} - \alpha_c^T M_{II} \alpha_c \tag{3.37}$$

$$\bar{M}_{NB} = \alpha_n^T (M_{IB} + M_{II} \alpha_c) \tag{3.38}$$

The third stage of the analysis deals with the complete structure. If, for example, the structure had been decomposed into two sub-structures, then there would be two sets of interior generalised coordinates q_n^1 and q_n^2 for each of the two sub-structures and a single set of boundary degrees of freedom u_B giving a complete set $U_c = \{q_n^1 \quad q_n^2 \quad u_B\}^T$ and a combined mass and stiffness matrix given by:

$$M_c = \begin{Bmatrix} \bar{M}_{nn}^1 & 0 & \bar{M}_{NB}^1 \\ 0 & \bar{M}_{NN}^2 & \bar{M}_{NB}^2 \\ \bar{M}_{NB}^{1T} & \bar{M}_{NB}^{2T} & \bar{M}_{BB}^1 + \bar{M}_{BB}^2 \end{Bmatrix} \quad K_c = \begin{Bmatrix} \bar{K}_{NN}^1 & 0 & 0 \\ 0 & \bar{K}_{NN}^2 & 0 \\ 0 & 0 & \bar{K}_{BB}^1 + \bar{K}_{BB}^2 \end{Bmatrix} \tag{3.39}$$

where the subscripts indicate the contributions from each sub-structure. The equation that now requires solution is:

$$M_c \ddot{U}_c + K_c U_c = 0 \tag{3.40}$$

which can be rendered into the standard eigenvalue problem for solution. The total set of displacements for the problem can then be found by using equation 3.31.

REFERENCES

1. Petyt, M., *Introduction to Finite Element Vibration Analysis*. 1990: Cambridge University Press, ISBN 0521266076.
2. Geradin, M., *Mechanical Vibrations: Theory and Applications to Structural Dynamics*. 2004: John Wiley & Sons, Ltd, ISBN 100470847859.
3. Kardestuncer, H., *Finite Element Handbook*. 1987: McGraw-Hill, ISBN 007033305.
4. Thompson, W.T., *Theory of Vibrations with Applications*. 2001: Nelson Thornes, ISBN 0748743804.
5. Rao, S.S., *Mechanical Vibration, 3rd edn. 1995: Addison-Wesley, ISBN 0201526867.*

4

What's Energy Got to Do with It?

4.1 INTRODUCTION

The user of the Finite Element Method assumes it is a convergent process so that increasing the number of elements provides solutions with increasing levels of accuracy. As discussed in Chapter 1 there are two meanings to the term accuracy: we can mean that the finite element solution is converging on an accurate description of the real-world problem or that it is converging to a correct solution of the mathematical problem represented by the finite element model. The two meanings of accuracy are often confused and taken to mean the same thing but this is not the case and the two relate to entirely different interpretations of accuracy. In this chapter the latter meaning is taken and the discussion focuses on the problem of determining whether a finite element solution converges to the solution of the mathematical problem as an increasing number of elements are employed.

The answer to this question is also not straightforward. In the case of a structure subjected to static loads, the purpose of a finite element analysis is to provide values for displacements, stresses, etc., at points distributed throughout the structure. If accuracy is taken to mean that a user can specify a point within a structure and be assured that all these parameters converge on the mathematical solution values, at that point, as more elements are employed, it is an unachievable request. This is because guaranteed point-wise convergence is not an attribute of the Finite Element Method. However, all is not lost as the method is assured to converge on the correct value for an energy functional for the elastic system being considered. In the case of a structure subjected to static loads, this energy functional is either the potential or complementary energy or appropriate developments of these two energy forms dependent upon the type of finite element being used. In the present case we are focusing on displacement finite elements so the appropriate energy form is the potential energy functional. Thus, the linear analysis of statically loaded structures using displacement finite elements exploits the principle of minimum potential energy to create an approximate computational analysis process that is guaranteed to converge. Convergence now means that the method will get closer and closer to the actual potential energy of the system being analysed as the number of elements is increased.

In the case where a structure is responding to dynamically applied loads, the situation is somewhat different. Here the energy functional is a combination of potential energy and kinetic energy, the potential energy being related to the elastic component of the structural response and the kinetic energy accounting for the inertia component. Once again the Finite Element Method converges towards a minimum energy state which, for the class of dynamic problems discussed in this book, can be interpreted as being related to a minimum kinetic energy state for the system. Although in both static and dynamic analysis cases the Finite Element Method is minimising an energy term, there is a significant difference between the two. In the case of a dynamic analysis, the primary terms often being sought by the analyst are the natural frequencies of the structure which are related to the structure's kinetic energy. In this situation the analyst has the advantage that the rate of convergence of the energy term is exactly the same as that of the design parameter.

The chapter uses two demonstration examples to illustrate the points raised. The first focuses on explaining the potential energy principle, showing how it is used to generate element stiffness matrices and how it is used to allow the allocation of distributed structural loads to element nodes. It then goes on to illustrate the convergence in energy for a simple bar load by a distributed force and shows how the shape of the energy term is very flat around the solution point. So, from an energy viewpoint, putting more and more elements into a structural model soon hits the law of diminishing returns. The problem also illustrates how the stresses in the elements converge more slowly and also how the stress discontinuity from element to element can be interpreted as a crude measure of point-wise accuracy. A results interpretation section exploits some of the methods for improving estimates of stress level within the bar structure to illustrate how improvement techniques work. All of these features of the Finite Element Method are exploited in later chapters of the book where a more detailed consideration is given to overall solution accuracy. The second is focused on demonstrating the kinetic energy principle as it applies to a structure undergoing a dynamic response and linking kinetic energy minimisation with convergence on natural frequencies. The same structure is used in the dynamic case as that employed in the static analysis example but without an applied load and targets the lowest natural frequencies of the structure.

This chapter takes a fairly simple approach to the topics considered; alternative approaches can be found in references [1], [2] and [3]. All the approaches are based on variational principles and an excellent explanation of this underpinning method can be found in the book by the late Kyuichiro Washizu [4].

4.2 STRAIN ENERGY

We start the process of developing the potential energy formulation with the definition of strain energy for an elastic system and begin by considering the strain energy of the simple loaded bar of length l and cross-sectional area a shown in Figure 4.1.

Its strain energy U is given by:

$$U = \int_0^u P du \text{ or } \int_0^{u/l} \frac{P}{a} a \frac{du}{l} l$$

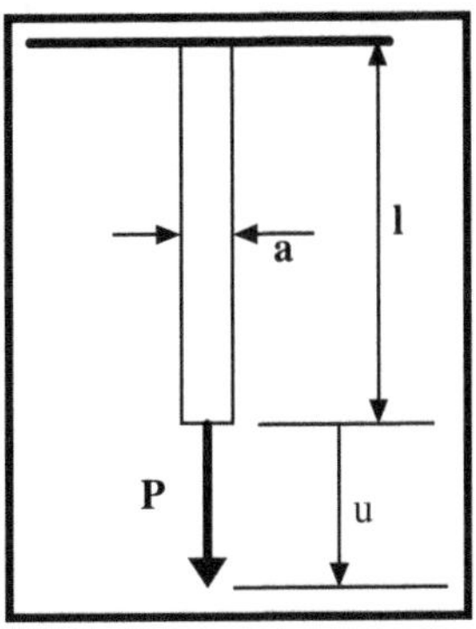

Figure 4.1 Single bar loaded by a single force.

Note that the strain in the bar $\varepsilon = u/l$, hence $d\varepsilon = du/l$ and the bar stress $\sigma = P/a$, therefore:

$$U = \int_0^\varepsilon \sigma \cdot d\varepsilon \cdot a \cdot l = \int_0^\varepsilon \sigma \cdot d\varepsilon \cdot vol$$

Where $vol = al$, i.e. the volume of the bar. Recalling that $\sigma = E\varepsilon$ where E is Young's modulus, then:

$$U = \int_0^\varepsilon \mathrm{E} \cdot \varepsilon \cdot d\varepsilon \cdot vol$$

which, after performing the integration, gives:

$$U = \frac{1}{2} \cdot \mathrm{E} \cdot \varepsilon^2 \cdot vol \tag{4.1}$$

This simple example can now be generalised to a three-dimensional body, but this introduces the complexity that the stress and strain terms are now tensors and they change as we move through the body – unlike the simple bar which has a constant strain state throughout each cross-section. With this in mind the generalised form for the strain energy is:

$$U = \iiint \frac{1}{2} \cdot \varepsilon^T \cdot \mathrm{D} \cdot \varepsilon \cdot d(vol) \tag{4.2}$$

The term D is the generalised form for Young's modulus E used in the development of the strain energy for the simple bar.

4.3 POTENTIAL ENERGY

The potential energy of an elastic system subject to static loads consists of two parts: one is the internal energy of the system – the strain energy just discussed; the other is the potential energy of the loads which is clearly external to the elastic system. Using the symbol π_p for the potential energy we have:

$$\pi_p = U + W_p$$

where W_p is the potential of the loads and can be replaced by the negative of the work done by the external loads W and hence:

$$\pi_p = U - W \tag{4.3}$$

4.4 SIMPLE BAR

Applying this formula to the simple bar for which we have already obtained an expression for the strain energy in equation 4.3 gives an expression for the potential energy:

$$\begin{aligned} \pi_p &= \frac{1}{2} E \cdot \varepsilon^2 \cdot vol - P \cdot u \\ &\text{or} \\ &= \frac{1}{2} E \cdot \varepsilon^2 \cdot a \cdot l - P \cdot u \end{aligned} \tag{4.4}$$

4.5 GENERAL CASE

In order to handle the general case we must recall that an elastic body has two types of boundary conditions:

1. One is where the displacements are specified to have given values, i.e. $u = u^*$, and this boundary is indicated by the use of the symbol S_u and is termed the displacement boundary or surface.
2. The other is where the stresses on the boundary are specified to have given values, i.e. $\sigma = \sigma^*$, and this boundary is indicated by the use of the symbol S_σ and is termed the traction boundary or surface.

The symbol S used above indicates that we are dealing with the surface of the elastic body.

The potential energy of a general three-dimensional body is still given by equation 4.3 so we can use the strain energy component as given by equation 4.2 and the remaining term to be calculated is the external work which is now given by:

$$\begin{aligned} W = &\iiint (F_x \cdot u_x + F_y \cdot u_y + F_z \cdot u_z) \cdot d(vol) \\ &+ \iint (T_x \cdot u_x + T_y \cdot u_y + T_z \cdot u_z) \cdot ds \end{aligned}$$

where F_x, F_y, F_z are body forces distributed throughout the elastic body (e.g. gravity forces) and T_x, T_y, T_z are the loads applied to the traction boundary of the body. The terms u_x etc. appearing under the triple integral and on the traction surface of the body under the double integral are the displacements throughout the body. There is no contribution from the displacement boundary as the displacements here are specified and, therefore, are not free to move and do work. If we write the work done by the body and surface forces in vector form, then we have:

$$W = \iiint F^T \cdot u \cdot d(vol) + \iint_{S_\sigma} T^T \cdot u \cdot ds$$

where S_σ is that portion of the surface of the elastic body where the traction forces are specified. Thus, the general form for the potential energy for a three-dimensional elastic body under static loads is given by:

$$\begin{aligned}\pi_p = &\iiint \frac{1}{2} \cdot \varepsilon^T \cdot D \cdot \varepsilon \cdot d(vol) - \iiint F^T \cdot u \cdot d(vol) \\ &- \iint_{S_\sigma} T^T \cdot u \cdot ds\end{aligned} \tag{4.5}$$

4.6 MINIMUM POTENTIAL ENERGY

Let us explore the properties of the potential energy to see how it applies to the development of the Finite Element Method and begin by seeing what happens when the potential energy is minimised using displacements as the optimising variables. Starting with the simplest case by returning to the bar shown in Figure 4.1 with the potential energy form given by equation 4.4 and recalling that $\varepsilon = u/l$:

$$\pi_p = \frac{E \cdot u^2 \cdot a}{2 \cdot l} - P \cdot u \tag{4.6}$$

In order to minimise this with respect to the variable u, π_p must be differentiated with respect to u and set equal to zero. Hence:

$$\begin{aligned}\frac{d\pi_p}{du} &= \frac{E \cdot u \cdot a}{l} - P = 0 \\ \text{or } u &= \frac{P \cdot l}{E \cdot a}\end{aligned} \tag{4.7}$$

Expression 4.7 is Hooke's Law and is the expression which answers the question 'What is the value of the displacement u of a bar of length l and cross-section a with a Young's modulus E when subjected to a static load P?'

In order to show a little more clearly what is actually happening, we take the following specific values for the constants:

- $a = 1$ unit
- $l = 10$ units
- $E = 100$ units
- $P = 10$ units

and inserting them into the expression for the potential energy 4.6 gives:

$$\pi_p = 5 \cdot u^2 - 10 \cdot u \tag{4.8}$$

In order to illustrate the fact that the solution to the problem of finding the equilibrium values for the end displacement of the loaded bar is directly related to the potential energy, Figure 4.2 shows the variation of π_p against u.

The required equilibrium value for the end displacement can be found by differentiating expression 4.8 and setting it to zero, which gives $u = 1.0$, and in Figure 4.2 the graph clearly shows that this is achieved at the point where the potential energy is at a minimum. This exercise has demonstrated one of the most important principles of structural mechanics discovered during the Napoleonic period in France by the

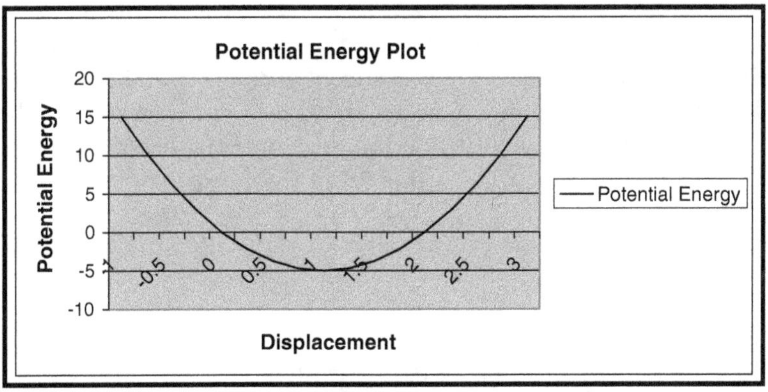

Figure 4.2 Variation of potential energy with bar end displacement.

Italian-born mathematician/engineer Joseph Lagrange and known as the *Principle of Minimum Potential Energy*:

> *For an elastic body subjected to static loads the Principle of Minimum Potential Energy states that from among all of the displacements satisfying the boundary conditions, the actual displacements are those which make the potential energy a minimum.*

As demonstrated, by example, in the next section the Finite Element Method is an application of the Principle of Minimum Potential Energy which means that it is guaranteed to converge on the correct potential energy for the structure being analysed as the number of elements used is increased. It should be noted, as already stated, that this is not a pointwise convergence principle so that it does not imply, for example, that the value of a stress at a specific point in the structure will get closer to the actual stress value as the number of elements increases.

4.7 THE PRINCIPLE OF MINIMUM POTENTIAL ENERGY APPLIED TO A SIMPLE FINITE ELEMENT PROBLEM

The numerical example used to demonstrate the minimum potential energy principle given above is, in fact, an application of the principle to a finite element problem but only to the simplest possible finite element problem, that of a single bar. The fact that this principle is used as the general derivation principle for all finite element systems is demonstrated by using a slightly more complex case employing a set of springs. Although still simple, this problem is truly representative of the entire class of finite element solutions for linear statics problems. Consider the loaded combination of springs originally employed in Chapter 2 and shown in Figure 4.3. As before, the springs are fixed at nodes 1 and 5 so that the displacements at these nodes, u_1 and u_2, have zero values ascribed to them.

The potential energy expression for this combination of springs and loads can be derived from equation 4.5 in the form:

$$\pi_p = \frac{1}{2}\cdot k_1\cdot\delta_1^2 + \frac{1}{2}\cdot k_2\cdot\delta_2^2 + \frac{1}{2}\cdot k_3\cdot\delta_3^2 + \frac{1}{2}\cdot k_4\cdot\delta_4^2 - F_3\cdot u_3 - F_4\cdot u_4 \quad (4.9)$$

where:

$$\delta_1 = u_2 - u_1 \quad \delta_2 = u_3 - u_2 \quad \delta_3 = u_4 - u_2 \quad \delta_4 = u_5 - u_4$$

The Principle of Minimum Potential Energy requires minimising equation 4.9 with respect to the optimising variables which, in this

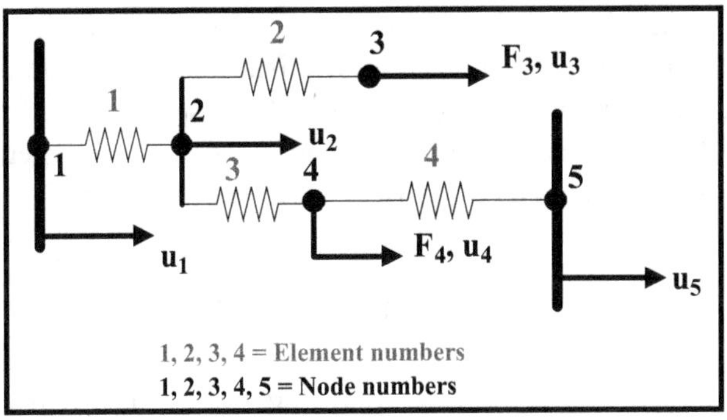

Figure 4.3 Loaded spring combination.

case, are the nodal displacements $u_1 \ldots u_5$. However, the two ends of the system are fixed and thus $u_1 = u_5 = 0$ and equation 4.9 can now be rewritten as:

$$\pi_p = \frac{1}{2} \cdot k_1 \cdot u_2^2 + \frac{1}{2} \cdot k_2 \cdot (u_3 - u_2)^2 + \frac{1}{2} \cdot k_3 \cdot (u_4 - u_2)^2 + \frac{1}{2} \cdot k_4 \cdot (-u_4)^2 - F_3 \cdot u_3 - F_4 \cdot u_4 \tag{4.10}$$

As stated above, it is necessary to differentiate this term with respect to the free optimising variables and set these equal to zero, hence:

$$\frac{\partial \pi_p}{\partial u_i} = 0 \quad \text{for} \quad i = 2 \ldots 4$$

giving:

$$\frac{\partial \pi_p}{\partial u_2} = k_1 \cdot u_2 - k_2 \cdot (u_3 - u_2) - k_3 \cdot (u_4 - u_2) = 0$$

$$\frac{\partial \pi_p}{\partial u_3} = k_2 \cdot (u_3 - u_2) - F_3 = 0$$

$$\frac{\partial \pi_p}{\partial u_4} = k_3 \cdot (u_4 - u_2) + k_4 \cdot u_4 - F_4 = 0$$

Taking the load terms F_1 and F_2 across the equals sign and putting the equations into matrix form gives:

$$\begin{Bmatrix} k_1 + k_2 + k_3 & -k_2 & -k_3 \\ -k_2 & k_2 & 0 \\ -k_3 & 0 & k_3 + k_4 \end{Bmatrix} \begin{Bmatrix} u_2 \\ u_3 \\ u_4 \end{Bmatrix} = \begin{Bmatrix} 0 \\ F_3 \\ F_4 \end{Bmatrix} \tag{4.11}$$

The matrix equation 4.11 is exactly the same as that which would have been generated had the direct approach been used. This demonstrates that the Finite Element Method is directly based on the Principle of Minimum Potential Energy and is, therefore, a convergent method (in energy) for the analysis of statically loaded elastic structures.

4.8 FINITE ELEMENT FORMULATION

Having considered a bar and a set of springs we can now turn to a more general application of the principle and see how it has other lessons for the Finite Element Method. Consider the four-noded element shown in Figure 4.4 where the displacement is a distributed function which stretches across the entire element and is, therefore, a function of x and y, i.e. $u = u(x, y)$. At each node of the element there are two nodal connection quantities d_i where $i = 1 \ldots 8$. Using these connection quantities the displacement at any point in the interior of the element can now be defined in terms of these quantities and a set of shape functions $N(x, y)$, thus:

$$u(x, y) = N(x, y) \cdot d \tag{4.12}$$

where d now denotes the vector containing the eight connection quantity terms $d_1 \ldots d_8$ as shown in Figure 4.4.

The general expression for the potential energy functional is given by equation 4.5; that is:

$$\pi_p = \iiint \frac{1}{2} \cdot \varepsilon^T \cdot D \cdot \varepsilon \cdot d(vol) - \iiint F^T \cdot u \cdot d(vol) - \iint_{S_\sigma} T^T \cdot u \cdot ds$$

Denoting the generalised relationship between strain and displacements as $\varepsilon = B \cdot u$ and using expression 4.12 gives the relationship between the

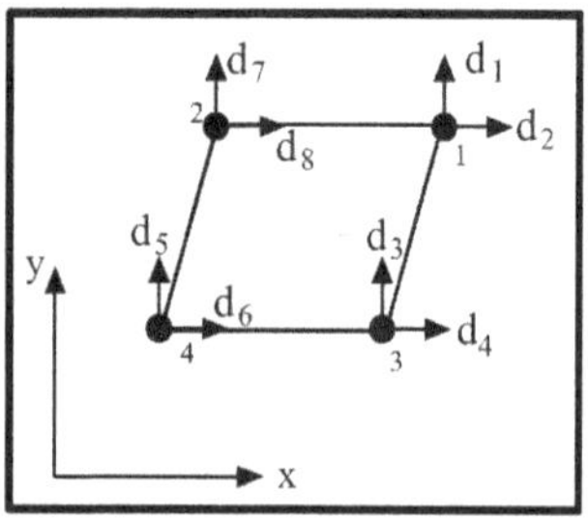

Figure 4.4 Four-noded finite element layout.

strains and the nodal connection quantities as $\varepsilon = B \cdot N \cdot d$. Using this expression and that for the potential energy, it is now possible to derive the element stiffness matrix and obtain an expression for calculating how a distributed load, which extends over the surface of the finite element, is represented by forces acting at the element nodes. In order to make life simple, the terms in the potential energy expression can be separated into the strain energy term and the load term. Taking the strain energy term first:

$$\iint \frac{1}{2} \cdot \varepsilon^T \cdot D \cdot \varepsilon \cdot dx \cdot dy = \frac{1}{2} \iint (B \cdot N \cdot d)^T \cdot D \cdot (B \cdot N \cdot d) \cdot dx \cdot dy$$
$$= \frac{1}{2} \cdot d^T \cdot \left\{ \iint (B \cdot N)^T \cdot D \cdot (B \cdot N) \cdot dx \cdot dy \right\} \cdot d \qquad (4.13)$$

The terms 'd' can be taken outside of the integration because, as nodal quantities, they are not distributed functions and do not exist in the interior of the element. The remaining term inside the curly brackets is the stiffness matrix for the element, thus:

$$k = \iint (B \cdot N)^T \cdot D \cdot (B \cdot N) \cdot dx \cdot dy \qquad (4.14)$$

Turning to the work terms in the second part of the potential energy expression:

$$\iiint F^T \cdot u \cdot d(vol) + \iint_{S_\sigma} T^T \cdot u \cdot ds$$
$$= \iint F^T \cdot (N \cdot d) \cdot dx \cdot dy + \int_{S_\sigma} T^T \cdot (N \cdot d) \cdot ds$$
$$= \iint d^T \cdot N^T \cdot F \cdot dx \cdot dy + \int_{S_\sigma} d^T \cdot N^T \cdot T \cdot ds$$

the term S_σ is associated with that part of the edge of the element which is subject to edge tractions as applied forces and the terms F relate to interior distributed forces such as gravity loads. As with the stiffness matrix derivation the terms 'd' do not exist on the interior of the element, nor along the element edge, but exist only at the nodes. Thus, the expression for calculating the work in terms of the nodal connection quantities is given by:

$$d^T \cdot \left\{ \iint N^T \cdot F \cdot dx \cdot dy + \int_{S_\sigma} N^T \cdot T \cdot ds \right\}$$

The terms inside the curly brackets are now generalised loads and are designated as P, thus:

$$P = \iint N^T \cdot F \cdot dx \cdot dy + \int_{S_\sigma} N^T \cdot T \cdot ds \tag{4.15}$$

This expression attributes a set of distributed body forces and applied surface tractions to the nodal connection quantities. Because this process uses the same shape functions as those used in the development of the stiffness matrix the loads in the vector P are known as consistent loads. Using these terms the potential energy function for this element can now be written in the form:

$$\pi_p = \frac{1}{2} \cdot d^T \cdot k \cdot d - d^T \cdot P$$

Minimising this potential energy expression for the new optimising variables 'd' gives:

$$\frac{\partial \pi_p}{\partial d} = k \cdot d - P = 0$$

$$\text{or} \qquad k \cdot d = P$$

4.9 DIRECT APPLICATION TO AN AXIAL BAR ELEMENT

The developments leading to the creation of a stiffness matrix and generalised load vector are now demonstrated using a simple axial bar element. The first step in the process is to compute the stiffness matrix using expression 4.14 for the element shown in Figure 4.5.

This element has two nodes and has, therefore, a linear variation of displacement along the length of the element which can be represented by the two shape functions shown in Figure 4.5, thus:

$$N = \{\, N_1(x, y) \quad N_2(x, y) \,\} = \{\, \left(\tfrac{l-x}{l}\right) \quad \left(\tfrac{x}{l}\right) \,\}$$

For this simple problem the relation between displacement and strain is given by:

$$\varepsilon = \frac{du}{dx} \qquad \text{hence} \qquad B = \frac{d}{dx}$$

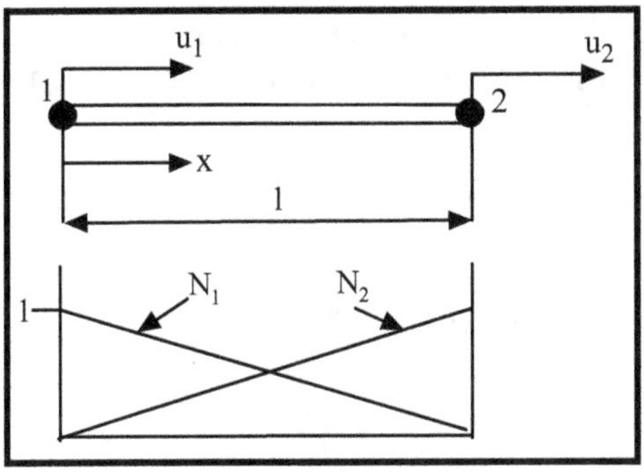

Figure 4.5 Axial element.

Operating on the shape functions with B gives:

$$B \cdot \mathrm{N} = \left\{ \frac{-1}{l} \ \frac{1}{l} \right\}$$

and for this simple case $D = E$ (Young's modulus) and the cross-section of the bar is A. Using these expressions the stiffness matrix is given by:

$$k = \int_0^l (B \cdot \mathrm{N})^T \cdot E \cdot (B \cdot \mathrm{N}) \cdot dx = A \int_0^l \left\{ \begin{matrix} \frac{-1}{l} \\ \frac{1}{l} \end{matrix} \right\} \left\{ \frac{-E}{l} \ \frac{E}{l} \right\} \cdot dx$$

$$= \frac{AE}{l} \left\{ \begin{matrix} 1 & -1 \\ -1 & 1 \end{matrix} \right\} \tag{4.16}$$

The second task is to calculate a set of generalised loads for a loaded axial element. The loads on the element are shown in Figure 4.6 and consist of a point load applied at $x = 2l/3$ of numerical value unity and a distributed traction along the length of the bar $q = 1/\text{unit length}$.

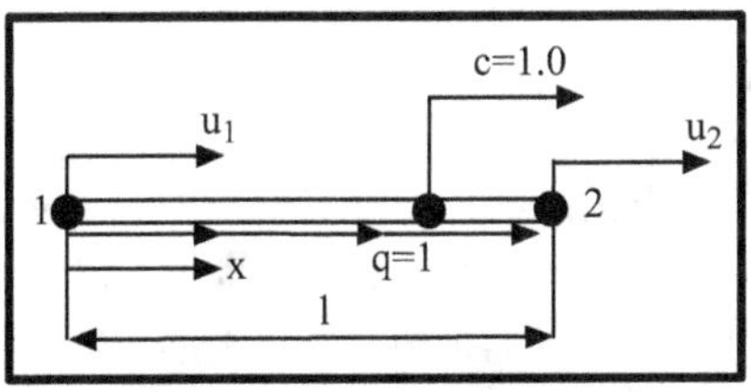

Figure 4.6 Loaded axial element.

There are no body forces applied to this element so there is only one of the generalised load components from equation 4.15, thus:

$$P = \int_{S_\sigma} N^T \cdot T \cdot ds$$

In this case, the expression becomes:

$$\begin{aligned} P &= \int_0^l N^T \cdot q(x) \cdot dx + N^T(x = 2/3) \cdot c \\ &= \int_0^l \begin{Bmatrix} (l-x)/l \\ x/l \end{Bmatrix} \cdot dx + \begin{Bmatrix} 1/3 \\ 2/3 \end{Bmatrix} \\ &= \begin{Bmatrix} l/2 \\ 1/2 \end{Bmatrix} + \begin{Bmatrix} 1/3 \\ 2/3 \end{Bmatrix} \end{aligned}$$

These are the consistent loads for this element for the load set shown in Figure 4.6.

4.10 CONVERGENCE IN ENERGY AND CONVERGENCE IN STRESS

So far we have demonstrated that the displacement Finite Element Method is based on the minimum potential energy principle and this assures us that the method will converge towards the correct value for the potential energy associated with the mathematical formulation of the structural analysis problem under review. This remark has to be treated with some caution. It assumes, for example, that the finite elements selected to model the problem have an appropriate formulation for the phenomena being addressed; if the problem involves the inextensional bending of a shell surface under applied load, then the elements chosen must be able to accommodate this type of bending action.

It may be argued that selecting the correct element is simply a matter of using common sense, but ensuring that the mesh does not give rise to elements that are so distorted as to inject significant error into the solution process is not so self-evident. We address these and other modelling issues in later chapters and link these to the errors associated with the idealisation process. For the moment these are overlooked in order to address a number of specific issues related to the convergence of the method in energy and in stress and having a direct relevance to error control when using the Finite Element Method:

- The first issue relates to the convergence in stress and although it is not possible to claim point-wise convergence in stress as the number of elements increases, it is possible to get some idea of the stress accuracy by looking at the stress jump as element boundaries are crossed.
- The second concerns the trajectory of the potential energy as the number of elements is increased – is the minimum a sharp point or a smooth curve and if the second, is the curve flat?
- Finally, the loss of applied load when using the consistent load method needs to be examined.

These are relevant to error control when the Finite Element Method is used in the analysis of a real-world structure and can be addressed by taking the simple problem shown in Figure 4.7 which illustrates a bar attached to a rigid wall at one end and subjected to a linear surface traction applied in the x-direction.

The problem involves a bar of length $L = 60\,\text{m}$, with cross-sectional area $A = 2\,\text{m}^2$, subject to an applied linear load $T(x) = 10.x$ newtons/m (N/m) and the bar material has a Young's modulus $E = 30 \times 10^6\,\text{N/m}^2$. In order to explore the three questions, in a demonstration mode, this problem is now solved using a sequence of elements starting with one element being employed and increasing the number up to 16 elements.

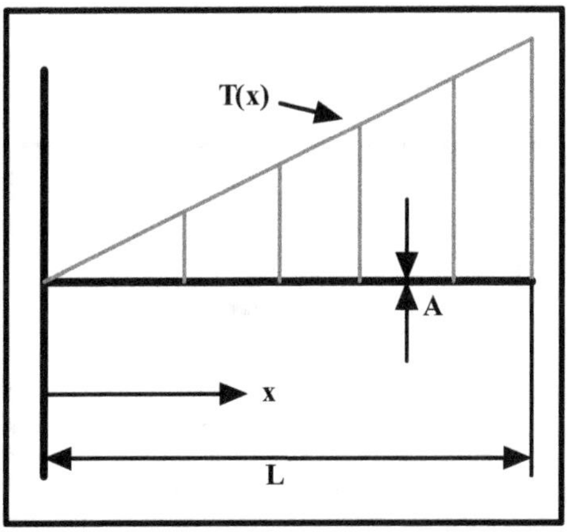

Figure 4.7 Loaded axial bar.

For reference purposes the following can be obtained from a direct mathematical analysis of the problem:

- the strain energy U of the system $= 86.4\,\mathrm{N \cdot m}$
- the work done by the external load W of the system $= 172.8\,\mathrm{N \cdot m}$
- the minimum potential energy π_p of the system $= -86.4\,\mathrm{N \cdot m}$
- the stress along the bar length $= 2.5(L^2 - x^2)$
- the displacement along the bar length $= 5x(L^2 - x^2/3)/EA$.

4.10.1 Single Bar Element Model

Starting with a single bar element modelling the system, the element layout is illustrated in Figure 4.8.

Using equation 4.16 the stiffness matrix for this element is given by:

$$k = \frac{EA}{L}\begin{Bmatrix} 1 & -1 \\ -1 & 1 \end{Bmatrix} = 10^6\begin{Bmatrix} 1 & -1 \\ -1 & 1 \end{Bmatrix}$$

Because there is only one element representing the problem, this is also the global stiffness matrix. The consistent load at the element level is obtained by employing equation 4.15:

$$\begin{Bmatrix} p_1 \\ p_2 \end{Bmatrix} = \int_0^{60} \begin{Bmatrix} \frac{(60-x)}{60} \\ \frac{x}{60} \end{Bmatrix} \cdot 10 \cdot x \cdot dx$$

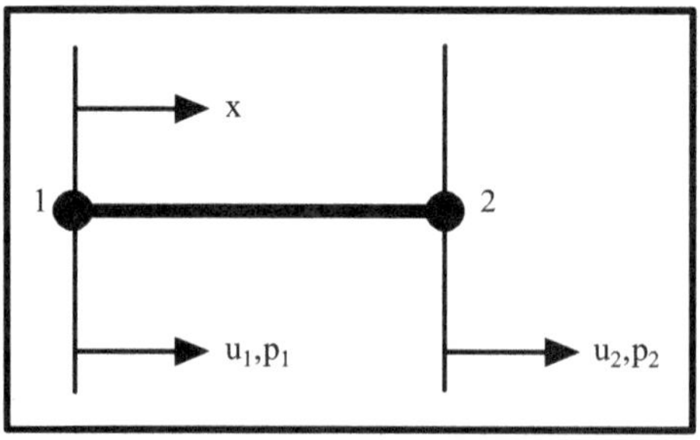

Figure 4.8 Single finite element model.

Evaluating this expression gives $p_1 = 6000\,\text{N}$ and $p_2 = 12,000\,\text{N}$ and, again because there is only one element, these are also the global values P_1 and P_2, thus the global finite element matrix equation is given by:

$$10^6 \begin{Bmatrix} 1 & -1 \\ -1 & 1 \end{Bmatrix} \begin{Bmatrix} u_1 \\ u_2 \end{Bmatrix} = \begin{Bmatrix} 6000 \\ 12,000 \end{Bmatrix} \tag{4.17}$$

When the boundary condition $u_1 = 0$ is applied, then the load at node 1 cannot play a role in the solution as the term P_1 is now the reaction and, thus, the 6000 N ascribed to this node by the consistent load process has vanished and is not 'seen' by the finite element system. Applying this boundary condition gives:

$$10^6 \cdot u_2 = 12,000$$

or:

$$u_2 = 0.012\,\text{m}$$

and the stress:

$$\sigma = 6000\,\text{N}$$

The reaction force can now be computed from:

$$10^6 \begin{Bmatrix} 1 & -1 \\ -1 & 1 \end{Bmatrix} \begin{Bmatrix} 0.0 \\ 0.012 \end{Bmatrix} = \begin{Bmatrix} R \\ 12,000 \end{Bmatrix}$$

giving:

$$R = -12,000\,\text{N}$$

This model gives a value of the strain energy for the system:

$$U = 72\,\text{N} \cdot \text{m}$$

The external work is:

$$W = 144\,\text{N} \cdot \text{m}$$

and the minimum potential energy generated by this finite element model is:

$$\pi_p = -72\,\text{N}\cdot\text{m}$$

> Results:
> Potential energy error $= -86.4 + 72 = -14.4\,\text{N}\cdot\text{m}$
> Stress values: single value for stress $= 6000\,\text{N/m}^2$
> Loss of load due to boundary condition $= 6000\,\text{N}$

4.10.2 2-Bar Element Model

Moving to a 2-element model, the resulting finite element model is illustrated in Figure 4.9.

In this case, the element stiffness matrix is given by:

$$k = \frac{EA}{30}\begin{Bmatrix} 1 & -1 \\ -1 & 1 \end{Bmatrix} = 2\times 10^6 \begin{Bmatrix} 1 & -1 \\ -1 & 1 \end{Bmatrix}$$

Hence the global stiffness matrix is:

$$K = 2\times 10^6 \begin{Bmatrix} 1 & -1 & 0 \\ -1 & 2 & -1 \\ 0 & -1 & 1 \end{Bmatrix}$$

the load vector for element 1 is given by:

$$\begin{Bmatrix} p_1^1 \\ p_2^1 \end{Bmatrix} = \int_0^{30} \begin{Bmatrix} \frac{(30-x)}{30} \\ \frac{x}{30} \end{Bmatrix} \cdot 10 \cdot x \cdot dx = \begin{Bmatrix} 1500 \\ 3000 \end{Bmatrix}$$

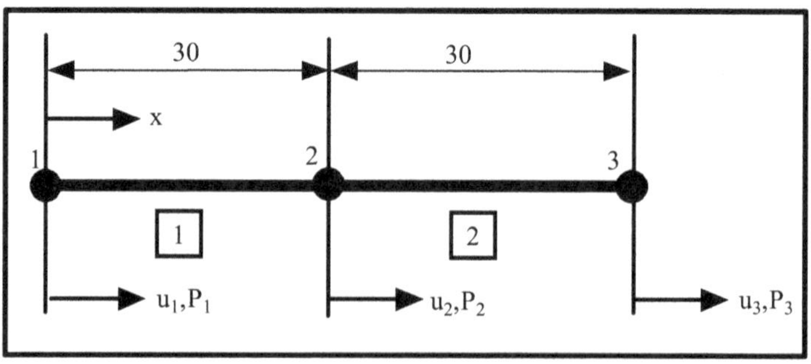

Figure 4.9 Double finite element model.

and the load vector for element 2 is given by:

$$\begin{Bmatrix} p_1^2 \\ p_2^2 \end{Bmatrix} = \int_{30}^{60} \begin{Bmatrix} \frac{(60-x)}{30} \\ \frac{(x-30)}{30} \end{Bmatrix} \cdot 10 \cdot x \cdot dx = \begin{Bmatrix} 6000 \\ 7500 \end{Bmatrix}$$

The assembled global load vector is therefore:

$$\begin{Bmatrix} P_1 \\ P_2 \\ P_3 \end{Bmatrix} = \begin{Bmatrix} p_1^1 \\ p_2^1 + p_1^2 \\ p_2^2 \end{Bmatrix} = \begin{Bmatrix} 1500 \\ 3000 + 6000 \\ 7500 \end{Bmatrix} = \begin{Bmatrix} 1500 \\ 9000 \\ 7500 \end{Bmatrix}$$

Hence the global system is:

$$2 \times 10^6 \begin{Bmatrix} 1 & -1 & 0 \\ -1 & 2 & -1 \\ 0 & -1 & 1 \end{Bmatrix} \begin{Bmatrix} u_1 = 0 \\ u_2 \\ u_3 \end{Bmatrix} = \begin{Bmatrix} 1500 \\ 9000 \\ 7500 \end{Bmatrix} = \begin{Bmatrix} R \\ 9000 \\ 7500 \end{Bmatrix}$$

R is the reaction because $u_1 = 0$ and hence:

$$2 \times 10^6 \begin{Bmatrix} 2 & -1 \\ -1 & 1 \end{Bmatrix} \begin{Bmatrix} u_2 \\ u_3 \end{Bmatrix} = \begin{Bmatrix} 9000 \\ 7500 \end{Bmatrix} \text{ giving } \begin{Bmatrix} u_2 \\ u_3 \end{Bmatrix} = \begin{Bmatrix} 0.00825 \\ 0.012 \end{Bmatrix}$$

The strains and stresses in the two elements are now found from back-substitutions and hence:

$$\text{for element 1, the strain} = \varepsilon_1 = \frac{1}{30}\{-1 \quad 1\}\begin{Bmatrix} 0 \\ 0.00825 \end{Bmatrix} = 2.75 \times 10^{-1}$$

$$\text{and the stress} = \sigma_1 = E \cdot \varepsilon_1 = 8250\,\text{N/m}^2$$

$$\text{for element 2, the strain} = \varepsilon_2 = \frac{1}{30}\{-1 \quad 1\}\begin{Bmatrix} 0.00825 \\ 0.012 \end{Bmatrix} = 1.25 \times 10^{-4}$$

$$\text{and the stress} = \sigma_2 = E \cdot \varepsilon_2 = 3750\,\text{N/m}^2$$

The reaction force can now be computed from the equation above giving:

$$R = -16,500\,\text{N}$$

This model gives a value of the strain energy for the system:

$$U = 82.125\,\text{N} \cdot \text{m}$$

The external work is:

$$W = 164.25\,\mathrm{N \cdot m}$$

and the minimum potential energy generated by this finite element model is:

$$\pi_p = -82.125\,\mathrm{N \cdot m}$$

Results:
Potential energy error $= -86.4 + 82.125 = -4.275\,\mathrm{N \cdot m}$
Stress values: element 1 stress $= 8250\,\mathrm{N/m^2}$
element 2 stress $= 3750\,\mathrm{N/m^2}$
Maximum stress jump across element junction $= 4500\,\mathrm{N/m^2}$
Loss of load due to boundary condition $= 1500\,\mathrm{N}$

4.10.3 4-Bar Element Model

Increasing the number of elements to four gives the finite element layout shown in Figure 4.10.

In this case, the element stiffness matrix is given by:

$$k = \frac{EA}{15}\begin{Bmatrix} 1 & -1 \\ -1 & 1 \end{Bmatrix} = 4 \times 10^6 \begin{Bmatrix} 1 & -1 \\ -1 & 1 \end{Bmatrix}$$

Hence the global stiffness matrix:

$$K = 4 \times 10^6 \begin{Bmatrix} 1 & -1 & 0 & 0 & 0 \\ -1 & 2 & -1 & 0 & 0 \\ 0 & -1 & 2 & -1 & 0 \\ 0 & 0 & -1 & 2 & -1 \\ 0 & 0 & 0 & -1 & 1 \end{Bmatrix}$$

the load vector for element 1 is given by:

$$\begin{Bmatrix} p_1^1 \\ p_2^1 \end{Bmatrix} = \int_0^{15} \begin{Bmatrix} \frac{(15-x)}{15} \\ \frac{x}{15} \end{Bmatrix} \cdot 10 \cdot x \cdot dx = \begin{Bmatrix} 375 \\ 750 \end{Bmatrix}$$

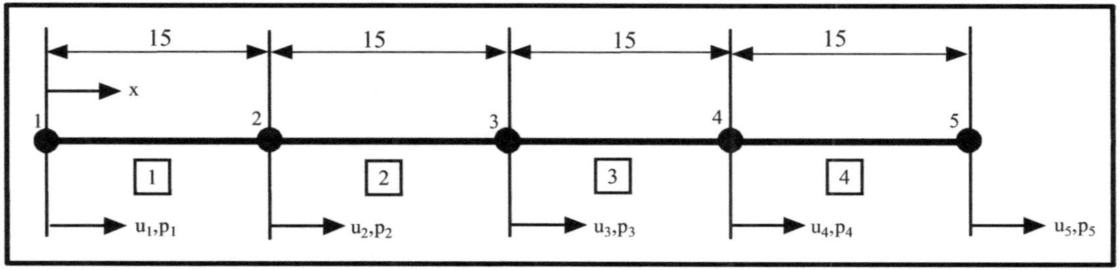

Figure 4.10 4-element model.

and the element load vectors for the other three elements are similarly calculated. This leads to an overall global finite element system given by:

$$4 \times 10^6 \begin{Bmatrix} 1 & -1 & 0 & 0 & 0 \\ -1 & 2 & -1 & 0 & 0 \\ 0 & -1 & 2 & -1 & 0 \\ 0 & 0 & -1 & 2 & -1 \\ 0 & 0 & 0 & -1 & 1 \end{Bmatrix} \begin{Bmatrix} u_1 = 0 \\ u_2 \\ u_3 \\ u_4 \\ u_5 \end{Bmatrix} = \begin{Bmatrix} R \\ 2250 \\ 4500 \\ 6750 \\ 4125 \end{Bmatrix}$$

Solving this matrix equation gives a displacement vector:

$$u = \begin{Bmatrix} u_1 \\ u_2 \\ u_3 \\ u_4 \\ u_5 \end{Bmatrix} = 10^{-3} \begin{Bmatrix} 0 \\ 4.406 \\ 8.25 \\ 11.0 \\ 12.0 \end{Bmatrix} m$$

and the individual element stresses are given by:

$$\sigma = \begin{Bmatrix} \sigma_1 \\ \sigma_2 \\ \sigma_3 \\ \sigma_4 \end{Bmatrix} = 10^3 \begin{Bmatrix} 8.813 \\ 7.688 \\ 5.437 \\ 2.063 \end{Bmatrix} \mathrm{N/m^2}$$

As before, the reaction force can now be computed giving:

$$R = -17{,}635\,\mathrm{N}$$

This model gives a value of the strain energy for the system:

$$U = 85.289\,\mathrm{N \cdot m}$$

The external work is:

$$W = 170.578\,\mathrm{N \cdot m}$$

and the minimum potential energy generated by this finite element model is:

$$\pi_p = -85.289\,\mathrm{N \cdot m}$$

Results:
Potential energy error $= -86.4 + 85.289 = -1.111\,\mathrm{N \cdot m}$
Stress values: element 1 stress $= 8813\,\mathrm{N/m^2}$
element 2 stress $= 7688\,\mathrm{N/m^2}$
element 3 stress $= 5437\,\mathrm{N/m^2}$
element 4 stress $= 2063\,\mathrm{N/m^2}$
Maximum stress jump across element junction $= 3374\,\mathrm{N/m^2}$
Loss of load due to boundary condition $= 375\,\mathrm{N}$

4.10.4 8-Bar Element Model

Moving to an element layout using 8-bar finite elements to model this structure and following the same procedure as before, it is found that the reaction force is now given by:

$$R = -17{,}906\,\mathrm{N}$$

This model gives a value of the strain energy for the system:

$$U = 86.12\,\mathrm{N \cdot m}$$

The external work is:

$$W = 172.239\,\mathrm{N \cdot m}$$

and the minimum potential energy generated by this finite element model is:

$$\pi_p = -86.12\,\mathrm{N \cdot m}$$

Results:
Potential energy error $= -86.4 + 86.12 = -0.28\,\mathrm{N \cdot m}$
Stress values: element 1 stress $= 8953\,\mathrm{N/m^2}$
element 2 stress $= 8672\,\mathrm{N/m^2}$
element 3 stress $= 8109\,\mathrm{N/m^2}$
element 4 stress $= 7266\,\mathrm{N/m^2}$
element 5 stress $= 6141\,\mathrm{N/m^2}$
element 6 stress $= 4734\,\mathrm{N/m^2}$

element 7 stress = 3047 N/m^2
element 8 stress = 1078 N/m^2
Maximum stress jump across element junction = 1969 N/m^2
Loss of load due to boundary condition = 94 N

4.10.5 16-Bar Element Model

Finally, moving to an element layout using 16-bar finite elements to model the structure and following the same procedure as before, it is found that the reaction force is now given by:

$$R = -17,977\,\text{N}$$

The model gives a value of the strain energy for the system:

$$U = 86.33\,\text{N} \cdot \text{m}$$

The external work is:

$$W = 172.625\,\text{N} \cdot \text{m}$$

and the minimum potential energy generated by this finite element model is:

$$\pi_p = -86.33\,\text{N} \cdot \text{m}$$

Results:
Potential energy error $= -86.4 + 86.33 = -0.07\,\text{N} \cdot \text{m}$
Stress values:
element 1 stress = 8988 N/m^2
element 2 stress = 8918 N/m^2
element 3 stress = 8777 N/m^2
element 4 stress = 8566 N/m^2
element 5 stress = 8285 N/m^2
element 6 stress = 7934 N/m^2
element 7 stress = 7512 N/m^2
element 8 stress = 7020 N/m^2
element 9 stress = 6457 N/m^2
element 10 stress = 5824 N/m^2

element 11 stress = 4348 N/m^2
element 12 stress = 3504 N/m^2
element 13 stress = 2590 N/m^2
element 14 stress = 2590 N/m^2
element 15 stress = 1605 N/m^2
element 16 stress = 550 N/m^2
Maximum stress jump across element junction = 1055 N/m^2
Loss of load due to boundary condition = 23 N

4.11 RESULTS INTERPRETATION

Now that the results from analysing the loaded bar with the aid of a number of finite element models are available, it is possible to draw some conclusions from the results and to exploit some of the properties of the Finite Element Method. This section begins by examining the convergence of the potential energy when an increasing number of finite elements are employed.

4.11.1 Potential Energy Convergence

Figure 4.11 illustrates the standard potential energy convergence progress for a displacement finite element model for this simple bar problem. It clearly shows that the curve depicting the variation of potential energy with an increasing number of elements is monotonic

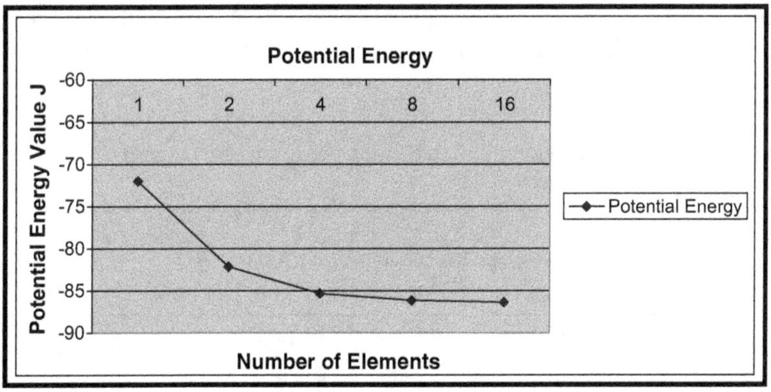

Figure 4.11 Potential energy convergence.

but relatively flat as the minimum value is approached. For example, the change in the value in moving from 8 to 16 elements, this doubling the number of elements, gives rise to a change in potential energy of 0.21 N·m. Although the convergence rate is illustrated by means of a very simple problem, the 'flatness' of the curve is a general property and occurs in all problems modelled by displacement elements. Referring to Section 4.10, the potential and strain energies vary at the same rate and the nature of the curve for the strain energy implies that for energy-based problems, such as the calculation of natural frequencies, the rate of convergence to the solution is likely to be rapid.

4.11.2 Stress Improvement

As demonstrated in Section 4.10 the maximum stress jump across element boundaries remains significant even when the number of elements employed is large and the potential energy is very close to the converged value. This would seem to imply that the predicted stresses are incorrect even for large numbers of elements. Although stress jumps are useful guides to the convergence of the stress values predicted by the finite element solution towards the mathematically correct values, they do not tell the complete story. The finite elements do their best to approximate the analytic values of the stresses according to the level of approximation available for predicting stress values. In the case of the simple bar elements employed here, these can only provide a single value for the stress in the entire element and, therefore, are constant stress elements. The first question to ask is 'How good are they are predicting the mean values for the stresses in each element used in the modelling process?' Let us answer this by examining a single case and take the model using four elements. In Table 4.1 a comparison is

Table 4.1 Finite element predicted mid-point stresses against analytically calculated mid-point stresses.

Element number	σ_{FE} Finite element stress value (N/m^2)	σ_{AN} Analytic mid-point stress value (N/m^2)	% Error $(\sigma_{AN} - \sigma_{FE}) \cdot 100/\sigma_{AN}$
1	8813	8859	0.52
2	7688	7734	0.59
3	5437	5484	0.86
4	2063	2100	1.76

made between the mean values predicted by theory and the values from the four-element model.

Table 4.1 clearly indicates that the finite elements are doing a good job in estimating the analytically obtained values of the stresses in the bar at the mid-points of the four elements. In order to exploit this ability to obtain relatively accurate values for stress at specific points not coincidental with the element mid-point, we could use an interpolation process.

As an example consider the question of employing interpolation as a means of improving the predicted values of stress at the element nodes. The stress jumps across element nodes were used earlier as a rough guide for estimating the convergence of the finite element estimated stresses in the bar. Staying with the four-element model, the stresses at node 3 can be estimated by fitting a separate linear interpolating curve through the mid-points of elements 1 and 2 and through the mid-points of elements 3 and 4. This gives two interpolating curves: one for $\sigma_L(x)$ that predicts the stress in the region to the left of node 3 and $\sigma_R(x)$ for the stress in the region to the right of the node:

$$\text{For elements 1 and 2} \quad \sigma_L(x) = 9376 - 75x$$
$$\text{For elements 3 and 4} \quad \sigma_R(x) = 13,870 - 224x$$

Now substituting $x = 30$ into each of these two equations gives:

$$f_L(30) = 7125\,\text{N/m}^2$$
$$f_R(30) = 7124\,\text{N/m}^2$$

Thus, the two equations predict almost identical values for the stress at the junction of elements 2 and 3, which compares well with the analytic value calculated from the analytic expression of 6750 N/m^2. Using these two values to estimate the nodal value also indicates that a set of stress estimates for the structural stresses can be obtained that are more accurate than is implied by comparing the stress jumps from the unmodified stress output from the finite element solution.

It is clear that it is possible to get more accurate values from the finite element output by using interpolation methods and this approach is used in many of the commercial finite element systems to create *a posteriori* (meaning after the event or in this case after the finite element analysis has been completed) error estimates. However, the simple bar problem used here for demonstration purposes is a very benign problem and caution needs to be exercised when applying interpolation and other stress improvement methods for complex structural analyses involving two- and three-dimensional analysis problems.

4.11.3 Displacement Convergence

Having observed the way that the stresses and potential energy converge as the number of finite elements used in solving the analysis problem is increased, we are in for a surprise when we examine the convergence of the displacements at the element nodes.

First, note that the predicted displacement at the end of the bar ($L = 60$) for all the finite element results is given as 12×10^{-3} metres. Second, for the finite element layouts with 2 and 4 elements, where the nodal displacements are given at the bar mid-point ($L = 30$), the finite element solutions give a value of 8.25×10^{-3} metres. Both of these finite element values agree with the analytic values. This agreement between the finite element and analytically predicted values of the displacements is not fortuitous because all the elements give the exact analytic value at the points where the models have nodes.

To illustrate this, consider the four-element model; if the displacement values at the element's four unconstrained nodes obtained from the finite element solution are compared with the computed values given by using the analytic expression, it will be seen that there is exact agreement. This agreement is due to the use of the consistent load method which, for a bar modelling by equal length two-noded elements, gives a solution in which the nodal displacement values from the finite element solution agree exactly with the analytic solutions of the mathematical model. If the elements used are not of equal length, the agreement is no longer exact but the rate of convergence for the finite element predicted displacements is still very fast. Although this rather nice property of exact agreement is not general for all element configurations, it does illustrate that using the consistent load method has some important advantages with respect to convergence to the correct analytic solution.

4.12 KINETIC ENERGY

In dealing with problems where the structure is responding to static loads, we have shown that the Finite Element Method minimises the potential energy of the system. If we move to structures where the response to the external environment is such that it gives rise to a dynamic behaviour, then the method has to take account of the kinetic energy involved. This requires that we add the kinetic energy to the potential energy discussed earlier so that a function is created, originally introduced into the mechanics of deformable systems by Lagrange and called after him the Lagrangian. This Lagrangian function[1] is constructed by combining the kinetic energy with the potential energy to form:

$$L = T - \pi_p$$

where π_p is the standard potential energy term and T is the kinetic energy of the system. In solving the structural analyses problems for statically loaded structures we noted that this is achieved by minimising the potential energy function; in the dynamics case there is a different minimising principle which states:

> *Of all possible time histories of the displacement states which satisfy the compatibility conditions and constraints on the kinematic boundary conditions and which satisfy the conditions at the initial and final times (t_1 and t_2) of the dynamic behaviour, the history which makes the time integral of the Lagrangian function a minimum corresponds to the actual solution.*

Hence the solution to the dynamic analysis problem is found by employing the following principle:

$$\delta \int_{t_1}^{t_2} L\delta t = 0 \tag{4.18}$$

known as Hamilton's Principle. Thus we have the same type of situation that we encountered in the static analysis case where we now have a specific function that must be minimised in order to generate the solution to a dynamic structural analysis problem.

Although the Lagrangian function can be complex for a very wide range of dynamic analyses problems, in the case of the free linear vibration of an undamped structure, the principle defined by equation 4.18 degenerates into the very simple proposition that the solution point

---

[1] It is worth noting that the Lagrangian used in structural dynamics is just one example of a Lagrangian; many other examples exist and are deployed in the solution of a wide range of optimisation and physical mechanics problems.

is reached when *the maximum value of the kinetic energy of the system is minimised* (we may recall from Chapter 3 that there is a reciprocal arrangement between the kinetic energy and the strain energy for a vibrating structure because of the exchange of energy taking place between these two energy types; thus we could equally well relate the solution to the value of the strain energy). This may seem to be a very restricted set of problems, but if we recall that the solution of a free vibration problem forms the building blocks for a very wide range of forced vibration problems, then we can see that examining the free vibration problem sheds light on the solution process in the design analysis for many, if not most, real-world structures.

In order to illustrate this minimisation process consider the bar problem used earlier in this chapter in Figure 4.7 for the static case but without the applied load. The vibration analysis is undertaken using the same physical dimensions and material properties as those employed in Section 4.10 and the same two-noded bar element. In order to compute the free vibration frequencies and modes, the element stiffness and mass matrices are required. The stiffness matrix is formulated in Section 4.10 as:

$$k = \frac{EA}{L}\begin{Bmatrix} 1 & -1 \\ 1 & -1 \end{Bmatrix}$$

and the consistent mass matrix as:

$$m = \frac{\rho A}{6}\begin{Bmatrix} 2 & 1 \\ 1 & 2 \end{Bmatrix}$$

where ρ is the density of the bar material which is taken as a typical value for aluminium at 2500 kg/m^2 and, as before, E is Young's modulus, L the element length and A the bar cross-section and we take their previously defined values. Taking the 2-element representation of the bar shown for the static case in Figure 4.1, the explicit form for equation 3.8 for this problem is:

$$\left\{\begin{Bmatrix} 2\times10^6 & -2\times10^6 & 0 \\ -2\times10^6 & 2\times10^6 & -2\times10^6 \\ 0 & -2\times10^6 & 2\times10^6 \end{Bmatrix} - \omega^2\begin{Bmatrix} 5\times10^4 & 2.5\times10^4 & 0 \\ 2.5\times10^4 & 10\times10^4 & 2.5\times10^4 \\ 0 & 2.5\times10^4 & 5\times10^4 \end{Bmatrix}\right\}\begin{Bmatrix} u_1 \\ u_2 \\ u_3 \end{Bmatrix} = 0$$

Table 4.2 Variation of kinetic energy with number of elements.

Number of finite elements	Kinetic energy N·m (J)
2	20
3	18
4	17.31
5	16.997
6	16.828
7	16.727
8	16.662
9	16.617
10	16.585
25	16.471

The boundary conditions have to be applied to take account of the fixed end condition at $x = 0$ which, for this simple problem, can be achieved by deleting the first rows and columns. The associated eigenvalue and eigenvector problems can then be solved to give estimates for the first and second natural frequencies and the kinetic energy for the bar of:

$$\text{First natural frequency } \omega_1 = 6.325 \text{ radians/second}$$
$$\text{and kinetic energy } KE_1 = 20 \text{ N}\cdot\text{m}$$
$$\text{Second natural frequency } \omega_2 = 12.647 \text{ radians/second}$$
$$\text{and kinetic energy } KE_2 = 80 \text{ N}\cdot\text{m}$$

If this process is repeated following the static example so that we have estimates for the natural frequencies for models employing 2 to 10 elements, then the values for the estimated kinetic energy associated with the lowest, first natural frequency are as shown in Table 4.2 with the 25-element result included as the converged value.

Alternatively, this can be shown graphically as displayed in Figure 4.12 where the variation of kinetic energy with respect to the number of elements used in the model from 2 to 10 elements is shown as a percentage of the final value achieved when 25 elements are employed.

Although presented in different forms, the graphs shown in Figure 4.11 and Figure 4.12 exhibit similar convergence rates for the potential energy associated with a static analysis problem and the kinetic energy associated with the first natural frequency for this bar structure. Thus, at the energy level, the finite element solution process has the same overall behaviour pattern for both static and dynamic structural problems. In addition to showing similarities between the

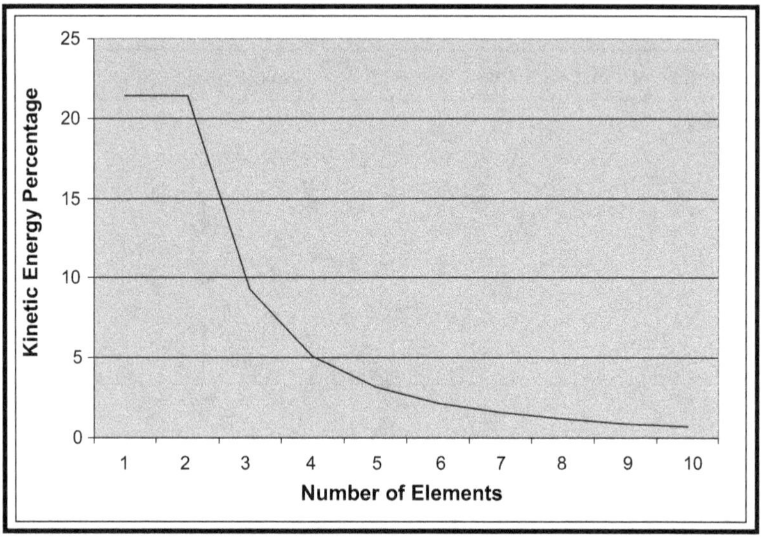

Figure 4.12 Kinetic energy convergence.

static and dynamic cases, they also exhibit a major difference in the convergence rates for some of the primary design variables in these two behaviour states. In order to explain this point note that the kinetic energy and the frequencies are directly connected so that as one converges so will the other in lock step (this also applies to the strain energy). This means that the rate of convergence shown in Figure 4.12 also applies to the first natural frequency.

In Section 4.10 it was shown that the stress estimates, even with some form of augmentation, converge to the final estimated value at a rate slower than that of the potential energy. The corollary of this observation is that the Finite Element Method converges more rapidly on the final value of natural frequencies than it does on the final value of the estimated stresses. In view of this the analyst, when confronted with a problem involving both static and dynamic cases, should give some thought as to whether it is better from an efficiency viewpoint to undertake separate static and dynamic analyses using different models.

Often there are powerful reasons why one analysis model only should be used in this situation, but if the same finite element models are used, then a reduction technique would normally be employed. Some care has to be employed when reduction techniques similar to that discussed in Chapter 3, Section 3.5.2, are used. This type of reduction technique requires that a set of equations based on static considerations are imposed and act as constraints on a dynamic analysis. Placing constraints on a structure undergoing elastic displacements can stiffen the structure. In order to provide a simple illustration the bar structure used above to illustrate convergence in kinetic energy can be employed. Consider the 2-bar element layout shown in Figure 4.9; if the reduction method from Chapter 3, Section 3.5.2, is invoked and node 2 taken as a slave, Table 4.2 shows that the kinetic energy for this configuration is 20 N·m with an associated first natural frequency of 6.325 radians/second. The reduced configuration also gives the same value for the natural frequency in conformity with the usual rule that this form of reduction generates the lowest order frequencies and for the kinetic energy. Moving to the 4-element layout shown in Figure 4.10 and taking nodes 2, 3 and 4 as slaves, the kinetic energy is once again 20 N·m, which is higher than that found in Table 4.2 for this element configuration. The reduction process has limited the vibration modelling capacity of the 4-element layout to that of the 2-element layout. In fact, using any number of elements for this problem and reducing the number of masters to just two end nodes produces the same result with respect to the first natural frequency. Making all the interior nodes slaves means that all the finite element models are restricted to behaving in the same way – this is not a particularly sensible slave selection when the number of elements is large, but it serves its purpose of illustration. If the normalised modes constructed from the reduction solutions are employed, then improved values for the first natural frequency are found that track the convergence process followed by the unreduced solutions but are not identical to them.

4.13 FINAL REMARKS

As stated in the opening section of this chapter, convergence discussed in this chapter relates to the convergence of the finite element solution towards the analytic solution of the problem that *it thinks it is solving*. If the model is incorrect in that it cannot represent the behaviour of the real-world structure, then we have an accurate solution to the *wrong problem*.

REFERENCES

1. Szabó, B. and Babuška, I., *Finite Element Analysis*. 1991: John Wiley & Sons, Ltd, ISBN 0471502731.

2. Tong, P. and Rossettos, J.N., *Finite Element Method: Basic Techniques and Implementation*. 1968: The MIT Press, ISBN 0262200325.
3. Zienkiewicz, O.C. and Taylor, R.L., *The Finite Element Method Set*. 2005: Butterworth–Heinemann, ISBN 100750664312.
4. Washizu, K., *Variational Methods in Elasticity and Plasticity*. 1981: Elsevier, ISBN 100080267238.

5

Preliminary Review of Errors and Error Control

5.1 INTRODUCTION

Chapters 2, 3 and 4 are concerned with the basics of the Finite Element Method focusing on how a computer goes about setting up and solving finite element analysis problems and emphasising the convergent nature of the process. These chapters show how the method can be viewed in a mechanical sense where data is input and results output. This point of view is reinforced by the ease of use of modern finite element and computer-aided design/analysis systems employing sophisticated graphical user interfaces that allow a finite element analysis to be set up and run by someone with little or no experience or understanding of the method. Similarly, the results of an analysis are presented in attractive forms employing coloured stress plots, animations, etc.

In essence, these systems are de-skilling the process of applying the Finite Element Method in the solution of complex structural analysis problems. Unfortunately, a lot can go wrong in a finite element analysis because there are many areas where errors can be generated or uncertainties appear and user experience and knowledge application are critical factors in their control. In addition to experience and expertise, an analyst or analysis team requires a methodology that can provide a set of procedures and processes that allow error and uncertainty sources to be identified and their influence quantified.

A Practical Guide to Reliable Finite Element Modelling A. Morris

The remaining chapters in this book focus on this last need and build up a process that allows an analyst to undertake an error-controlled finite element analysis to provide reliable results. By reliable we mean that the output from an analysis accurately predicts the behaviour of a real-world structure, after it has been entered into service, to within defined tolerance levels. The process is endeavouring to create the equivalent of a validated finite element structural analysis model and is achieved using a combination of user experience, specific finite element analyses and test data. The full process requires an understanding of the nature of the error sources, their potential influence on the finite element analysis results and the generation of methods whereby these can be identified, treated and controlled.

This chapter starts the process of creating a methodology that is able to locate errors or uncertainties within a proposed finite element analysis and provide techniques for treating or controlling their influence on the eventual results from an analysis. It also points out that the overall nature and type of analysis require early assessment so that the analysis requirements can be linked to the experience base of the analyst or analysis team.

Although the methodology developed in the next chapters is targeting error and uncertainty identification and controlling their impact, there is the implication that it must also include some aspects relating to consistency. For a set of analyses to be consistent implies that every analysis is treated in an identical manner so that there is no apparent contradiction at any given stage in going from one analysis to another. This requires that the analysis process itself can be decomposed into a logical and coherent set of sub-processes with associated models or representations. The introduction of error control methods within these sub-processes creates consistency between analyses ensuring that an identical analysis problem solved by different teams would yield the same results.

5.2 THE FINITE ELEMENT PROCESS

The overall finite element analysis process is illustrated in Figure 5.1 where it is shown as a series of discrete sub-processes starting with a real-world structure and ending with a set of results that purport to model its actual in-service behaviour as described in outline in Chapter 1. These sub-processes create a sequence of models or representations of the structural analysis problem that have to be assessed from an error

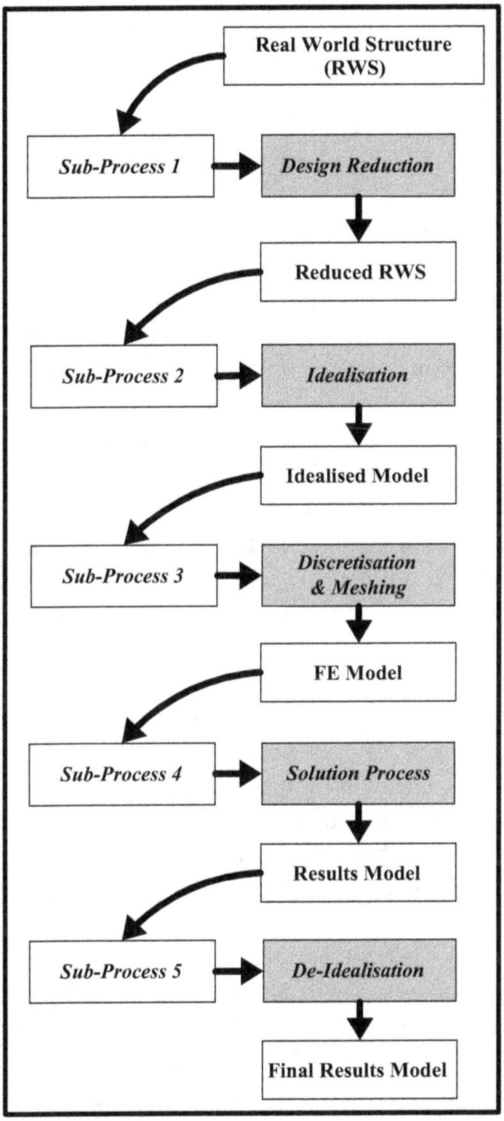

Figure 5.1 The finite element analysis process.

viewpoint. Although presented in Figure 5.1 as a single path through a logical and coherent sequence, the process does admit feedback loops. This is particularly important in an error and uncertainty control methodology which must permit assumptions to be made early in the analysis process and then revisited and changed as required.

The control methodology advanced in this text recognises the fact that, as an analyst progresses through these models or representations, errors and uncertainties must be identified at each stage. These may be different in character and type, dependent on the particular sub-process and the type of model being constructed, reference [1]. Despite these differences, the method developed in this book allows an analyst to take account of the total error accumulation process resulting from cascading through the complete sequence of sub-processes operating with a number of radically different models.

All analyses commence with a requirement to provide data to support the design and entry into service of a structure within a specified design environment. The structure, the loads applied to it and the constraints placed on the design are called the real-world structure and shown in Figure 5.1 as the 'Real World Structure' or RWS. The problem with the RWS is that, normally, it does not exist in a form amenable to analysis and in this situation the analyst(s) must perform the first sub-task that reduces the RWS to a problem designated as the Reduced RWS or RRWS. Essentially this limits the scope of the design problem to be analysed; it defines the domain limits and identifies the loads and loading actions that will be applied during the analysis. This is the first point where the analyst(s) encounters qualification and certification requirements and, if lucky, this will reduce the scope and the complexity of the analysis by requiring that it is performed using predefined or stylised loads that the structure must be demonstrated to be able to sustain without experiencing some form of failure. If the analysis has a low level of novelty, the creation of the RRWS may be relatively straightforward, but if it has a high level of novelty, the reduction process is far from simple and can introduce significant error into the analysis process. In this situation the scoping process may require a number of initial, limited size, finite element analyses as indicated in Chapter 9 and which are also discussed in Chapter 8 where the question of how to assess and quantify errors and uncertainties is addressed.

When the 'Design Reduction' sub-process 1 has been successfully concluded, the critical sub-process of creating an idealised model is undertaken in sub-process 2. This takes the RRWS that is defined in terms of actual structural properties with its discontinuities, joints, welds, rivets, etc., and generates what is essentially a mathematical definition of the structure and its loads that can be modelled by the finite element process. The error and uncertainty types and sources that are encountered during this process are defined and characterised in Chapter 7. Once the process of generating an idealised model or models has been completed, the tasks associated with selecting elements and mesh creation are undertaken in sub-process 3, here called 'Discretisation and Meshing'. Advice on how to get the best out of the models created during this sub-process is given partially in Chapter 4 but mainly in Chapter 6.

When all the decisions are made with respect to finite elements, models, loads, boundary conditions, response types, etc., the resulting finite element data set can be solved in sub-process 4, the 'Solution Process'. Of course there is ample opportunity to create numerical errors during the solution process but methods for the identification and control of this type of error are adequately covered in a range of authoritative texts on numerical analysis and are, therefore, not covered in this book.

Finally, sub-process 5 takes the results obtained from a finite element analysis and interprets them in terms of the RWS that is to be designed. This provides the designer with the necessary information to continue with the design process leading to the creation of a structure that fits the requirement and is fit for purpose. Because this process is very design specific it is not covered in this book.

Looking at the total process outlined in Figure 5.1 it is clear that a dividing line occurs as the analyst moves down the sub-process chain. This line occurs at sub-process 3 where the discretised and meshed model is created. The sub-processes 3 and 4 are concerned with selecting appropriate finite elements, meshing the structure and offering up a model to the computer for numerical solution. It is convenient to consider these sub-processes as 'internal' to the Finite Element Method as they are concerned with manipulating finite element entities only. Sub-processes 1 and 2 are 'external' as they are concerned with interpreting the behaviour of a real-world environment so that it is in a form to which the Finite Element Method can be applied.

5.3 ERROR AND UNCERTAINTY

Thus far no real attempt has been made to define what is meant by error, nor how it relates to the uncertainty that is inevitably encountered in engineering design. If we take a quantity 'q' to represent a characteristic property of a structure, this could be a stress in a structural member, a displacement, a vibration frequency, etc. Taking q_{is} as the value of this quantity in the in-service structure and q_{fe} as the value given by a finite element analysis, we have:

$$q_{is} = q_{fe} \pm \delta q \tag{5.1}$$

This means that the prediction given by the finite element analysis will differ from that found in the structure when in operational service by an amount $+\delta q$ or $-\delta q$ where δq is the error. Defined in this way, error is the measured difference between a finite element model and the real world. As stated in Chapter 1 an appeal to the Church–Turing Theorem establishes that it is, in principle, reasonable to ask that a computer program can perfectly simulate the behaviour of a physical system, in this case a structure responding to external loads. It implies that it should be possible to drive terms such as δq to zero. However, this is not realisable but the corollary to the theorem is that a process must exist whereby the term δq can be controlled and its impact bounded: constructing such a process is our present task.

Normally error, used in an engineering sense, has the technical meaning that makes it synonymous with uncertainty as discussed in Taylor's excellent text, reference [2]. Whilst this is true in the case of finite element analysis, there are situations where the use of the term uncertainty would not be applicable. It is, therefore, appropriate in this chapter to use the term error as the basic term applying to all the factors that cause the structural behaviour, predicted by a finite element analysis, to differ from that found to be the case when the structure is introduced into service. In this context error has two components, generated error and uncertainty, and the total error is the addition of these two components, thus:

$$\delta q = \delta q_{ge} + \delta q_{un} \tag{5.2}$$

The term δq_{ge} is the generated error and relates to the errors that are introduced into an analysis by the action of the analyst(s) by, for example, employing inappropriate finite element meshes or elements. This type of error for the most part relates to the 'internal' sub-processes identified and defined in Chapter 6, Section 6.2. Errors of this type are introduced into the analysis through mistakes made because the individual or team undertaking the analysis does not have sufficient knowledge of finite element theory and practice. Chapter 6 endeavours to provide sufficient help and background knowledge to allow this type of error to be removed or controlled.

The term δq_{un} is the 'error' due to the presence of uncertainty found in the two 'external' sub-processes 1 and 2 defined in Section 5.2. In principle, there are two main types of uncertainty that give rise to errors in a finite element analysis:

1. Random variability in a parameter or measurable quantity so that it cannot be fully quantified.
2. Lack of knowledge on the analyst's part about:
   (a) the completeness or applicability of models selected for an analysis;
   (b) the precision of the variables used in an analysis;
   (c) the model's ability to predict correctly the final behaviour of the structure.

As an example of the uncertainties defined in (a) consider the loads experienced by an airliner during the course of its service life. While the designers may know the type of loading that the aircraft will be subjected to during the course of its active life, i.e. that it will experience buffeting due to turbulence, there is no way of knowing in advance the frequency and load intensity to be experienced. Normally this type of uncertainty is handled through the use of representative load functions coupled with the application of appropriate safety factors that usually ensure the resulting design is not optimum but is, at least, safe.

There is often a linkage between the three types of error cited in definition 2, particularly between (b) and (c). To illustrate this point, consider the finite element model to be used in the analysis of a plate-like structure. If this structure genuinely exhibits a behaviour pattern that should be modelled by classical plate-bending theory, then the analysis should be undertaken using plate finite elements. However, if the analyst has an incomplete understanding of the structural behaviour, it is possible that he or she will select a membrane element for this finite element analysis. Incomplete knowledge has led to the use of a structural model that is not applicable and the results achieved by such an analysis will not correctly predict the behaviour of the structure. An example of (b) is the accuracy or otherwise of the material properties to be used in an analysis. Material properties for many complex materials, carbon fibre reinforced plastics, for example, are derived from coupon tests and this generates uncertainties for the analyst who many not know the size of the coupon used or the type of test undertaken to define the specific property required for an analysis. It should be noted that lack of knowledge in this section does not imply ignorance of the subject matter on the part of the analyst, nor inexperience; it simply means that the required information is not directly available.

The above description shows that uncertainties in our error definition are *random* as they relate to uncertainties in measured quantities being fed through as structural properties, support properties, applied loads, uncertainties in build quality, etc. Clearly a complete finite element analysis has *systematic errors*, for example, relating to the accuracy and stability of the numerical algorithms used in the solution process. However, although these do not form part of the set of the errors and uncertainties considered in Chapter 7 and Chapter 8, a very comprehensive description of this type of error can be found in the book by Higham [3].

Using this interpretation of error the results from a finite element analysis can be considered as measurements of the performance of an RWS and the differences between these and the values obtained from measurements made of the actual performance in service are caused by the presence of random uncertainties. The aim of this chapter is to start the process of identifying the sources of these uncertainties and to complete the process in Chapter 7 so that Chapter 8 can provide methods for 'measuring' their magnitude or, at least, controlling their influence.

5.4 NOVELTY, COMPLEXITY AND EXPERIENCE

In addition to the specific sources discussed in other parts of the book there are a number of more general factors that can make the process of undertaking a finite element analysis either more or less prone to generated error and these relate to the analyst's experience, the degree of complexity of the analysis or the degree of novelty of the design. These are now examined separately.

5.4.1 Analysis Novelty

The degree of novelty in a design is often directly linked to the industrial sector. In certain industries, economic factors or the need to ensure a low-risk design encourages the designer to retain as much as possible from previous designs. In other industrial sectors the opposite is the case and the success of a design in breaking into the marketplace or outperforming the competition is linked to a high degree of novelty. The decision to design a highly novel product does not necessarily mean that the required structural analysis has a high degree of novelty as the new product may simply involve the novel use of existing components. Nevertheless, a novel design often implies that a novel finite element analysis is required. To be clear, the novelty addressed in this section relates only to the novelty of the finite element analysis. Below we attempt to give some broad definitions for the level of novelty associated with a specific analysis, but these are not meant to be definitive as individual companies will have their own ideas on novelty levels. It should also be noted that the definition of novelty is somewhat subjective and is directly connected to the experience and maturity of those undertaking the finite element analysis. However, assessing the level of novelty and relating that to the experience of the team must be part of any error control process.

5.4.1.1 Low level

A new analysis with a low level of novelty is one that is, to a very large measure, similar to one already performed by the individual or team tasked with undertaking the analysis. Often it reuses an existing analysis with minor modifications. Thus the structural form, layout and materials should be the same as those used in the earlier analysis. The load cases applied to the new structure should be identical in all significant aspects, though the value of the loads may change. However, the changes in the load values should not change the nature of the structural responses; if the earlier analysis was linear static, then the new analysis should be linear static and not, for example, require a non-linear analysis. Similarly, if the first analysis is a static case, the new one should not involve dynamic loads and in a dynamic case should not migrate from a low-energy to a high-energy case.

These conditions, for considering an analysis as significantly similar to an earlier analysis, relate to the overall aspects of the analysis process. Conditions also apply at the model and element level. Thus a low level of novelty implies that the structural models used in the earlier analysis remain unchanged in the new analysis. If the models are to remain unchanged, so should the elements selected with respect to their broad representational characteristics; thus a membrane element used in one analysis should not be changed to a plate element in the next, but it is perfectly reasonable to change the number of nodes from (say) a four-noded membrane to an eight-noded element without changing the level of novelty. Implicit in this discussion is that the underlying mathematical models used to model the structural behaviour within the finite element formulation are not changed. However, the size of the analysis with respect to the mesh densities can be changed without having to raise the level of novelty, but there are clearly limits to the expansion of the model size.

5.4.1.2 Medium level

In this case the finite element analysis being undertaken is based on an existing analysis but has been subjected to modification because the earlier analysis does not meet the new specification. This level of novelty occurs when an existing product or component is developed to meet an enhanced or more complex requirement. The overall structural layout should remain similar to the earlier analysis but major components may have changed. This can result in the need for a different mathematical model. For example, a component's thickness has changed and in the earlier analysis was modelled as a membrane, but this has now to be modelled as a thin plate. In the case of a dynamically loaded structure, a significant change in the number and type of vibration modes could be tolerated within this level of analysis novelty.

In both static and dynamic cases medium, level novelty implies that there is no major change in the predicted structural responses from, for example, linear to non-linear. A change in material properties can be accepted, but a change from the use of an isotropic to anisotropic material might require the analysis to be moved to a high novelty level state. The load cases and the loading types can be changed within this novelty level providing this does not result in a major change in the structural responses.

5.4.1.3 High level

A high level of analysis novelty is usually encountered when a new product or component is being designed for which there is only limited previous experience. One example would be the move from designing a sub-sonic civil passenger-carrying airliner to designing a super-sonic civil airliner. The new aircraft would involve analyses where thermal heating becomes a new major feature, with significantly different loads, and, in order to create an aerodynamically efficient aircraft, the structural shape and layout would be completely different.

An analysis with a high level of novelty is radically different from previous analyses undertaken by the analyst or analysis team. The structural form and layout will be new, requiring new structural models with significantly different behavioural patterns from those previously encountered. The loads will give rise to a different set of structural phenomena from those within the analysis experience base. Similarly, a move from isotropic to highly anisotropic materials would also indicate a need to categorise the analysis as having high-level novelty. A long list of factors indicating that an analysis has a high level of novelty could be easily drawn up but the key factor is that the analyst or analysis team is encountering a major set or sets of structural analysis modelling problems that have never been addressed before.

5.4.2 Degree of Complexity

It may be tempting to consider complexity and novelty as different sides of the same coin but this is not the case. An analysis could be very complex, such as the analyses associated with a major gas turbine engine to power a civil airliner but may not have a high level of novelty if it is analysed by a very experienced analysis team and does not include any design innovations. While it is possible to attempt to define, at least roughly, levels of complexity as done in Section 5.4.1 for novelty, defining the degree of complexity of an analysis problem is more difficult because most engineering designs are complex and the associated analysis is also complex. Nevertheless, some assessment of the degree of complexity needs to be undertaken as a very complex analysis offers untold possibilities for the generation of errors through simple mistakes or the impact of uncontrolled uncertainty.

Although no specific degrees of complexity are expounded here, it is important that a 'level of complexity' table is created by those responsible for undertaking finite element analyses. As an aid to defining levels of complexity we discuss a number of categories that should be considered in any complexity table.

Depth of analysis: With the exception of very trivial problems, all finite element analyses have a certain 'depth' relating to the way that the internal structure is constructed. For example, an aircraft wing has an outer skin within which lies a load-carrying structure with spars, ribs, stingers, etc., and represents the first level of decomposition or analysis depth. The second might include the attachments that connect the moving surfaces (flaps, etc.) to the main wingbox and attachments connecting the wingbox to the fuselage. This process would continue 'downwards' until all the structural details have been included. As a second example, take the case of an analysis of a single span bridge. The first level would include the major features of the structure: the deck, the girders and the bracing beams. The second level might focus on the details of the components making up these main features: the fact that certain girders are I-beams in certain locations and stiffened plates in others. At a lower level still one might now include the bearings that support the underside of the girders etc. The process of evaluating the analysis depth requires:

- Identifying the number and type of major and minor component interfaces.
- Identifying the number and type of joints that connect components; these can be components at the same 'depth' level in a structure or those in a lower level connecting to a higher level component.
- Identifying the number and type of attachments; in many modern design environments the most important attachment is that linking the analysis under the control of one analyst or team with that of a separate analyst or team – distributed design with separate teams working with separate analyses is common in most major engineering industries.

Response types: In the case of linear static or linear dynamic analyses it would seem reasonable to identify these as having a low level of analysis complexity. Analyses involving material or geometric non-linearities ought to be considered as more complex. A further level of complexity arises when a multi-disciplinary finite element analysis is undertaken where, for example, fluid–structure coupling is involved. Possibly the most complex analysis case occurs when a multi-disciplinary analysis also includes multiphase.

Materials: In the materials category the least complex analysis cases are those involving a single isotropic material. Next in line are analyses with a single anisotropic material such as carbon fibre reinforced plastics, or an analysis involving a number of different isotropic materials requiring the creation of finite element models involving material boundaries. Finally, we have analyses with multiple anisotropic materials; here the main difficulties may arise during the results interpretation process.

Qualification and certification requirements: As discussed in the Introduction, qualification and certification requirements can simplify an analysis. However, in a safety-critical environment where structural failure can result in very serious loss of life, satisfying the safety case can significantly increase the level of analysis complexity. Once again complexity may lie both in the analysis itself and in the results interpretation process.

Design complexity: The complexity of the design specification has a direct impact on the level of complexity associated with the required analysis. All of the above four categories for assessing the level of complexity of the analysis can be directly associated to the complexity of the design specification.

5.4.3 Experience

The primary method for creating a starting point for the majority of analysts is to employ past experience. While this is an effective way to make progress, it is often not done in a systematic manner which allows a logical connection from a current problem to one previously encountered. Engineers rely on intuitive knowledge in deciding that one structure is sufficiently close to a second example to allow the modelling procedures used in the one to be repeated in the second. Many years of experience in solving problems using finite element analysis methods are a very valuable commodity in solving new structural analysis problems. Nevertheless, it does not provide an infallible process for achieving an error bounded analysis method which can be used *ab initio* for all new structural design problems. Nor does it provide a logical method which can be used by the less experienced engineers who are fresh to the analysis process or by experienced engineers operating in a completely new analysis area.

The approach put forward in this book for controlling the influence of error on the results from a finite element analysis requires a systematic application of experience in support of the idealisation process. Experience is used at a number of levels within the qualification process reflected in the way that it is incorporated into the analysis tasks. Upper levels relate to the way that the qualification criteria influence the setting up of the analysis process. Factors such as the qualification criteria themselves, the load type, the analysis type, etc., are part of this upper level. Lower levels are concerned with the way that experience is connected to the more detailed aspects of the analysis process through which features can be identified.

A key concept in the exploitation of experience is the establishment of an experience database consisting of a set of finite element models which have been evaluated against real-world performance with errors assessed and bounded, reference [4]. Such models provide a basic data source for the construction of a finite element model for a new structural analysis problem. Models from this database can provide both the starting point for modelling a new structural analysis problem and a link with the exterior real world. If properly constructed, the datum is a way round the paradox established by Gödel's theorems since it relates the finite element model to a knowledge base which has not been obtained from the finite element model itself. The use of the datum model concept is not new, reference [5], since all companies which routinely use finite element analysis have available to their analysis teams past analysis models. However, these past analyses have not usually been treated to a comprehensive error treatment or control process, nor can they be reliably compared to a new structure in a logical and coherent manner.

The question of how to relate one structure to another through a logical connection requires the establishment of similarity rules. These, in turn, require that a set of parameters be identified which uniquely define a given structure. In complex analyses a range of parameters may be required to identify fully a specific structural problem which may not have equal weight. Thus strong and weak parameters are needed, the strong being absolute in nature while the application of the weak may be more relaxed. The identification of appropriate comparison parameters opens the way for the establishment of similarity rules which allow two structures to be compared and their relative similarity defined. A new structure can then be directly compared to an established datum and an initial model built up by using the datum model supplemented by the treatment processes described in Chapter 9. Unfortunately, at the present time, there is no strong theory on similarity rules to support the finite element analyst in logically comparing analyses. As a result, it is not possible to put forward a well-defined set of rules for establishing the similarity of one analysis model to another. Nevertheless, categorising analyses by load type, response type, element and mesh layouts, etc., can provide basic information to suggest that analyses are either similar or dissimilar. It is recommended that analysts or analysis teams establish model databases and that each model is accompanied by a categorisation that allows it to be logically compared with later analyses using this type of information.

A specific datum model for a given problem may be either a complete model for a comparable problem or a model for a major sub-component. In the case of a complex structural design the error treatment process may, therefore, require the use of several such models employed in a hierarchical sequence. In this scheme a datum model could be used for the first time deep within the modelling process where model improvement procedures come into play. Any model which is being employed as a datum must have certain properties. First, it must be a finite element meshed model with identified errors which are controlled and bounded. This implies that all the procedures introduced in Chapter 6 have been employed and that the resulting meshed model has been fully optimised with respect to the elements selected, the mesh parameters, etc. Second, it must be fully characterised with respect to the set of the comparison parameters. Finally, the similarity rules which link it, in an error-controlled manner, to the new structure presented for analysis must be available.

Lack of experience is possibly the greatest source of error in a finite element analysis. Modern finite element systems provided by the many reputable suppliers have easy-to-use front ends that allow very inexperienced users rapidly to become expert manipulators of these systems. Being able to drive a finite element system is not the same as understanding how to use it. A good manipulator is not equivalent to a good finite element analyst! While experience is essential, it is the level and scope of the experience that are important and these link to the level of novelty and degree of complexity of the analysis.

The link between experience, novelty and complexity is illustrated in Figure 5.2. This figure illustrates the common-sense notion that

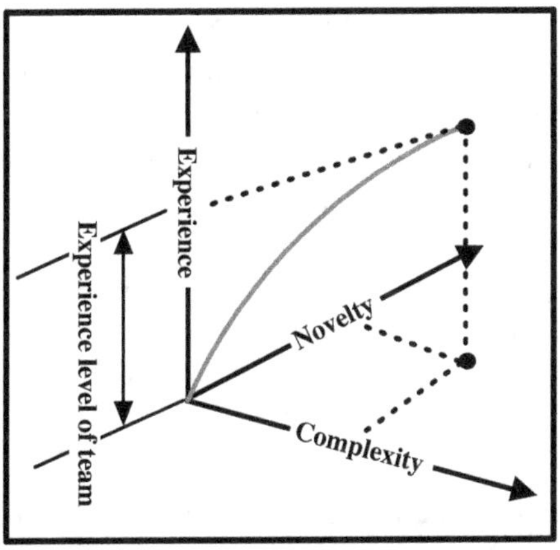

Figure 5.2 Relationship between experience, complexity and novelty.

experience is built up by moving along the complexity and novelty axes. It indicates that the team or person in question has an experience profile that has developed in both handling complexity and novelty as typified by the curved grey line. One may note that it appears, in this case, that there is more experience in dealing with novelty than complexity. In Chapter 8 we return to the question of taking into account the experience level of an individual analyst or analysis team. Here we are simply remarking that assessing experience may not be all that simple when a new analysis problem is being confronted.

5.5 ROLE OF TESTING

Within the methodology and processes being introduced in this book the position and role of testing have changed. Normally testing is performed to give additional assurance, or possibly as the main assurance, that a structure will perform as required to satisfy the qualification regulations. Here its role is to provide assurance that the assumptions identified in the finite element modelling process are appropriate so that the errors implicit in the process are controlled and bounded. As pointed out in Chapter 1, testing is now subservient to analysis and it is the responsibility of the analyst to define the test parameters in order that appropriate error information is furnished. This approach has implications for the test itself since its results must be validated with respect to the actual behaviour of the real world. As with analysis, the results from a test must now be validated and the errors associated with it assessed in a manner similar to that for analysis results. These test errors are not the usual experimental errors, since these can be controlled by a competent test house, but are the errors associated with the deviation of the test from the requirement to replicate accurately the real structural behaviour. They represent the difference between the behaviour assumed by the analyst for modelling purposes and the real-world behaviour. This requires using some process which provides an assurance that the test matches the real-world situation. It contrasts with the usual assumption that a test always replicates the real world and that the only errors which need be considered are the traditional experimental errors associated with uncertainty in measurement.

As an example of the process we might consider the analysis of an aircraft structure involving a complex joint. The analyst may feel that an appropriate model is provided by assuming the joint behaves as a beam. If the specific joint has not been modelled before, no datum is available and if the assumption cannot be adequately treated by the methods of Chapter 9, then the analyst must have recourse to testing. The requirements for the test are clear. First, does the joint actually behave in a way which conforms to the assumptions of beam theory? If the answer is broadly affirmative, then the analyst needs to know how much deviation might occur in order to provide an error bound. Second, the analyst must be informed on the limits to the assumption, i.e. how much deformation can the model predict before the basic beam assumption is invalid? Finally, if the answer is negative, then the analyst would wish to know in what aspect this modelling assumption is inadequate. For example, if a thin beam has been assumed and is inadequate, the subsidiary question would be 'Is a thick beam model more appropriate or is the structure behaving in a far more complex mode?'

5.6 INITIAL STEPS

One of the aims of this book is to assist analysts in formulating or creating a finite element analysis methodology that ensures a structure is fit for purpose and, *in extremis*, provide the foundations for a legal defence in the event of a structural failure associated with finite element results that do not match the actual real-world structural behaviour. The process of ensuring a product is fit for purpose is normally achieved by embedding the design activity within a well-established quality system. Such a system should have a process that establishes design or structural behaviour parameters through which the structure can be qualified or certified and places limits on them. The type of parameters employed depends on the analysis and, typically, may be stresses, displacements, frequencies, etc. The aim of the analysis is to find values for these parameters as they have now become the primary targets for the analysis. Once the analysis targets have been selected, the analyst must set about defining the limits on the acceptable levels of error or uncertainty for each target. The first step is to place a finite element analysis within a qualification process.

5.6.1 Qualification Process

Structures are required to be designed and manufactured with a level of integrity appropriate to their intended use and to the possible consequences in the event of a structural failure. The required level of integrity is industry dependent with very high levels of integrity necessary for civil aircraft or nuclear power stations and lower levels for household appliances. In the case of many industries, where the structure is to operate in a safety-critical environment, the qualification rules and requirements are set by regulatory bodies. In other cases there may be industry-based qualification rules and, in the absence of externally applied codes, companies must devise their own internal codes in order to develop an effective quality control process. In the development of an internal code, a risk assessment is required in order to quantify the consequences of a structural failure both at the overall product level and at the component level. Even where there is an externally applied set of qualification requirements, a detailed risk assessment should still be undertaken as the primary role of the qualification code is to protect the user community and not the company. In the case of aircraft, for example, the certification process is there to ensure that aircraft do not suffer failures that could lead to the loss of life, but is disinterested in any impact that a structural failure might have on the manufacturing company's financial viability.

The qualification of a design within its operating environment normally has a scope that goes well beyond structural analysis issues. It uses rules and codes of practice that place limits on the magnitude of specified parameters or on operational performance criteria often linking these with supporting tests at both component and full-scale level. In the case of a finite element analysis, the focus is on that part of the qualification process relating to the parameters selected to assess the structural performance criteria. There are a large number of ways to measure structural performance to create qualification criteria. These can be drawn together into three broad categories, but this should not be taken as being comprehensive.

Class 1: This category pre-dates modern computer-based finite element analysis and is very much based on *empirical rules* based on an accumulation of past experience. Often these rules are concerned with the specification of geometric parameters or require that the structure is shown to be able to sustain an idealised load case. Although it is logical to assume that the introduction of advanced computer-based codes would have rendered this type of empirical rule-based approach obsolete, this is not the case as many certification codes still retain empirical rules. These are, nevertheless, useful in providing independent, though somewhat conservative, corroboration of structural integrity. However, Class 1 qualification codes often require the application of rules that do not represent the real world and the analyst, in this situation, should recognise that the 'real world' is now the 'certification world'.

Class 2: The second class is concerned with the use of *permissible states*, usually permissible stress, but may also be permissible strain or other single parameters. The specified states are calculated for the structure and compared with 'allowable values'. In certain cases these 'allowables' can be based on experience but often are derived from standard structural test data or from environmental tests. If based on test data, the tests will have been performed using coupon test samples and the size of the coupon is a factor in making decisions on the lower limit on the element size. It is worth noting that this latter comment is not usually understood by analysts and, therefore, is often ignored, resulting in costly detailed finite element models that have dubious value.

Class 3: One of the objections to the permissible state qualification process is that it does not permit the analyst to work with the actual failure mechanism and does not allow advantageous 'post-failure' behaviour to be taken into account. This is because Class 2 criteria are usually based on the linear theory of elasticity and in many cases a structure can enter the non-linear regime without structural failure occurring. A second objection is that a redundant structure may have considerable capacity beyond the first indication of failure that is not recognised by the Class 2 rules. As an example, the aircraft industry now exploits redundancy in creating fail-safe structures rather than exclusively safe-life structures. This third class is termed *limit state* qualification and is based on the concept that a structure can sustain various loadings with a number of failure states at their limiting states. The limit state is a limiting condition beyond which the structure is considered to be unsafe. This might occur because the structure has reached a limiting physical state, such as stress, or a serviceability limit state has been reached. The first of these implies that the applied loads have reached values that will cause the permanent failure of the structure. The second assumes that a condition has been reached that requires the structure to be withdrawn from service. For example, in the civil aircraft area, the certification authorities lay down a rule that states the point at which an aircraft must be withdrawn from service due to the presence of a crack of specified length. In terms of a risk analysis it is clear that serviceability limit states should avoid the danger that life will be lost but hitting a serviceability limit could result in serious financial loss.

Because Class 3 is a qualification process designed for the analysis of complex structures subject to multiple sets of applied loads, many finite element analyses may be required. The fact that there are many analyses requires that these are drawn together and the manner in which this is done depends on the specific qualification process adopted by the company or industry. However, the final stage is always the same in that the finite element outputs, i.e. the structural responses, are compared to the allowable values. In some cases the qualification code may require that non-linear analyses are undertaken in order to ascertain the post-failure behaviour of the structure either to exploit the remaining strength of the structure or to assist in defining the safety factor. This latter could apply in a case where the allowable represents a buckling state so that the security of the design can be evaluated should the allowable be exceeded in service.

The application of this approach to qualification does not imply that the selected limit state parameter for a specific application is the same for all applications. As an example take the case of analyses used in a fatigue life assessment. In the case of an aircraft with a fail-safe component, the analysis is linked to a requirement that it is possible to instigate an inspection process that minimises the risk of catastrophic failure. The associated allowable relates to the crack growth rate and is focused on the time that the crack is nucleated and the crack propagates until failure results in a complete failure of the structure. In another industry the fatigue limit state could relate to peak stress and characteristic frequency. Thus, in one case the required analysis may limit the value of (say) the stress intensity factor, while in the other it will require forced response analysis.

Summary. Through the application of a qualification process the structural parameters or behaviour characteristics will, or at least should, be fully identified for the analyst. These must now become the subject or targets for the error and uncertainty control procedures discussed in the rest of this chapter.

5.6.2 Acceptable Magnitude of Error or Uncertainty

Section 5.3 introduced the concept that the value of a specific structural parameter, such as stress, obtained from a finite element analysis will not be identical to the measured value when the structure is in operational service as defined by equation 5.1. This has now been refined as the parameters defined in equation 5.2 are not selected by the analyst but result from the qualification process. In this modified situation it is the error or uncertainty in the qualification parameters that must be subjected to control processes. The terms q_i are now the values of the qualification parameters and δq_i the differences between computed values of q_i and those occurring when the structure is in service; these differences are the 'error' between the actual and calculated values. While in practical applications it is impossible to reduce to zero the errors occurring in the solution of a structural analysis problem using the Finite Element Method, it should be possible to provide bounds on δq.

The ideal would be one in which hard error bounds can be introduced in equation 5.1 so that for all the 'n' parameters of interest in the structure q_i, defined by the qualification criteria, with a total analysis uncertainty or error specified by δq_i, there exists a set of positive numbers N_i such that:

$$N_i \geq |\delta q_i|$$

with the N_i having values significantly smaller than the values of the quantities of interest q_i. It is worth re-emphasising the point that the terms N_i are bounds on the differences between the real world and the finite element world and are not bounds on the differences between one finite element formulation and another. Discrepancies between the real-world performance of a structure and the finite element predictions of its performance are not differences between two sets of physically measured quantities so it may not always be possible to find or create accurate numbers that can be used for N_i. As a result a slightly different approach is adopted which uses a reference value for the qualification parameter q_i(ref). To explain the meaning of these reference values consider the measurement of a physical quantity. Normally we would directly measure the value of a real-world quantity and then look at the error sources and work out the probable error in the measurement. In analysing the performance of a structure in the design phase there is nothing to measure as the structure does not yet exist. What does exist is the analysis model which acts as a reference structure; that is, it is a point of reference for assessing how the in-service structure will perform. The reference values are analysis-based estimates, hopefully 'best' estimates, of the actual in-service values for the qualification parameters. If, for example, the analysis model is indeed a finite element model, this will provide estimated reference values for the qualification parameters which can then be compared with the qualification limits.

The terms q_i(ref) are now the values for the qualification parameters obtained from the reference finite element analysis. The error terms δq_i are now the difference between the qualification parameters q_i seen by the in-service structure and the reference values. The error expression can now be written as:

$$q_i = q_i(\text{ref}) \pm \delta q_i$$

Taking this new form for the error term, the qualification parameter can now be related to the reference values by:

$$q_i = q_i(\text{ref})\{1 \pm \delta q_i/|q_i(\text{ref})|\}$$

The term $\delta q_i/|q_i(\text{ref})|$ is the fractional uncertainty or error and this new term requires bounding by a new set of positive numbers M_i such that:

$$M_i \geq |\delta q_i/q_i(\text{ref})| \qquad (5.3)$$

It is worth noting that reference values for the qualification parameters lying outside the acceptable range imply that the design does not satisfy the qualification requirements and is not fit for purpose and requires re-designing. However, it is possible that the finite element model used to generate the reference values is not sufficiently accurate. As an example, consider the case of a free vibration analysis where there is one qualification parameter q which is the lowest natural frequency and the qualification criterion is that this parameter is lower than a specified value of 10 Hz. If a finite element analysis estimates (or predicts) that the lowest frequency is 10.5 Hz, then $q(\text{ref}) > 10$ Hz and the structure appears to violate the qualification criterion but the answer might be, taking account of the discussion in Chapter 4, that the model is over-stiff due to the use of insufficient elements. The analyst would need to give serious thought to running a more refined model which would constitute a new reference model that might lead to the situation where $q(\text{ref}) < 10$ Hz. In this new situation the analyst could then progress to exploiting the methods displayed in Chapter 8 to generate values for the bounding parameter M in expression 5.3.

Although the example used above to describe the reference state is a finite element analysis, the methodology described and deployed in Chapter 9 does not exclusively use finite elements to obtain reference values. Within this methodology the analyst can employ a variety of methods to obtain these values which might include taking 'best estimates' obtained from employing simple engineering formulae or be based on experience, or it may mean values taken from a number of analyses or by other methods. Despite the problems associated with generating these numbers the analyst must decide how much error or uncertainty is acceptable in the estimated values of parameters q_i within the qualification process.

The very first stage in any analysis before any thought is given to the type of analysis, or the elements to be used, or the mesh type, etc., is the process of generating or estimating values for all the M_i. In essence the analyst is deciding how much tolerance is acceptable with respect to variations in the qualification parameters. Essentially, this part of the analysis process is defining the margins or acceptable differences between the values for the qualification parameters predicted by the finite element analysis and those that will occur in the structure when in operational service. In most structural designs these are not decisions that the analyst or analysis team can make alone and these have to be made working with the designer or design team as they are the basis for establishing safety margins for the design. Only when this task has been completed can the real task of undertaking an analysis begin.

The process of creating an initial set of bounds on the qualification parameters may seem an obvious first task as it is not really possible to undertake an analysis without knowing the permitted tolerances. However, it is common practice for analyses to be started and go all the way through to completion without any consideration being given to the required level of accuracy.

5.7 ANALYSIS VALIDATION PLAN (AVP)

The purpose of an AVP is to describe how the analysis will be shown to have been performed in a manner which ensures that it adequately predicts the in-service structural behaviour of the end product. The AVP normally forms one component in a much wider document that demonstrates the product is fit for purpose as a whole and this extends beyond demonstrating the accuracy of results obtained from a finite element analysis. In certain industrial sectors the construction of a validation plan is complex and may be one of the key documents presented in public inquiries. It is not, therefore, our intention to provide a comprehensive description of how to create a validation plan but to highlight the key features required in an AVP that is to be integrated into the methodology advanced in Chapter 9.

Although focused on validating a finite element analysis, the AVP normally draws on factors identified in a main validation plan that influence the structural analysis. For example, in safety-critical design environments complex risk assessments can be required by the main validation document that takes into account likely structural failure modes. A key component of these assessments will involve the impact that uncertainties in build quality, material variability, etc., have on these failure modes. In this situation the certification bodies might require that major structural tests are undertaken to demonstrate the integrity of a structure, irrespective of the quality of the finite element analysis. These statutory tests can be incorporated into the analysis AVP and be exploited to support the bounding processes applied to the finite element solutions in the Chapter 9 methodology.

Within this broad context the AVP must contain all the relevant information to allow the accuracy of the finite element analysis to be fully assessed and to provide guidance to those undertaking the analysis and those required to review it from a validation viewpoint. The AVP needs to have a reasonably well-developed structure and should contain, as a minimum, the points in the following outline (note that this is not a template but an outline guide to the creation of an AVP):

1. Introduction
   - States who is responsible for ensuring that the AVP is adequate for the design task and who is responsible for ensuring that it is implemented.
   - For multi-company analysis teams the AVP must ensure that the qualification codes or criteria are coherent across the groups involved.

2. Initial considerations
   - State the objectives and requirements of the qualification code as applied to the design problem.
   - Interpretation of these objectives and requirements in terms of structural aspects; critical components, failure types and modes which can then be reinterpreted in terms of qualification parameters such as maximum allowable stress, maximum displacements, vibration limitations, etc.
   - Link these to the analysis type: isotropic, anisotropic, linear, non-linear, etc.

3. Validation criterion

   - Acceptable levels of error or uncertainty for qualification parameters defined, see Section 5.5.3.
   - Processes to be employed to quantify the impact or extent of errors present in a finite element analysis on qualification parameters. One is based on processes internal to the analysis, the others external:

     - *Internal*:

       - Methods for identifying sources of error and uncertainty (Chapter 7).
       - Methods for calculating their influence (Chapter 8).
       - Incorporation of these two methods within a measurement and control methodology (Chapter 9).

     - *External*:

       - Other analytical models: These can be finite element analyses carried out by other analysis teams or analysis employing alternative solution processes, e.g. using finite difference methods.
       - Experience: This is data within the experience base of an analyst, analysis team and accompanying database. It could include existing Quality Reports (see Chapter 9), data sheets, design codes, reference books, etc.
       - Experimental data: In this category it is important to distinguish between tests undertaken before an analysis is started and those commissioned during the course of the error control process introduced in Chapter 9. Pre-existing tests would include tests undertaken on similar structures with similar loads or environments that can be meaningfully employed to assist in the validation process. They would also include those tests necessary to supply data that allows a finite element model to be constructed, i.e. material property coupon tests. Tests commissioned to support the analysis while models are being built up are called into play when it is clear that the internal methods fail to deliver a bound on specific qualification parameters. These, however, cannot be detailed at the start of the analysis process as the requirement for such tests becomes apparent as the process outlined in Chapter 9 unfolds.

- Process to be employed to control errors or uncertainties from multi-team involvement in the analyses.

4. Revision procedure

- Explanation of probable evolution of the AVP – in the case of simple analysis problems the AVP may not require any modifications but for most analyses it will change as the analysis process unfolds.
- Identify those responsible for revision process.
- Process for initiating and progressing revisions.

5.8 APPLIED COMMON SENSE

When the bounds on errors have been defined and the AVP drawn up, the process of undertaking the actual analysis can begin. As has been emphasised throughout this book many decisions have to be made to ensure that the results obtained from an analysis accurately predict the behaviour of the structure once it has been constructed and entered into service. This cannot be undertaken 'blind'; the analyst needs to have some understanding of how the structure is likely to behave. Selecting an appropriate set of structural behavioural states on which to base an element selection requires giving consideration to the likely behaviour modes that the structure will experience when acted upon by the loads. This requires an initial, possibly closed-form, analysis to be performed that allows the analyst to form a view as to what is the expected structural behaviour that will be corroborated from the finite element analysis results. This is important and allows the analyst to interpret the finite element results. If the stresses at specific points, for example, turn out to be twice as large as anticipated from the initial analysis, this may be flagging up the fact that incorrect loads have been applied or incorrect boundary conditions applied. Hopefully the methods advanced in this book will ensure that, should the finite element results appear to be counter-intuitive, this is not due to the inclusion of errors.

Before proceeding further the problem needs to be assessed for its level of complexity and novelty using the principles discussed in Section 5.3. The degree of novelty and/or complexity defines the maturity required by the individual or team doing the work which is then matched against the experience of the analyst or the analysis team. If the experience of those doing the analysis exceeds the level of novelty and complexity, then the analysis process can be launched. If not, then additional action is required in the form of back-up expert help to augment the experience base of the analyst or analysis team either as permanent or temporary support. The specific areas where expert help is required need to be fully identified by the analyst(s) together with the expert resource(s) to be employed. Waiting until problems arise before seeking outside expert help is not wise for obvious reasons.

From time to time analysis teams are confronted with a highly novel design of a type not seen before and for which there is no available expertise, i.e. a first-time analysis. In this situation it is essential to create a comprehensive plan that fully identifies those aspects of the analysis that are novel. The plan should have a full description of the approach that will be used to allow the analyst or analysis team fully to control errors and uncertainties associated with the finite element analysis. The greater the level of novelty or complexity, the greater the need for an effective error and uncertainty control methodology.

In addition to explicitly defining how an error- and uncertainty-controlled analysis is to be undertaken, the analysis plan has to cover a wide range of topics starting with the qualification criteria and finishing with an evaluation of the output. The analysis plan links the control process into an analysis scheme that allows the analyst to draw firm conclusions on the validity of an analysis and make judgements on the quality of the final results that purport to represent the behaviour of the in-service structure and link it to the qualification criteria. Chapter 9 shows that a well-organised analysis plan is underpinned by a well-organised error or uncertainty control methodology. This can be done by using the ideas advanced in this book or by following the SAFESA process, reference [4] (or from NAFEMS, see Preface), or by using a reliable in-house approach – whichever path is followed, it is essential that an analysis is undertaken in line with the requirements of an integrated plan.

The final application of 'common sense' is to time. The analysis plan must be associated with an effective process for estimating the time required for an analysis to be undertaken and completed with reasonable accuracy. The analyst or analysis team must always ensure that the finite element analysis job is allocated sufficient time. It is surprising how often this aspect of the analysis plan is given scant consideration. Remember: to do an analysis in haste is to regret the outcome at leisure, possibly faced with a lawyer seeking excessive compensation following a structural failure caused by an inadequate analysis!

5.9 THE PROCESS

This chapter has introduced some of the basic concepts and activities that have to be undertaken when a finite element analysis is to be employed within a qualification-controlled environment where the impact of errors must be controlled and quantified. These form part of a set of building blocks that allow error-controlled analyses to be undertaken. The full process is described in Chapter 9 where the material contained in this chapter is brought into a wider and more comprehensive system. Methods for assessing and bounding the impact that uncertainties (until this point included under the broad term error) have on the results of a finite element analysis are introduced and discussed in Chapter 8.

Chapter 7 describes where and how uncertainties can be introduced into a finite element analysis. This chapter emphasises that uncertainties mainly result from decisions that the analyst must make during the reduction of the real-world problem and, particularly, in the idealisation processes that are illustrated in Figure 5.1.

As already indicated, generated errors are the result of misunderstandings or lack of knowledge at the stage where the analyst has reached the point that the model is being meshed and individual element types selected. Information on how to control this part of the analysis process is contained in Chapter 6. It should be noted that the full process will normally require that the methods and techniques described in Chapter 6 will be used repeatedly as the finite element process passes through the Chapter 9 methodology.

Essentially the total process for controlling the impact of errors and uncertainties has information being built up by the application of a variety of methods and techniques that often require feedback loops. Unless the problem is extremely simple the process will require a number of different analyses that build towards a final model. As the models used in the build-up process may provide information that implies that some of the initial assumptions are incorrect, this may require changes to the validation plan as unforeseen sources of error or uncertainty enter the domain of analysis.

Let us now move on and begin looking at the details of the methodology, starting with elements and meshing that may be the source of user-generated errors.

REFERENCES

1. Bechkoum, K. and Vignjvec, R., *Report on Error Classification.* 1992: SAFESA Report 9034/TR/CIT/2001/0.0/30.9.92, Cranfield University.
2. Taylor, J.R., *An Introduction to Error Analysis, 2nd edn.* 1997: University Science Books, ISBN 093570275X.
3. Higham, N.J., *Accuracy and Stability of Numerical Algorithms.* 1996: SIAM, ISBN 0898713552.
4. Fox, M.J.H., *SAFESA Classification and the Use of Experience.* 1994: SAFESA Report 9034/TN/NE/3051/0.0/18.2.94, Nuclear Electric.
5. Fox, M.J.H., *Uncertainty, Error and Datum Points.* 1994: SAFESA Technical Note 9034/TC/WSA/4061/1.0/04.02.94, Nuclear Electric.
6. Morris, A.J. and Vignjvec, R., Consistent finite element analysis and error control. *Comput. Methods Appl. Mech. Eng.*, 1997. **140**: 87–108.

6

Discretisation: Elements and Meshes or Some Ways to Avoid Generated Error*

6.1 INTRODUCTION

Finite element analyses are undertaken to provide information on the fitness for purpose of a structure or structural component. This creates a set of numbers characterising the behaviour of a real-world structure, i.e. the analysis is to predict accurately such factors as the stress levels in the structure, the displacements, etc. This chapter assumes the analyst has gone through the stage of creating an idealised model of sufficient accuracy and the remaining question is one of meshing the model and selecting the appropriate elements so that there are no errors introduced into the analysis through inappropriate meshes or element selection. The question of how the analyst creates an error-controlled idealised model is the subject matter of Chapters 7, 8 and 9. However, the generation of an error-controlled idealised model cannot be totally separated from element selection and the creation of high-quality meshes. Generating a high-quality finite element model with appropriate loads, boundary

* This chapter draws very heavily on notes written by John Barlow for a lecture on finite element analysis at Cranfield University. At the time when the notes were prepared John was a structural analyst at Rolls Royce PLC at Derby, undertaking finite element analyses on aircraft gas turbine engines. He was also a founder member of the NAFEMS organisation set up to promote the safe and effective use of the Finite Element Method in industry.

A Practical Guide to Reliable Finite Element Modelling A. Morris

conditions, etc., is not a single pass process but for complex analyses one in which the analyst has to pass back and forth using outputs from finite element analyses to guide the error control process. This is made clear in Chapters 7, 8 and 9 where the process of building a finite element model for a structural analysis problem is presented as an integrated activity in which the subject matter of the present chapter is incorporated into a general methodology.

This chapter provides assessment techniques that allow an analyst to make informed judgements when selecting elements and assessing mesh quality. It avoids detailed mathematical derivations as these are not required in order to explain basic principles necessary for deciding on what specific finite elements are able to deliver, nor what makes good or bad meshes. The original work upon which this chapter is based can be found in references [1], [2] and [3]. A more detailed and comprehensive discussion of the topics in this chapter is given by Richard MacNeal, who developed NASTRAN and founded MSC Inc., in his book, reference [4]. Readers with a sound understanding of finite element theory may elect to move directly to other chapters; however, even an experienced analyst would profit from a review of this referenced material.

In order to decide on the type and number of elements needed for a specific analysis, it is necessary to know what an element can deliver. In the case of static analysis or in the case of a dynamic analysis where dynamic stresses are sought, this usually means answering the question 'What stresses can the element deliver?' A second question addressed concerns the error injected into the stress predicted by an element due to element distortion. The answers to these questions are critical to the overall accuracy of the finite element analysis results in describing the actual behavioural responses of the real-world structure whose performance we are attempting to quantify. Because the majority of commercially available finite element packages use displacement finite elements, the discussion on element evaluation covered in this chapter relates to this type of finite element only. The section begins with the question of element delivery and then moves on to consider the effects of mesh distortion on element accuracy.

6.2 ELEMENT DELIVERY

Because the first question raised above relates to stress output from an element, this section provides a very rapid method for assessing the number of stress states that two-dimensional (2-D) and three-dimensional (3-D) finite elements can deliver and by implication the number that linear 1-D bar-type elements deliver. The method is not foolproof but is conservative so the approach will not suggest that an element can deliver higher order stress states than is the case in reality. It assumes that an analyst wants to know what the elements within the vast list supplied by the major finite element systems can deliver in terms of stress output. Do they provide just a single constant value for the stress predicted across the element or accommodate a linear, quadratic or higher order stress variations across the area or volume making up the interior of the element? With this information rational judgements can be made during the process of selecting element types, the number of elements to be used and the layout for a specific analysis.

6.2.1 Two-Dimensional Elements

The method we are going to use was created by John Barlow and it is convenient to begin with 2-D membrane elements. We start with a simple 4-noded membrane element shown in Figure 6.1 and want to know what the order of the stresses is that this element can deliver. Figure 6.1 has an illustrative description of the stress regime within the element where it is shown that the complete stress set for the element has three components: two direct stresses and the shear stress. We shall shortly show that all elements, including that shown in Figure 6.1, can represent constant stress, but can this and other elements represent higher order stress states? In order to assess this, it is necessary to introduce the Barlow Method that allows an evaluation to be made of the number of complete stress sets that an element can accommodate. This is done herein by inference.

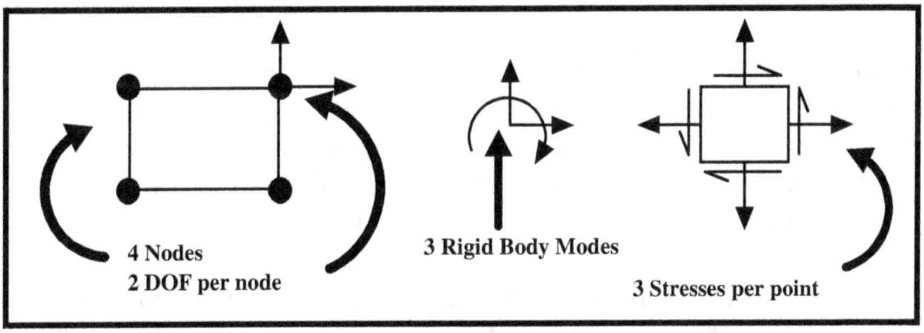

Figure 6.1 2-D 4-node membrane finite element.

Consider a general polynomial representation for the stress in a 2-D domain:

$$\sigma = a_1 + a_2x + a_3y + a_4xy + a_5x^2 + a_6y^2 + \ldots \tag{6.1}$$

This equation tells us that to represent a constant stress state we need to have available one constant a_1; thus, to argue that an element can accommodate a state of constant stress it must be demonstrated that it can represent one complete stress state.

In order to see how the relevant information is extracted from the basic element data consider the 4-noded element shown in Figure 6.1. The procedure follows four steps:

1. Compute the number of available displacement degrees of freedom (dof) which is the number of nodes (4) $\times$ number of dof per node (2) $=$ 8 dof.
2. Reduce the available dof by subtracting the number required to accommodate rigid body movement $= 8 - 3 = 5$.
3. Compute the number of complete stress states that can be accommodated; noting from the figure that this has three components, hence the number of stress states available for use by the element $= 5/3$. Since the process is identifying the number of complete stress states in order to decide how many constants are available for use in equation 6.1, only integer values can be employed.
4. Step 3 shows that this 4-noded element can accommodate *one* stress state + a bit and because all the stress components are required we conclude the element can deliver *constant stress only*.

Of course it can be argued that there will be situations where the neglected non-integer part of the stress computation, i.e. 2/3, can be exploited. However, this concerns a special state of stress which may or may not occur in an analysis; hence we accept the integer part component only of the number representing the number of stress states in the element. Although described as a four-stage process, in essence, it is two-stage only:

1. Calculate the available dof per element = displacement dof − the number of required rigid body modes (zero-energy modes).

2. Calculate the number of complete stress sets = integer part after the available dof per element are divided by the number of required stresses per point in the element.

The stress sets computed in this manner are known as the *Barlow stress sets* and provide estimates of what an element can deliver with respect to stress representation without having to go through an elaborate and complex element derivation process.

In order to see what is required if higher order elements are being evaluated, consider, once again, equation 6.1. If three complete stress sets were available these could be considered as playing the role of the three constants a_1, a_2 and a_3 so that a linear stress variation could be assumed across the area of the element. Thus an element capable of providing a linear stress variation must be able to exhibit three Barlow stress sets. Similarly, if six stress sets are available, representing the six coefficients $a_1, a_2, \ldots, a_6$, then a quadratic stress state can be modelled by the element.

To illustrate a higher order element consider the 8-noded serendipity shown in Figure 6.2.

For this element the algorithm is:

1. Number of available dof $= 8 \times 2 - 3 = 13$
2. Number of Barlow stress sets $= 13/3 = 4$ (+ a bit)

This element has provided a number of stress sets that are larger than the three sets required for linear stress variation but significantly less than the six required for a quadratic variation. Thus, the conservative approach indicates that it is safe to assume the element can deliver linear stresses but nothing higher. While this is the normal inference to draw

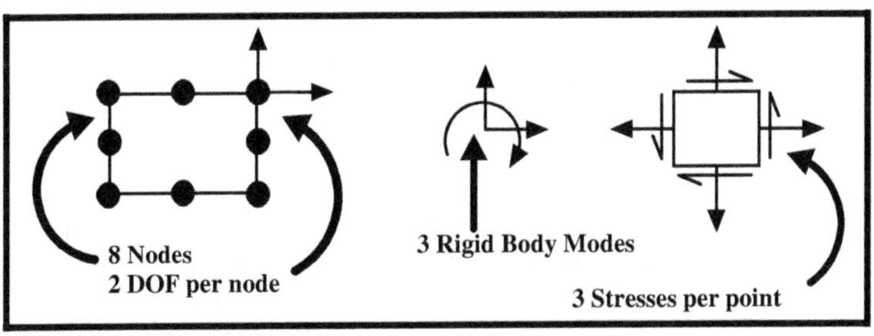

Figure 6.2 2-D 8-node membrane finite element.

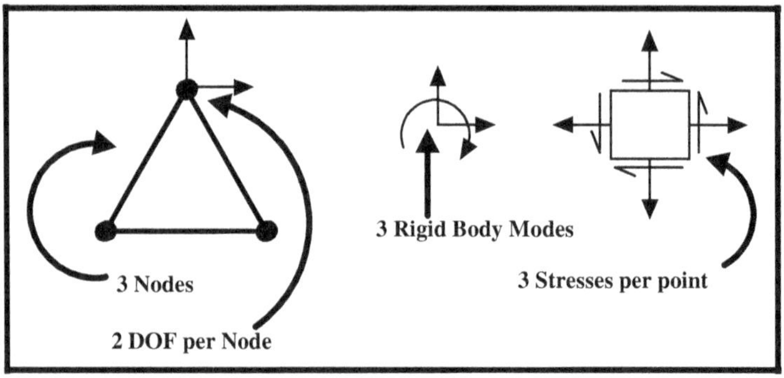

Figure 6.3 2-D 3-node membrane finite element.

from this number of Barlow sets it will be shown later in the section that, for this type of element, it is possible to use the extra terms above and beyond the integer 3 value to gain a higher order prediction for the stress – but for the moment we consider the element as linear in conformity with the Barlow prediction.

Turning now to triangular elements, Figure 6.3 illustrates the simple 3-noded element with 2 dof at each node. Applying the algorithm to this element gives:

1. Number of available dof $= 3 \times 2 - 3 = 3$
2. Number of Barlow stress sets $= 3/3 = 1$

Thus, this element has one Barlow set without any additional 'bits' and is, therefore, a pure constant stress element.

The addition of side nodes to this element gives rise to the 6-noded membrane element shown in Figure 6.4. The reader may easily verify that this element has three Barlow stress sets, enough for linear stresses exactly.

6.2.2 Three-Dimensional Elements

Extending the argument to account for 3-D elements requires re-examining the representational equation 6.1 so that it becomes:

$$\sigma = a_1 + a_2x + a_3y + a_4z + a_5xy + a_6xz + a_7zy + a_8x^2 + a_9y^2 + a_{10}z^2 + \cdots \quad (6.2)$$

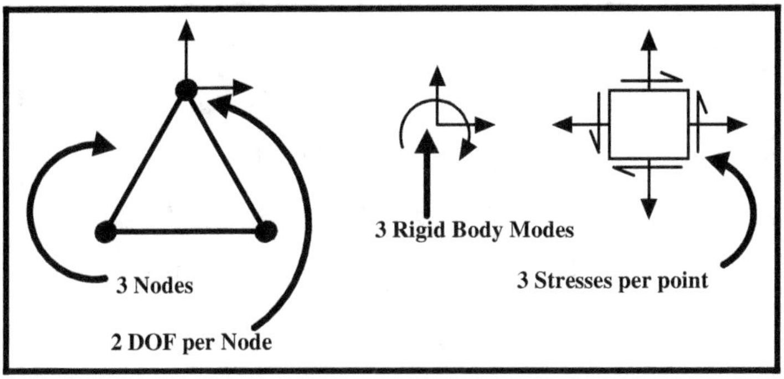

Figure 6.4 2-D 6-node membrane finite element.

Once again, an element having one Barlow stress set, equivalent to demonstrating the existence of the coefficient a_1 in the above equation, is a constant stress element. For a linear stress element there is now a requirement for the existence of four Barlow stress sets implying the existence of the four coefficients a_1, a_2, a_3 and a_4. Similarly for an element to be able to model a quadratic variation of stress, the equation indicates that 10 Barlow stress sets are required, equivalent to the 10 coefficients $a_1 \ldots a_{10}$.

Following the pattern used for the 2-D elements we begin by evaluating the 8-noded hexahedral (brick) element shown in Figure 6.5. It will be noted that the number of rigid body modes has also risen and is now the usual 6 dof required to capture the translating and rotating motion of a 3-D body.

In evaluating this element note has to be taken of both the increased number of rigid body modes and that the number of required stresses needed to create one complete set of Barlow stresses has increased to six as shown in Figure 6.5. Taking these facts into account the algorithm used to evaluate the stress delivery of this element is given by:

1. Number of available dof $= 8 \times 3 - 6 = 18$
2. Number of Barlow stress sets $= 18/6 = (\text{integer})3$

This integer value is more than 1 but less than 4 so one can conclude that the element delivers constant stress + a bit. Since it is not possible to decide satisfactorily what the 'bit' can do, it is better to take the conservative conclusion that the element is a constant stress element.

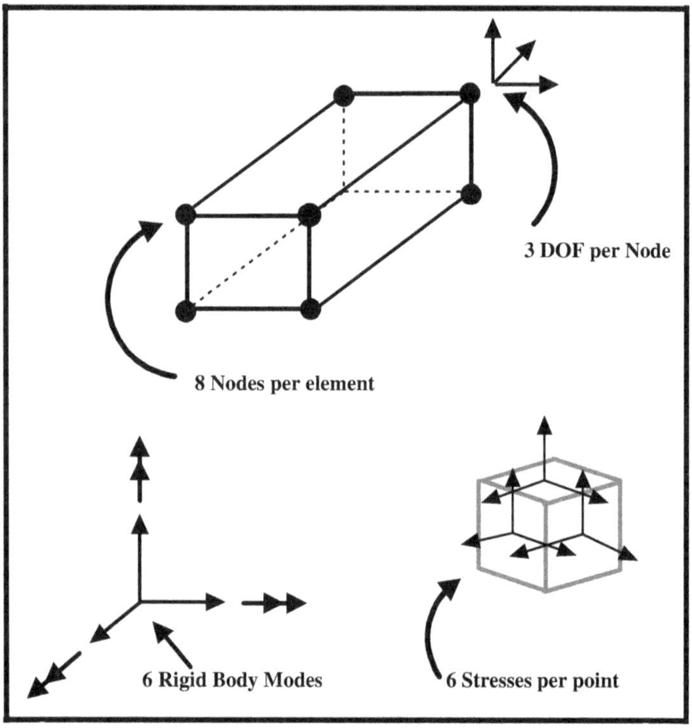

Figure 6.5 3-D 8-node hexahedral finite element.

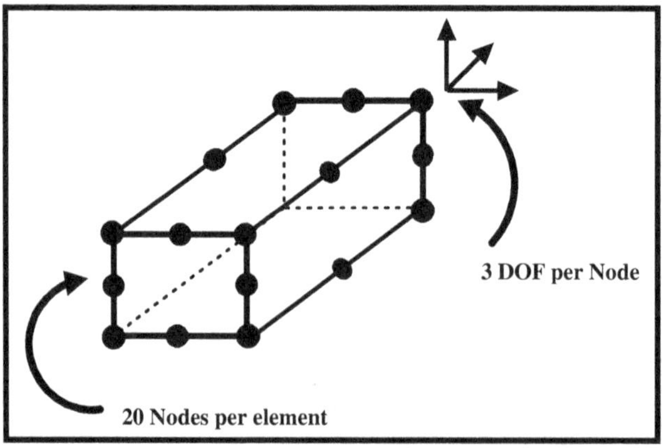

Figure 6.6 3-D 20-node hexahedral finite element.

Figure 6.6 illustrates the 20-noded hexahedral (brick) finite element and applying our algorithm to this element gives:

1. Number of available dof $= 20 \times 3 - 6 = 54$
2. Number of Barlow stress sets $= 54/6 =$ (integer)9

As with the 8-noded element above, the number of Barlow sets lies between the 4 required for providing a linear stress variation and the 10 required for quadratic variation. As a consequence it is assumed, for the moment, that this element can yield linear stresses only.

The reader might like to work out the Barlow stress sets for the 4-noded and 10-noded hexahedral elements and show that these are a constant stress element and linear stress element respectively. It is also instructive to perform these calculations for plate and shell elements.

6.2.3 Why Do This?

Analysing the behaviour of a structure is a complex activity involving a number of important steps and one of these is the selection of a combination of finite elements able to meet the requirements laid down by the analyst responsible for creating a finite element model that is fit for purpose. As is fully demonstrated later in this book, one of the requirements is to select appropriate elements for the task in hand: 'horses for courses'. A lot of the information needed to make decisions relating to element performance is 'hidden' in the finite element formulation or code and is not in the manual. Nevertheless, the analyst needs to have an understanding of what an element can deliver and the method provided above, while not foolproof, is effective in providing a relatively conservative procedure for getting at the 'hidden' information without going into a complex derivation of each and every element in a commercial system. A professional analyst should go through the manual of the finite element system to be used in commercial analyses and do the sums described above for all the elements used. This takes time but only has to be done once and, thereafter, this information can be regularly used to make the appropriate decisions when the question of element selection arises for a new finite element analysis.

6.2.4 Optimal Stress Points and Making the Most of Them

Sections 6.2.1 and 6.2.2 concentrated on the use of a simple method to evaluate the stress performance of a specific finite element. It is a very useful method and very effective in providing the analyst with a procedure for evaluating element performance without having to go into the details of the element's mathematical formulation. However, the method does not tell the whole story and is only indicative of what an element can deliver. Certain elements perform better than others. For example, experience shows that 2-D 8-noded quadrilateral membrane finite elements perform better than 6-noded triangular elements and, equivalently, 3-D hexahedral finite elements perform better than tetrahedral elements. To understand why we need to examine these two element types in a little more detail.

As demonstrated in Section 6.2.1 the 6-noded triangular finite element has just enough available degrees of freedom to accommodate a linear stress variation across the element. The same section also demonstrated that the 8-noded membrane finite element had more than the required three Barlow stress sets to model linear stress but less than the six sets required for quadratic stresses. It is natural to ask of what use are these extra bits of information which lie between the 'linear 3' and 'quadratic 6' Barlow stress sets? The answer is quite subtle!

Although the 8-noded membrane or 20-noded hexahedral finite elements cannot represent all quadratic stress, they have sufficient information locked in these 'extra bits' to detect the presence of such higher order stresses. The element cannot follow a quadratic stress variation across the entire element but there are specific points at which the difference between these quadratic stresses and the stress predicted by the element are zero.

In order to understand what is going on we need to recall that these elements employ numerical integration with a set of specific points being used as the sample points for the Gaussian integration normally used in finite element systems. These are illustrated in the case of the 8-noded membrane and 20-noded 3-D element in Figure 6.6. Although numerical integration can employ any number of sample points, it is normal in the case of the 8-noded membrane element to use four sample points, known as Gauss points, equivalent to a 2×2 integration as shown in Figure 6.6. For the 20-noded hexahedral element this same order of integration in this 3-D situation requires $2 \times 2 \times 2$ Gauss points located at equivalent points to the 2-D case, illustrated in Figure 6.7. If the element stresses are evaluated at these numerical integration sample points then, at these specific points, the difference between the quadratic stress state and the element stress values is zero. These numerical integration points are, therefore, often called the optimal stress sample points!

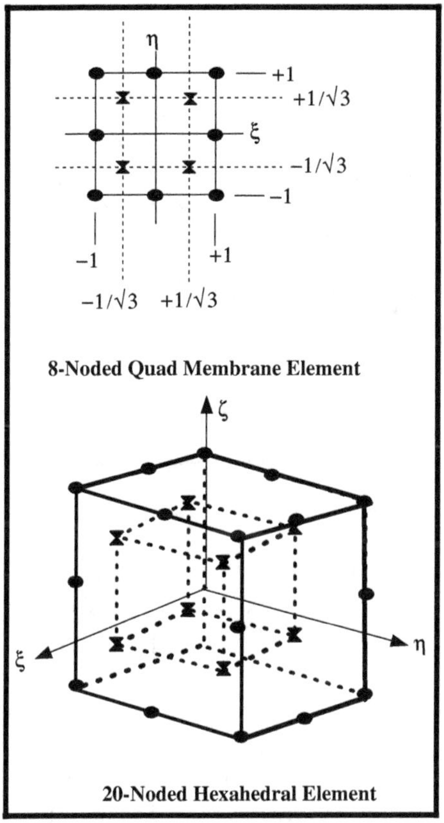

Figure 6.7 Gaussian integration points.

This happy situation arises because of the nature of the numerical integration that implicitly employs polynomials as approximations to the actual stress profile and for the elements being discussed here this polynomial is a quadratic function. By evaluating the stress at the optimal stress sample points, the element can exploit the 'extra bits' to match a quadratic stress variation at the optimal sample points. This situation is shown diagrammatically in Figure 6.8 where the element stress profile and the actual quadratic stress profile can be seen to coincide at the two Gauss points. At these points the elements produce stress predictions of exceptional accuracy. In fact, the stresses at the optimal points have the same order of accuracy as the nodal displacements and have moved the element up the accuracy 'ladder'. This 'ladder' is discussed in the next section. This attribute is sometimes referred to as 'superconvergence' and the message is clear: *always*

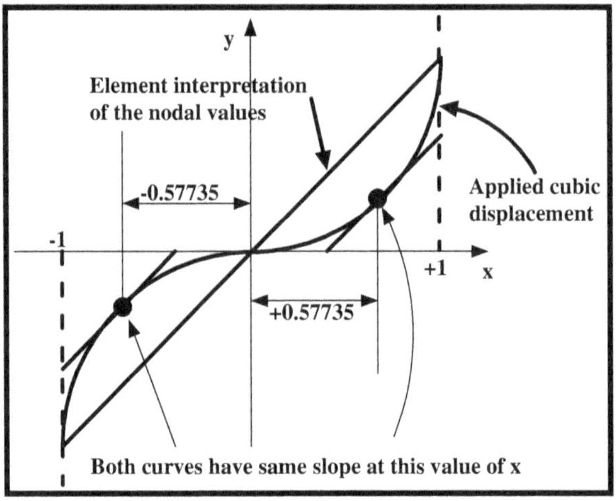

Figure 6.8 Optimal stress points.

sample stresses at the optimal points. The position of the optimal sample points coinciding with the Gauss points is fortuitous for the class of element discussed above and the actual optimal sample points are known as Barlow points for higher order elements. For more complex finite elements, the Barlow points are near to the Gauss points but not actually coincidental with them. For most purposes the use of Gauss points for sampling will improve the accuracy of the stress output and should be used for sampling stresses. But if your software supplier offers Barlow points as the sample stress points, take them!

The 6-noded triangular membrane and the equivalent 10-noded 3-D tetrahedral elements do not have this information and when higher order stresses are encountered, the stress accuracy deteriorates when compared to the above 8- and 20-noded elements. Hence for the 2-D triangular and 3-D tetrahedral elements one needs to make the mesh size sufficiently small to ensure that the stresses are sensibly linear. In consequence fewer quads are required for equivalent stress accuracy than 'triangle pairs' (and the 3-D analogue), provided that stresses are sampled at the optimal stress points.

It should be emphasised that good finite element codes evaluate stresses at the optimal points in the case of the 8-noded membrane and 20-noded brick elements, but in some cases this may not be obvious as these stresses may be output as though they are nodal stresses or the nodal stresses may be extrapolated from the Gauss point stresses. As noted in Chapter 4, extrapolating stresses may improve the accuracy and be advantageous when higher order accuracy is being sought from the finite element system. The user must always be alert when employing a computer package to ensure that the most accurate stresses are being provided by the system so that the results are of high quality.

6.3 MESH GRADING AND MESH DISTORTION

So far we have discussed how to use information about an element's ability to model the stresses within the structure being analysed. However, we have made the implicit assumption that mesh layout within which an element is located has no influence on the element's performance – this is not the case. The mesh defines the shape of an element and the shape has a significant impact on the ability of an element to produce accurate predictions of the stress levels within the structure. Inaccurate stresses mean that the stress levels predicted by the analysis will be in error when compared to those that occur in the structure once manufactured and introduced into service. Since controlling errors is the primary purpose of this book, then we must open the door to mesh-based errors and endeavour to explain some of the problems that can arise and how to try to circumvent them!

Meshing errors and the effects of mesh distortion have been topics of research interest for a considerable time and a major body of theoretical work has appeared in the research journals and in books associated with the theory of finite element modelling of mathematically defined structural problems. While this is very important in providing a mathematical foundation to finite element theory, it is not essential to follow the detailed mathematical arguments advanced in this theoretical work in order to avoid some of the more significant error sources associated with the misuse of distorted meshes. In this section, therefore, an illustrative approach to explaining the problems associated with mesh distortion is taken, leaving it to the reader to take the matter further, from reference [4], if a more profound understanding is required.

6.3.1 Mesh Grading

High stresses inevitably occur at a surface between two dissimilar materials. These materials might be metal and air at the 'free surface' of a structure or between dissimilar metals or between carbon and the reinforcing fibres within a CFRP structure, etc. In order to illustrate the point to be discussed in this sub-section, consider the case of the stresses near to the free surface of a structure as shown in Figure 6.9.

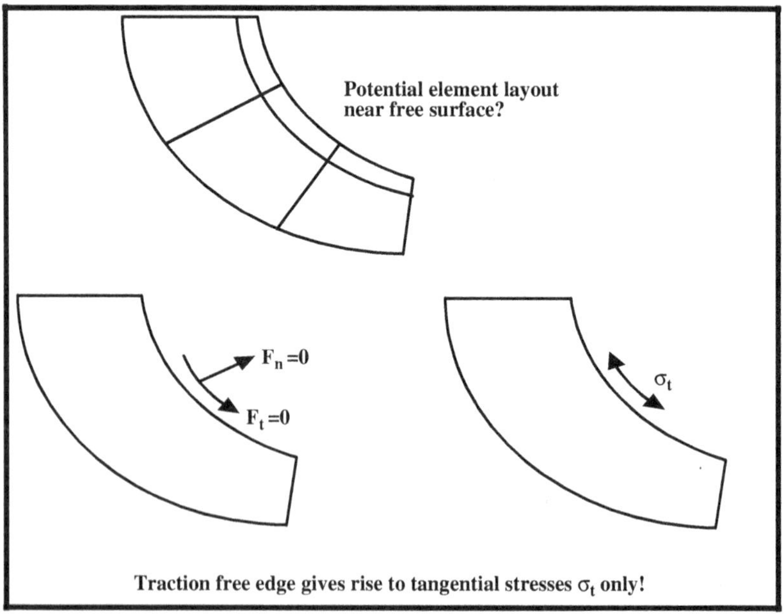

Figure 6.9 Free surface metal–air with no applied surface tractions.

In order to recover the stresses near to the surface there is a temptation to put a single thin element at the surface as shown in Figure 6.9. This does not work and it is important to understand the reason for this!

Near a surface the applied surface tractions F_n and F_t are zero; thus, the tangential strains/stresses dominate as illustrated in Figure 6.9. Displacement finite elements have complete strain compatibility in the tangential direction and 'stick' together very well in that mode as illustrated in Figure 6.10, but not in other directions. In this illustrative

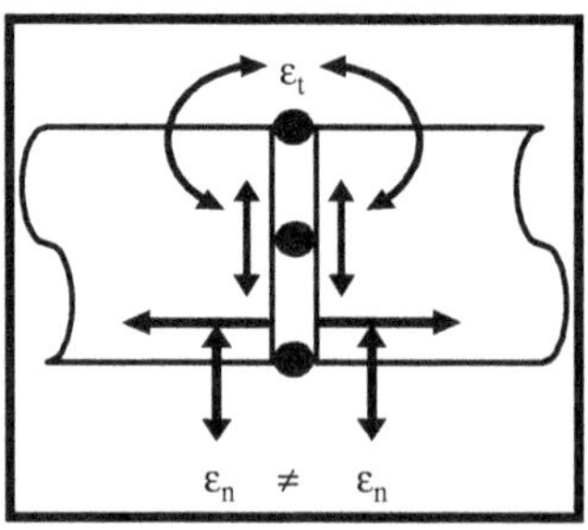

Figure 6.10 Tangential ε_t strains are compatible; normal strains ε_n are not!

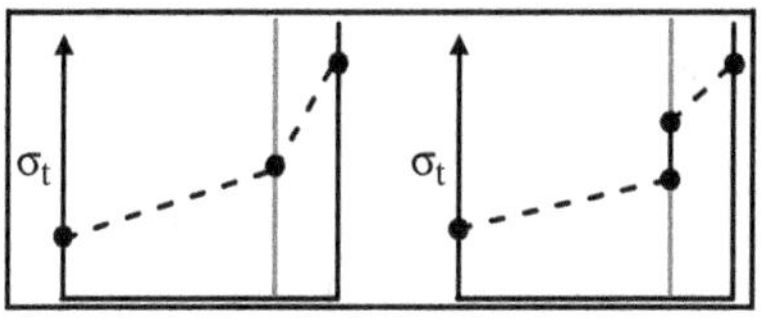

Figure 6.11 Actual stress state compared to required stress state.

example the tangential strains in going from one element to the next are shown as compatible while the normal strains across the element boundary are not compatible. However, elements rely on some incompatibility in order to change the stress in moving from element to element and the compatibility in the dominant strain field, in this case the tangential strain, gives rise to a particular problem illustrated in Figure 6.11.

The left-hand side of Figure 6.11 is showing that there is, essentially, no 'stress jump' across the element boundary in moving from the surface element to the adjacent element. This is because the dominant stress σ_t is the only meaningful stress present and is forced into being in equilibrium across the element interface. The required situation is that shown on the right-hand side of Figure 6.11 where this single dominant stress is allowed to be discontinuous across the element interface. It is also worth noting that the majority of the strain energy is in the big element and as illustrated in Chapter 4 the best results are obtained when there is an equal distribution of this energy among the elements participating in the analysis of a component. Taking the strain energy consideration into account, it would be better to have used two elements of equivalent size to model the domain.

If the elements being employed are required to allow some stress variation from element to element and permit a more equitable distribution of strain energy, then the use of a 'graded' mesh layout is indicated. For this simple illustrative problem an appropriate 'graded' mesh is demonstrated in Figure 6.12.

The lesson from this simple illustrative problem is clear and general – *avoid sudden changes in element size normal to surface stress concentrations.* Although a simple free surface problem has been used in this section, in fact the principle of using graded meshes should apply to all problems where there is a rapid change in stress in a specific direction. *Thus grading should be used whenever there is a change in material properties in a structure or wherever a stress raiser is encountered.*

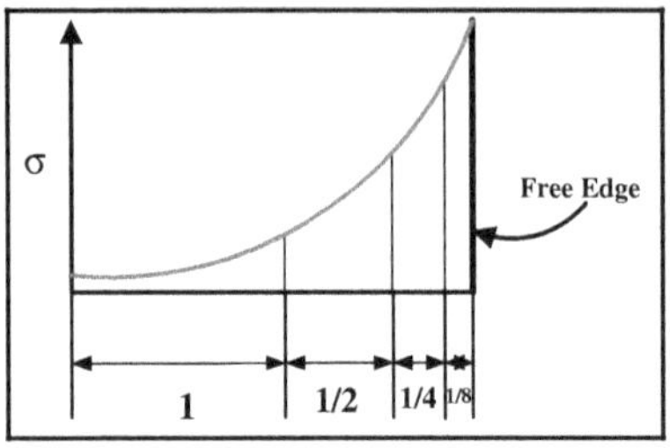

Figure 6.12 Graded meshes work best.

6.3.2 Element Distortions

Almost all elements used in modern displacement-based finite element packages are mapped elements and each element employed in an analysis is mapped onto an underlying regular shape where all the computation leading to the creation of an element stiffness matrix is undertaken. Thus, the 4- and 8-noded and similar shaped higher order membrane elements are built using a perfect square shape and the 20-noded brick and associated higher order 3-D elements are based on the perfect cube. If an element in an analysis has the same shape as the base shape, then the mathematical process of creating the element stiffness matrix is straightforward. If, as is usual, the element does not conform to the base shape, then the mathematical mapping that has to be undertaken by the computer in creating the element stiffness matrix becomes increasingly severe as the element becomes more distorted from its basic shape. The question is 'Does this distortion have any effect on the accuracy of the stress output from an element?' The answer is yes, but depends on the type of distortion injected into the element shape. It also depends on whether the structure is displacement or load dominated. For load-dominated structures the integration order is significant. In this section an examination of what is allowable and what to avoid is presented using quads and hexahedral elements but the conclusions also apply to all elements in both two and three dimensions.

It may be argued that modern pre-processing CAD-based systems automatically avoid using finite elements that display several of the distorted shapes or the inappropriate nodal positions shown in this section. But some of the distortions, contained in this section, can find their way into a mesh/element configuration and it is important that the analyst recognises when a system has created a configuration that is unacceptable. In addition to recognising that a problem exists, the analyst must have sufficient knowledge to be able to modify the computer-generated configuration to achieve acceptable results from the finite element solution. It is also worth noting that a professional analyst should have a clear understanding of why the finite element systems are programmed to do what they do – for a professional analyst, nothing is worse than having to answer 'I don't know' when asked why systems do not tolerate certain element shapes or distortions.

In order to illustrate the points being made, this section uses the 8-noded membrane element but the arguments apply equally to all 2- and 3-D 'square' or 'cubic' shaped elements and therefore apply also to the equivalent plate and shell elements. Also parallel arguments apply to triangular and higher order tetrahedral elements.

6.3.2.1 Group 1 distortions (linear distortions)

This type of distortion is illustrated in Figure 6.13 where the element has undergone a linear distortion in the direction '*L*'. Such distortions do not make any difference to the element performance independently of whether the structure is displacement or load dominated but does increase the effective element size and this influences the representation of the stress field. This is clear when consideration is given to the fact that the same degree of polynomial representation is being used to

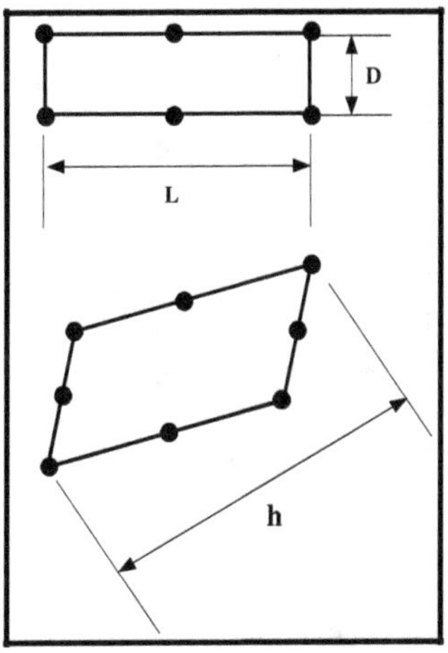

Figure 6.13 Group 1 distortions.

model the displacement along the length '*L*' as well as along the length '*D*'. The influence of this difference in dimensions feeds through to the stress representation. The same arguments apply if the distortion is applied across a diagonal '*h*' as also shown in Figure 6.13.

This type of distortion can be exploited if there is a significant difference in the loads in different directions in a structure but its use is not recommended and most of the major finite element systems will alert the user to the fact that elements having this type of distortion exist in the model.

6.3.2.2 Group 2 distortions (bi-linear distortions)

Bi-linear distortions are better known as 'taper' and the illustrative element used in this section is shown in Figure 6.14. This figure shows that taper is equivalent to keeping all the edges of the element straight but moving the centre node. In this case the node has been moved by an amount '*e*' that can be interpreted as an error input. Taking this as an error input, the question that needs to be answered is 'How much of the "error" finds its way into the stress output when a linear stress field is being modelled and optimal sample points used?'

In fact, the effect of 'taper error' *e* on the resulting stress is relatively small and of $O(e^2)$,[1] a conclusion that applies to structures that are subjected to both displacement and load-dominated fields. This implies that the analyst can accept a reasonable amount of element taper before needing to become concerned.

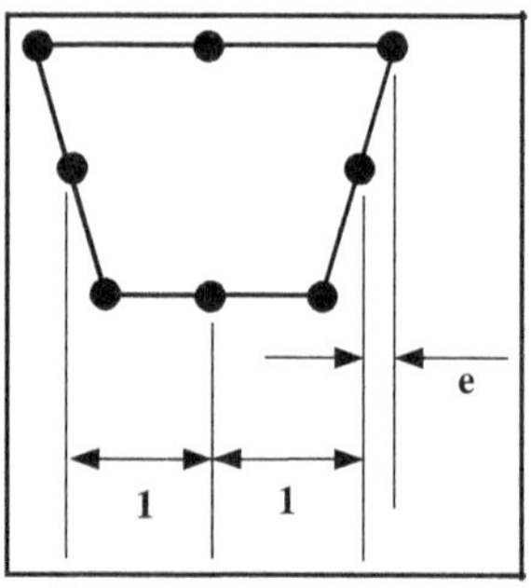

Figure 6.14 Tapered 8-noded membrane element.

[1] The term O means 'of the order of'; thus $O(e^2)$ means that the error in the stresses predicted by the element is of the order of the error squared. Since $1 > e$, see Figure 6.14, then the input distortion $O(e) > O(e^2)$.

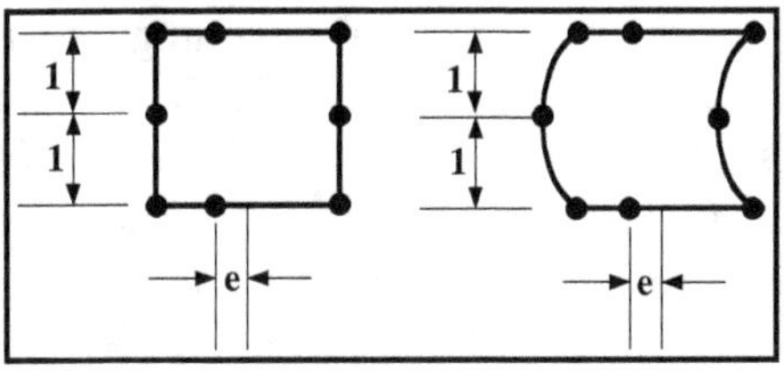

Figure 6.15 Examples of quadratic distortions.

6.3.3 Group 3 Distortions (Quadratic Distortions)

The third group of distortion errors is illustrated in Figure 6.15 and is equivalent to moving two of the element's side nodes together by an amount e. Figure 6.15 displays two elements with group 3 distortions but only the square element on the left side of the figure actually incurs errors from this type of distortion. The reason why the element on the right side of the figure does not incur an error for this distortion class is discussed below and illustrated in Figure 6.16.

The effect on the stresses predicted by elements distorted in this manner is more severe, in principle, than that introduced by tapering the element (group 2 distortions) in displacement-dominated situations with an error of $O(e)$. However, the effect of sampling at the optimal sample points is to reduce the distortion error from $O(e)$ to $O(e^2)$ in the representation of linear stresses.

For load-dominated conditions the situation is more complex. If the element employs the 2×2 Gaussian integration scheme shown in Figure 6.7 and optimal sampling is also employed, the errors in the element stress output now change to $O(e)$. This change of characteristic error occurs because of the presence of spurious null-energy modes that restrict the performance of the element. Although the finite element model maintains global equilibrium, not all of the load gets into the distorted

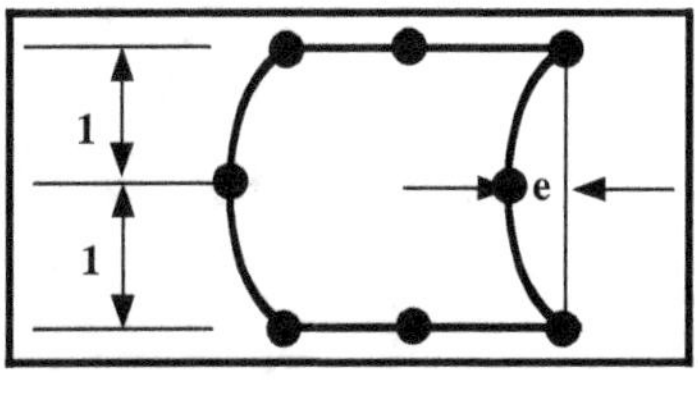

Figure 6.16 A benign distortion.

element and has to be reacted by the surrounding structure. It is possible to circumvent this load truncation problem by employing a 3×3 integration rule with stress sampling at the usual 2 × 2 points. However, many of the commercially available systems do not offer such an option.

Turning to the case illustrated in Figure 6.16 for displacement-dominated structures, because this distortion is quadratic and the element can model this type of distortion at the optimal sample points, it does not 'see' this particular distortion as an error. Thus the stresses at the optimal stress points are not influenced and, therefore, the error is zero resulting in a benign and totally acceptable distortion. Providing, of course, that e does not become too large, a good rule of thumb is to keep e less than 0.2 for elements of length 2 as shown in Figure 6.16.

6.3.4 Group 4 Distortions (Cubic and Higher Order Distortions)

Finally, we come to a set of distortions that are best avoided and these are illustrated in Figure 6.17. These give rise to errors in the element stresses of $O(e)$ in linear stresses which optimal point sampling does not cure. For this type of distortion there is little difference in the order of the error for displacement or load-dominated structures. However, some improvement can be achieved for the load-dominated situation by using a 3 × 3 integration rule with the usual 2 × 2 stress sampling procedure. If this type of distortion cannot be avoided, it is clearly best to keep it small!

6.3.5 Distortions of Other Element Types

For convenience, the discussion relating to the various distortion groups has been restricted to 2-D elements. However, the same results apply

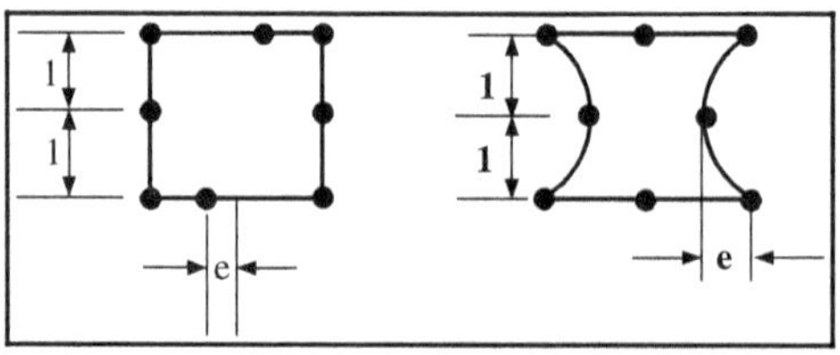

Figure 6.17 Cubic distortions.

when these distortion patterns are applied to the hexahedral (brick-type) elements. The deformation patterns are a little more difficult to visualise in three dimensions but simply consist of constructing a deformed cube from the shapes given in the above sections of this chapter.

In the case of 2-D triangular elements, the distorted shapes are created by cutting the distorted shapes for the square elements given above along a diagonal. However, these elements do not have the advantage of optimal stress sampling and, therefore, the effect of group 3 distortions is more severe in the case of this class of element. As in the case of the hexahedral element, the equivalent distortion shapes for the tetrahedral elements can be constructed from the 2-D triangular cases.

6.3.6 General Principles with Respect to Element Distortions

The discussion on the effects of element distortions given above is not intended to be comprehensive but to provide basic information to guide the analyst when a major analysis is being undertaken. The resulting concepts can be distilled down into a number of broad general principles:

- Keep the edges of the elements as straight as possible commensurate with the requirements of the analysis.
- Keep the mid-side nodes of the element equally spaced between corners.
- Moderate taper (< 25%) is acceptable.
- Match curvatures and offsets on opposite edges and faces of an element, as shown in Figure 6.18, and use optimal sample points where appropriate.

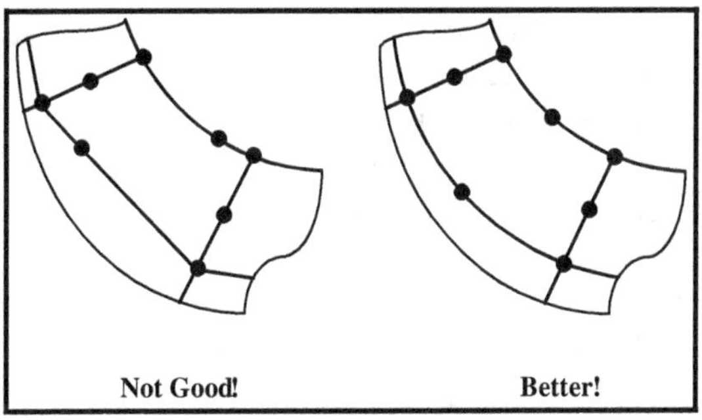

Figure 6.18 Matching curvatures and node positions.

- If possible, use a 3×3 Gaussian integration with stress sampling at the 2×2 optimal points.

These broad principles should be applied with full account taken of the rest of the information in this chapter. As discussed in the next section, these principles still apply even when automatic mesh generation and adaptive methods are employed!

6.4 THE ACCURACY LADDER

The above sections have concentrated on a number of aspects relating to what elements can deliver from the viewpoint of stress accuracy, the effect on accuracy of element distortions and the need to ensure that meshes are not subject to inappropriate grading. Taking advantage of this information allows the analyst to exploit a ladder where certain actions improve the stress estimates provided by an analysis.

6.4.1 The Stress Ladder

When the Barlow stress sets were computed, it was noted that the ability of elements to deliver quality stress estimates is directly related to the number of nodes and, in the case of certain elements, is further augmented by using optimal stress sample points. Thus the stress 'accuracy' is improved by increasing the stress recovery potential of the element by increasing the number of nodes and using optimal stress points and this movement towards increased accuracy is one component of the 'accuracy ladder'. Element stress delivery needs to be given thoughtful consideration by the analyst when setting up an analysis approach[2] to a complex problem. In certain industries, constant stress elements, such as the 4-noded membrane or the 8-noded brick elements, are popular but many of them are needed to give an adequate stress representation in the presence of complex or rapidly changing stress levels; the analyst would be much better advised to move up the accuracy ladder. A further step up the ladder of accuracy can be made by exploiting extrapolation methods where the results from a number of sample points are used to create an extrapolating polynomial as illustrated in Chapter 4.

---

[2] The analysis approach is addressed in greater detail in Chapters 8 and 9.

6.4.2 The Mesh Ladder

A further move up the accuracy ladder can be achieved by noting the influence of element distortions on element performance and ensuring that the element shapes are appropriate. The list of distortion errors in Section 6.3 and their degradation on the element's performance clearly indicate that some distortions can bring the analysis down the accuracy ladder. The fact that many structural analysis problems require the Finite Element Method for designs with complex shapes means that distorted elements will inevitably be needed. In this situation, if there is no escape from having distorted elements that cause concern, then the analyst should try to ensure that these elements are located at positions in the finite element model where both stresses and stress gradients are relatively low. The word 'if' used in the preceding sentence is meant to indicate that an analyst in the situation of being presented with a design with shapes and shape changes that cause serious distortion to the mesh shapes should challenge the need for these shapes. Often designs are initially generated by CAD specialists who have a different set of criteria from those of the analyst and the people in question will probably have little or no experience of using the Finite Element Method. A conversation on the need for the offending shapes or shape change gradients can lead to modifications that make the analysis a whole lot simpler. As proposed in later chapters, it is better to have the discussion on the analysis approach and the development of the analysis itself as a joint debate involving the design and the analysis teams early in the development of the product.

6.4.3 Automatically Moving up the Accuracy Ladder

One way to move up the accuracy ladder in a relatively easy manner is to employ finite element codes that either automatically increase the number of elements used in an analysis or increase the order of the polynomial employed in the elements. Increasing the number of elements within a domain without changing the order of the polynomial used to approximate the displacements within the element automatically is known as h-adaption. This adaption process is illustrated in Figure 6.19 where a 2-D 4-noded membrane element is used to illustrate the process. In practice, it is more usual to employ triangular elements in 2-D analyses and tetrahedral elements for 3-D problems, as these as easier to use by programmers developing h-adaptive codes to model automatically

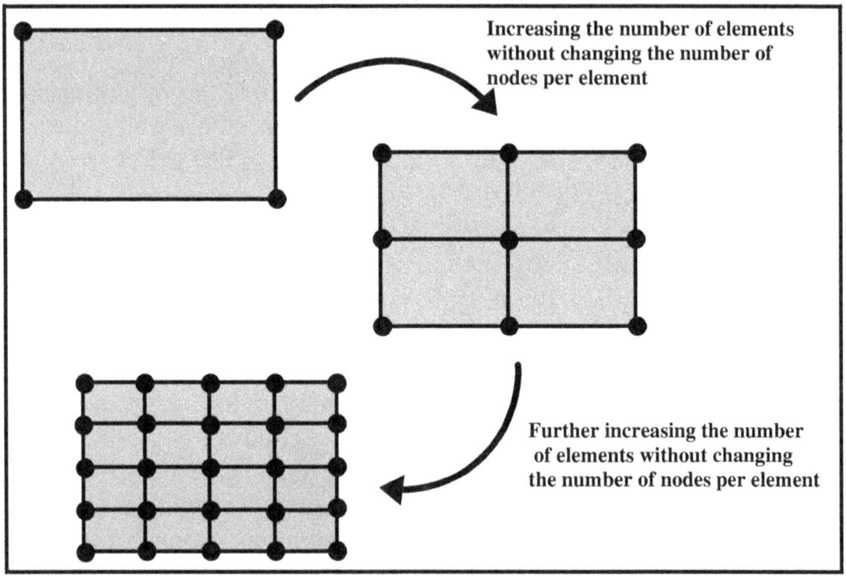

Figure 6.19 Stylised h-adaption employing 4-noded membrane elements.

structures that have curved surfaces or edges. Employing this type of code, in principle, drives the user up the accuracy ladder by increasing the number of elements within a domain so that the average stresses predicted by an individual element more accurately match those of the underlying mathematical model.

An alternative to employing more elements is to move up the accuracy ladder by increasing the order of the polynomial used within the element to model the displacement field, and this process is known as p-adaption. The user sees no change in the number of elements being employed but the number of nodes per element increases. Again this is demonstrated using a very simple stylised 2-D model in Figure 6.20.

Both h- and p-adaption codes are automatic in that they assess the accuracy of the solution being generated by a current number of elements, in the case of h-adaption, or the current order of the polynomials employed in the case of p-adaption, and then decide on whether to put in more elements (h-adaption) or higher order polynomials (p-adaption). It is important to understand how these codes decide to increase the number of elements or the order of the approximating polynomials within the elements being used. The codes use *a posteriori* error estimates to make this decision. There are many papers and books that explain the development of this type of error estimator and the more

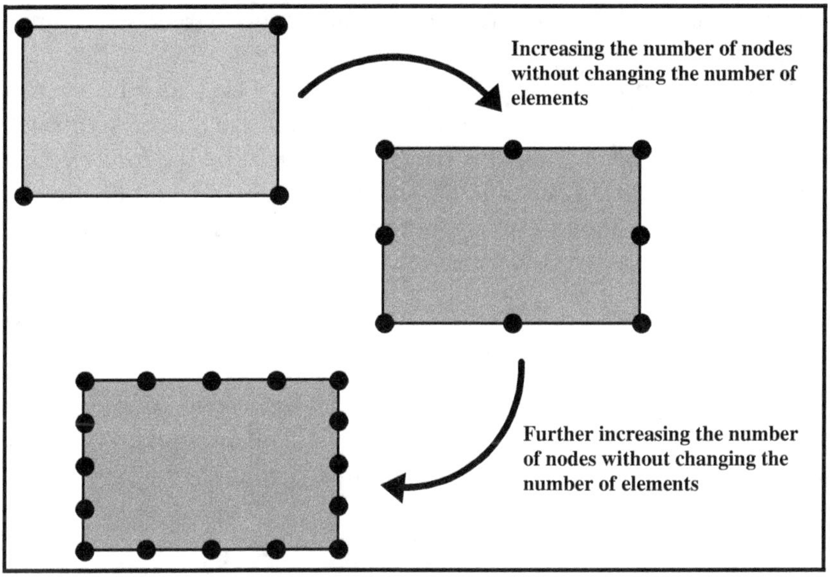

Figure 6.20 Stylised p-adaption employing a single 2-D membrane element.

mathematically inclined reader may wish to consult this literature in order to gain a firmer grasp of the basic theory associated with these error estimators; references [5] and [6] provide useful starting points. Although the development of these estimators appears complex, the underlying principles are easy to understand and have already been alluded to in Chapter 4. In order to start our explanation of the process used by adaption methods, consider the element layout shown in Figure 6.21. In this figure, four elements are displayed and these could be either the entire

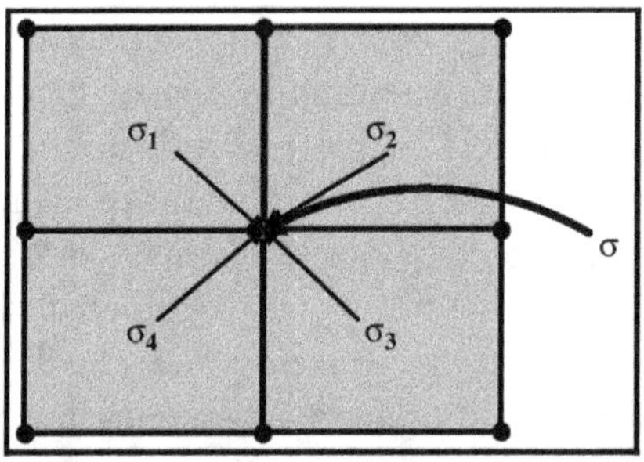

Figure 6.21 4-element layout with a common node.

finite element model or a small portion of a much larger set of finite elements.

At the centre node there is a specific stress σ, which may be a direct stress component or a stress invariant, but each of the four elements will predict a different value for this stress as indicated by the four element values σ_1, σ_2, σ_3, σ_4. The value for the nodal stress σ could be directly obtained from the finite element output or can be taken as the mean average of the four stress values:

$$4\sigma = \sum \sigma_1 + \sigma_2 + \sigma_3 + \sigma_4 \tag{6.3}$$

The error in the stresses generated by the finite element solution could then be estimated by comparing the four individual stresses with the average value from the above equation, thus giving:

$$\begin{aligned} e_1 &= \frac{|\sigma_1 - \sigma|}{|\sigma|} \\ e_2 &= \frac{|\sigma_2 - \sigma|}{|\sigma|} \\ e_3 &= \frac{|\sigma_3 - \sigma|}{|\sigma|} \\ e_4 &= \frac{|\sigma_4 - \sigma|}{|\sigma|} \end{aligned} \tag{6.4}$$

and an error estimate can be calculated by taking the largest of the four values e_1 to e_4 as the approximate error in the stress predicted by this element layout. By using these four 'error' values, the analyst can gain some understanding of the quality of the stresses predicted by the element.

The use of these measures by an analyst to gain a measure of the quality of the results from a finite element analysis means applying a simple procedure. It requires zooming in on one or more regions where the output is giving relatively high stress values and picking a node in each region that lies at the vertex of a number of elements. The value of the vertex stress predicted by each element is then obtained, the equivalent of the stresses σ_1, σ_2, σ_3, σ_4, together with the nodal value, the equivalent of stress σ in equation 6.3. Such nodal values are normally provided by all finite element systems but the analyst is still at liberty to perform the calculation. With this information the terms given in equation 6.4 can be computed to provide a simple measure of

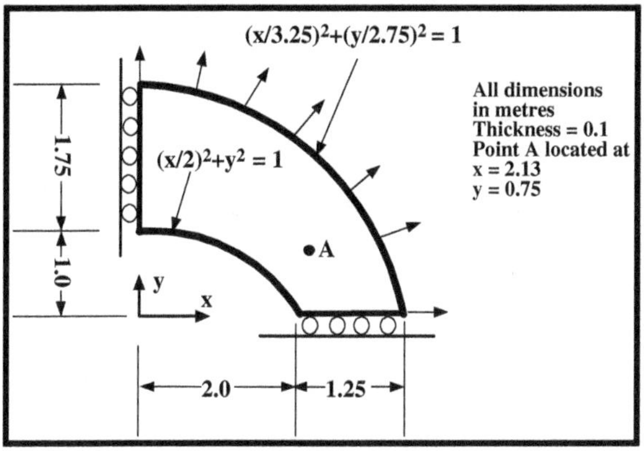

Figure 6.22 Elliptic membrane.

the overall quality of the results. This is a simple task to undertake and it is advisable always to undertake this type of examination before the results from a finite element analysis are accepted.

In order to take an illustrative example, consider the problem shown in Figure 6.22 that has been put forward by NAFEMS as a test example for finite element systems. The analysis problem is represented by an elliptic membrane subject to a uniform pressure of 10 MPa on the outer edge with the inner edge unloaded. The boundary conditions are that a symmetry condition exists along the edge that lies on the x-axis and along the y-axis as shown. The material is isotropic with an elasticity modulus $E = 210 \times 10^3$ MPa and Poisson's ratio $\nu = 0.3$. The analysis has employed 4-noded membrane elements and we shall see what happens to the stress jumps for four elements that coincide at point A.

The analysis starts with 6 elements as shown in Figure 6.23, moves on to consider a 24-element layout shown in Figure 6.24 and then 96- and 1536-element layouts which are not shown.

All the element layouts have been constructed so that there is always a node at the point A in Figure 6.22 and there are four elements with a common node at this point. This allows us to obtain values for the predicted stresses at point A and to construct estimates of element analysis errors at this point using equation 6.4. In Figure 6.23 node 6 coincides with point A and in Figure 6.24 node 17 is the coincident node.

Table 6.1 shows the results from applying equation 6.4 to the four separate meshes used in this analysis and using the two direct stresses σ_{xx} and σ_{yy}. The use of 4-noded membrane elements implies that

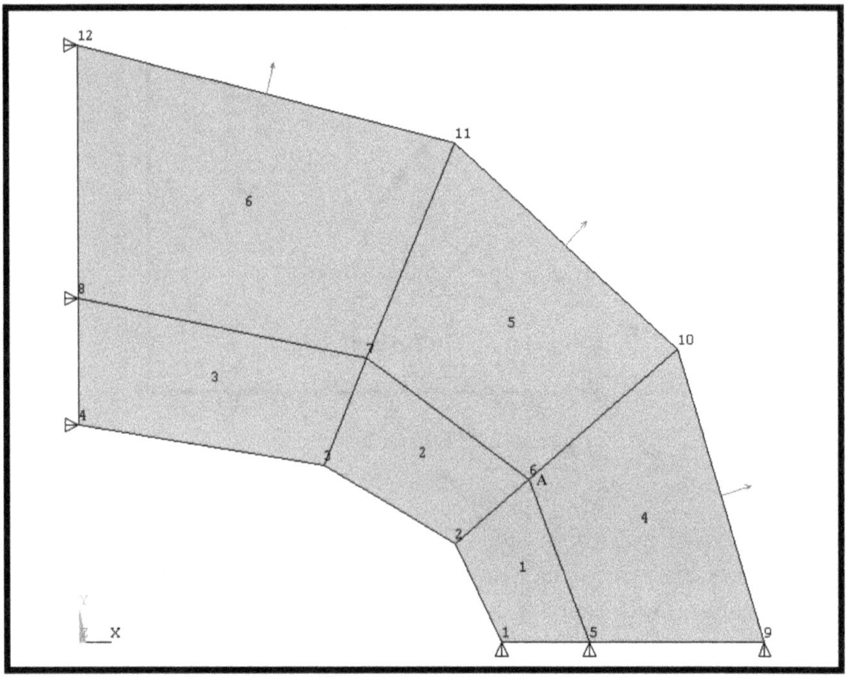

Figure 6.23 6-element model.

convergence to an acceptable answer is slow and this is evidenced by the slow convergence of the stresses shown in the first two columns of Table 6.1. The most interesting results are the values of the error where the direct stress in the x-direction actually displays an increase in error as the number of elements increases. This is due to the elements having straight sides so that the initial meshes are modelling an ellipse with faceted outer and inner surfaces; once a sufficient number of elements is employed, then convergence towards a curved outer and inner surface ellipse takes place. The results from Table 6.1 now allow an analyst to make a decision on the number of elements that are required to reach a reasonably accurate value for the stress at point A.

This process is used by the adaptive codes to decide if more elements (h-adaption) or higher order polynomials (p-adaption) are required in order to achieve a required level of internal accuracy. However, the simple estimates from equation 6.4 are not sufficiently reliable and recourse is made to more extended measures of error that combine the results from a number of elements, some of which are not adjacent to the sample point. This is illustrated in principle in Chapter 4 where it is

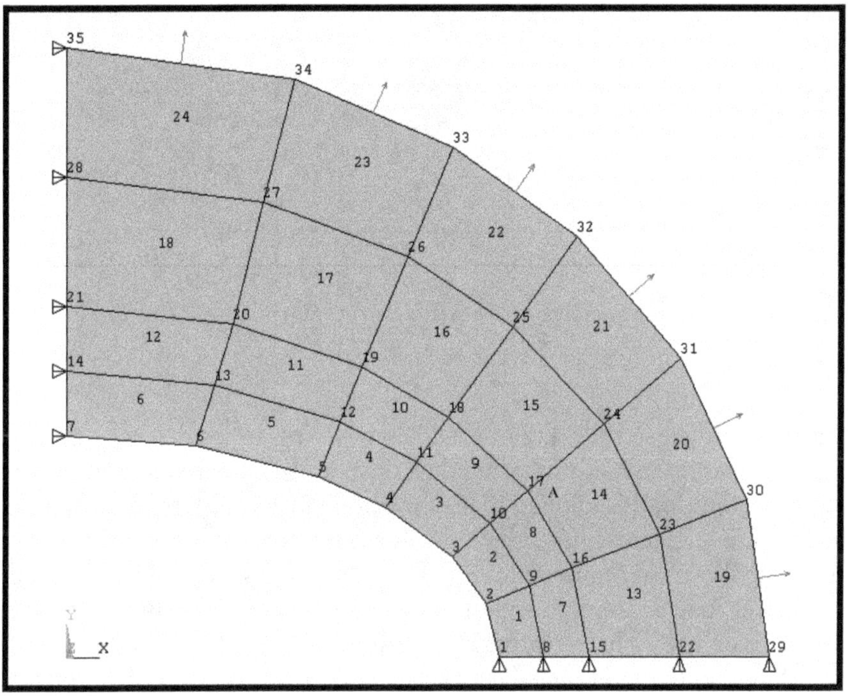

Figure 6.24 24-element model.

shown that much better estimates of the stress at a particular point can be achieved by using interpolated values when the results from a number of elements are taken into account.

However, using adaptive procedures can give rise to a different set of errors that arise from the adaption procedure itself: h-adaption can give rise to misshapen elements and both h- and p-adaption methods or

Table 6.1 'Error estimation' at point A.

Number of elements in the model	Mean value for σ_{xx} (MPa)	Mean value for σ_{yy} (MPa)	Maximum error in σ_{xx} from 5.4	Maximum error in σ_{yy} from 5.4
6	9.00	22.19	0.42	0.44
24	7.26	26.33	0.45	0.10
96	6.99	27.65	0.25	0.04
1536	6.77	27.39	0.06	0.07

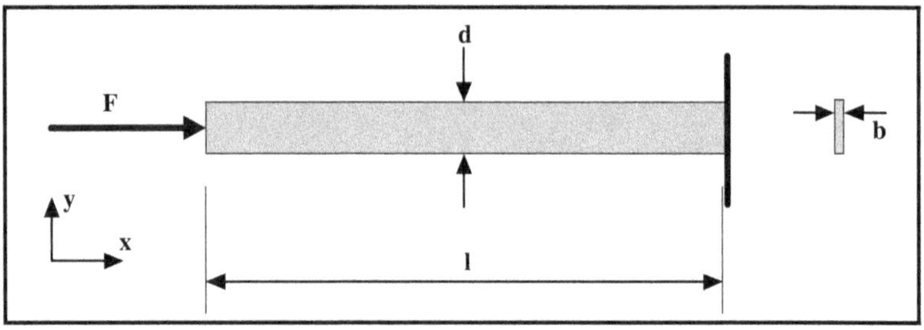

Figure 6.25 Membrane example.

systems can chase their tails by endlessly trying to model increasingly high stresses if the analyst has made a mistake and created a singularity, e.g. if a point load is applied to a single node in a structure modelled by membrane elements. In order to illustrate this point consider the problem shown in Figure 6.25 where a membrane structure is subjected to a point load at the free end of value $F = 200$ kN and is fully encastré at the fixed end. The structural dimensions are $l = 2$ m, $d = 0.2$ m and $b = 0.006$ m, the modulus of elasticity $= 200\times 10^3$ N/m^2 with Poisson's ratio $= 0.3$; the structure is modelled using a regular array of Quad 8 elements and solved using the Strand 7 finite element system. Three illustrative cases are used involving 40, 200 and 4000 elements respectively and the results are shown in Table 6.2, which clearly illustrates the fact that we have created a singularity by applying a load at a single node.

The presence of this singularity is illustrated by the stress plot shown in Figure 6.26 and the tabulated results show that increasing the number of elements merely accentuates the influence of the singularity and would encourage an adaption process to continue increasing the number of elements adjacent to the loaded node. Figure 6.26 also shows that, in

Table 6.2 Variation of stress in the x-direction with increasing number of Quad 8 elements.

Number of Quad 8 elements in the x-direction	Number of Quad 8 elements in the y-direction	Stress in MPa at point of load application in the x-direction
20	2	−609
50	4	−873
200	20	−5549

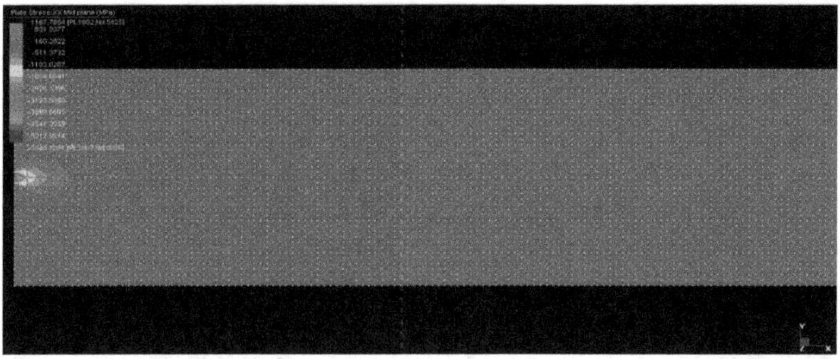

Figure 6.26 Plot of direct stresses in the x-direction for a model with 4000 elements. (See also Plate 1.)

conformity with Saint-Venant's Principle, the elements correctly balance the applied load in regions sufficiently distant from the load point.

Similarly, a displacement-generated singularity would occur if the problem defined in Figure 6.25 were changed so that it resembles a cantilever beam fixed at $x = l$ with the point load removed and replaced by a vertical pressure applied on the top surface and with an additional single node at (say) $x = 0.5l$ having fixed displacements. As with the point load case, we would find that a singularity exists at the fixed node at $x = 0.5l$ so that the apparent stresses at that point increase in proportion to the number of Quad 8 elements employed. Had a standard beam finite element layout been used, this singularity would not have been created.

As we can see, there are many snares to catch the unwary when adaptive meshing is employed, but even when properly employed the use of these h- and p-adaption automatic methods does not negate the application of the principles and tests introduced in the above sections and in later chapters. In fact, it makes them more important because the system has created the mesh without a capability to apply all the lessons from these sections – a post-mesh-generation examination, where the analyst applies the criteria for assessing the quality of the mesh and the elements employed, is an essential task before proceeding to the solution stage in the computational process.

Although we have concentrated on treating h- and p-adaption methods separately, they have their individual strengths and weaknesses and code developers have attempted to get the best of both worlds by combining both approaches with h- and p-adaptive codes.

Because these adaptive codes are clearly using 'interior' error measures to decide on how to adapt an analysis in solving a specific structural problem, there are a number of points to note with respect to the calculation of these errors. The first is the observation made in Chapter 4 that the Finite Element Method does not yield point-wise accurate results, only integrated accurate results. As a result, there is no absolute guarantee that the accuracy in stresses at a specific point in the analysis model will increase in proportion to the indicated reduction in the error measures. Second, as pointed out before, these estimating methods are self-referencing and are not connected to the external world. Thus the most they can say is that the solution is getting closer and closer to the solution of the mathematical problem that the finite element analysis 'thinks' it is solving – *whether this represents a convergence to the solution of the real-world problem cannot be estimated.* Salespeople often tell the purchasers of these codes that they reduce or even eradicate the error in the finite element solution, but this, as we have seen, is not necessarily or even normally the case!

REFERENCES

1. Barlow, J., Optimal stress locations in finite element models. *Int. J. Numer. Methods Eng.*, 1976. **10**: 197–209.
2. Barlow, J., More on optimal stress points – reduced integration, element distortions and error estimation. *Int. J. Numer. Methods Eng.*, 1990. **28**: 1457–1501.
3. Barlow, J., *Critical Tests for Element Geometric Distortions*. In *6th World Congress on Finite Element Analysis*. 1990: Robinson & Associates.
4. MacNeal, R.H., *Finite Elements: Their Design and Performance*. 1993: Marcel Dekker, ISBN 0824791622.
5. Szabó, B. and Babuška, I., *Finite Element Analysis*. 1991: John Wiley & Sons, Ltd. ISBN 0471502731.
6. Babuška, I. and Strouboulis, T., *The Finite Element Method and Reliability*. 2001: Clarendon Press, ISBN 0198502761.

7

Idealisation Error Types and Sources

7.1 DESIGN REDUCTION AND IDEALISATION ERRORS

The breakdown of the finite element analysis process described in Chapter 5 shows that it is possible to identify two broad error categories associated with the analysis of a real-world structure using the Finite Element Method. One of these may be considered as 'internal' to the analysis and is concerned with the analysis tasks associated with the lower levels in the total finite element process identified in Figure 5.1 as sub-processes 3 and 4 where errors may be generated in the creation of meshes, the selection of elements, etc., and in the numerical solution of the resulting model. These errors often arise because the user lacks the necessary experience or knowledge of the Finite Element Method or from mistakes made by an experienced analyst who is undertaking an analysis without following an approved quality control process. The methods and procedures for controlling this type of error are presented in Chapter 6. The information contained in that chapter guides the user through the process of building a finite element model prior to invoking the solution sub-process and focuses on the selection of appropriate elements and meshes. However, to undertake this part of the analysis and control the impact of errors by following the procedures laid down

A Practical Guide to Reliable Finite Element Modelling A. Morris

in Chapter 6, it is necessary to know what the set of elements and meshes selected by the analyst is supposed to be modelling!

In order to demonstrate the point being made, consider the case where an analyst is presented with a plate-like structure. If it can be established that the characteristic behaviour of the real-world structure is adequately represented by a thin plate mathematical model, then thin plate finite elements can be selected and subjected to the processes and procedures described in Chapter 6. This requires that a one-to-one correspondence is demonstrated which directly links the characteristic behaviour of the real-world structure to that of the mathematical model, in this case the thin plate theory that purports to replicate it. This link is set up through the sub-processes 1 and 2 in Figure 5.1; that is, in the design reduction and idealisation tasks. These two sub-processes are endeavouring to characterise the behaviour of the real-world structure in such a manner that it can be represented by an appropriate finite element model. Errors associated with these sub-processes constitute the other broad error category and can be considered as 'external' to the finite element analysis process as usually understood by the typical user.

We now move to this second set of error sources which, following the discussion in Chapter 5, are more properly defined as uncertainties which occur when a real-world structure is interpreted in terms that allow it to be represented by a finite element analysis. These uncertainties arise because the analyst has to make decisions about the behaviour of a structure often on the basis of limited or incomplete information. Clearly if the structure is very simple, such as that outlined above, which involves a linear static analysis of a thin plate and this is constructed from an isotropic homogeneous material with unambiguously defined and understood loading and support environments, then there may be little difficulty in devising an ad-hoc method for limiting the impact of uncertainties and, thereby, to create a satisfactory finite element model for the analysis problem. This type of simple analysis situation rarely, if ever, occurs in practice as most structures presented for analysis are multi-component, subject to a complex loading environment and have support and boundary conditions that are difficult to characterise. A multi-component structure normally gives rise to a multi-model configuration for the finite element analysis with internal boundaries, where components are attached to each other or come into contact during the course of a loading procedure. In this chapter we endeavour to categorise and explain the many sources of error or uncertainty which can be encountered in sub-processes 1 and 2 in Figure 5.1 during a complex analysis; Chapter 8 proposes methods for quantifying them as required by the discussion in Chapter 5 and Chapter 9 draws the outputs of these two chapters into a methodology.

There are several important and relatively discrete error or uncertainty sources associated with the design reduction and idealisation processes that make up sub-processes 1 and 2 as defined in Chapter 5. In the following sections these are categorised and presented in a chronological order which reflects the procedure recommended to the analyst in Chapter 9. In order to identify the uncertainty sources in these two sub-processes the first concept to take on board is that a domain of analysis has to be defined and this draws a boundary around the analysis. The process of selecting an appropriate domain is an activity that arises in the sub-process 1 that leads to the creation of the RRWS and is an important component in the analysis task. It can also occur as an activity in the idealisation process as certain parts of the structure may be given a reduced role to play. Choosing a domain of analysis brings with it uncertainties in the boundary conditions with respect to both type and influence.

Having constructed a domain of analysis, attention then moves to considerations relating to the behaviour of the structure within this domain. At this juncture uncertainties arise relating to the structural performance that is to be characterised by the final finite element model and include a wide range of considerations which might involve asking 'Is the structural behaviour linear, non-linear, static, dynamic, etc.?' Then the analyst must select the correct level of abstraction so that the resulting mathematical model used in the analysis adequately represents the RRWS's actual or required behaviour. There are potential uncertainties associated with the assumed (or selected) material properties, the loads or load paths, the boundary conditions at the edges internal to the structure under analysis, etc. This total set of analysis uncertainties is now considered in detailed in the remaining sections of the chapter.

Although this chapter attempts to list the major sources of uncertainty in the reduction and idealisation processes, it cannot be completely comprehensive. In working towards the creation of an adequate idealised model that allows the process of selecting element types and mesh layouts, the analyst is operating more like a detective than an engineer. The process begins with an initial set of assumptions about the nature of the analysis and the characteristics that require modelling with an uncertainty control mechanism in place. However, as the analysis proceeds, it often becomes clear that this initial set is incomplete or even incorrect. This may require a reconsideration of the initial assumptions in the light of new information extracted from a close analysis of the results obtained as the analysis proceeds. We have now entered a realm where a finite element analysis is really a series of analyses that might require a complete reconsideration of all the information used to control the impact of uncertainties in an analysis. This cyclic process is discussed in Chapter 9 where the process of evaluating the effect of sources of uncertainty and error is linked to the definition of these sources discussed in this chapter and methods for quantifying their impact on the analysis results described in Chapter 8.

The concepts and ideas presented in this chapter follow a common-sense approach but it may be argued that when the discussion is focused on selecting an appropriate level of abstraction in Section 7.4 common sense is abandoned. This section describes a process for assisting the analyst in selecting a finite element model or sets of models that adequately represent the actual behaviour of the in-service structure when subjected to service loads. However, worrying about the selection of elements best suited to a particular application seems to fly in the face of current popular 'common-sense' practice where 3-D solid elements are chosen as the first and only modelling option on the basis that modern computers have sufficient power to cope with any large finite element model. In addition, the availability of adaptive meshing programs means that no thought need be given to the number and grading elements in regions where high-definition models are required. This is a tempting policy but there are a number of reasons why this approach should be avoided which are presented in Section 7.4.

7.2 ANALYSIS FEATURES

The word 'feature' is commonly used in finite element books and texts and is used to describe entities resulting from the process of decomposing an analysis into a discrete set of separate structural objects. In the present text the word is used in a slightly more precise manner when employed within the error and uncertainty control process that forms the rationale for this chapter and the following two chapters. It is still used to describe an entity comprising an object or set of objects that have a specific location within a structure and a specific set of properties but is now linked to the presence of error or uncertainty sources within this structure. A feature can be a single component within the structure or a collection of components making up a sub-structure that exhibit the same behaviour or have the same physical characteristics or attributes.

In this situation a key aspect of a feature is that it has uniform characteristics or attributes. These characteristics or attributes might include geometric shape, material properties, response to external loads. In a static analysis the defining property for a feature may be that it carries a particular load path or paths.

For complex structures with a high level of depth involving many structural levels, as described in Section 5.4.2, there is likely to be a cascade effect in which lower level features are contained within higher level features. However, this should not be considered a hierarchy! At first sight a set of components within a multi-component structure may be considered as a single feature, but as the uncertainty review proceeds it may become clear that these components have separate uncertainty sources and that their interactions at internal boundaries also have associated uncertainties. In this situation the number of features has to increase as the uncertainty review progresses to reflect the increasing number of uncertainty sources that have to be characterised and, through the methods of Chapter 8, quantified.

A simple example is shown later in this chapter when the structure shown in Figure 7.5 is decomposed into a set of features in Figure 7.13. In this example a panel with stringer reinforcements is shown broken down into a set of features that are linked to the selection of a specific set of models for analysing the structure and depend on decisions made by the analyst. The process is driven by the analyst's view of the potential sources of uncertainty that lie within the modelling process. Although the feature creation process described in Section 7.4 considers the main panel and the reinforcing stringers as separate features, it would be possible to combine the panel and stringers into a single feature by modelling these as an orthotropic plate. If this approach had been taken, then uncertainties in the properties of the stringers and panel would have to be introduced into the orthotropic model. This would be an example of a number of features being drawn into a single feature that encapsulates all the potential uncertainties or errors within one feature.

The process of continuously breaking the structure into ever finer features in order to isolate an error source was introduced in the SAFESA project, references [1], [2] and [3], where the lowest level feature was called a primitive. We feel that the introduction of the term primitive does not add anything significant to the process of controlling error or uncertainty propagation in a finite element analysis. We now proceed to expand this brief discussion of error and uncertainty sources by examining the important sources in detail.

7.3 THE DOMAIN

7.3.1 The Domain of Analysis

Since no structural analysis problem extends to infinity the analyst must make a decision on the extent of the domain to be analysed. In certain cases this is straightforward if there is a complete separation between the structure that is the focus of the analysis and the surrounding domain. For example, a car is connected to the road through relatively flexible tyres while the road is significantly stiffer and in this situation could be considered as infinitely stiff. Thus the domain of analysis can, in principle, be limited to the car and its tyres and this complete vehicle can be considered as being supported on an infinitely rigid plane.[1] A bridge analysis is entirely different as the main supporting columns are located below ground level and the nature of the foundation support depends upon the nature of the sub-soil or the rock strata. Thus, in most cases, some assumptions have to be made relating to the limits of the analysis domain. Decisions on this domain can be taken early in the analysis process as part of a scoping exercise or later as the complexities of the selected analysis problem become manifest. This decision is not independent of the nature of the analysis and, therefore, is linked to the novelty of the analysis and the experience of the team.

In order to illustrate the points being discussed consider the simple analysis problem shown in Figure 7.1 which illustrates a thin beam subjected to a distributed load that is supported at one end by an elastic wall. At a point distant from the wall the beam is resting on a roller that is supported on an elastic foundation. There are a number of possibilities with respect to the domain of analysis for this simple problem and these depend on the relative flexibilities of the beam, roller and the foundation support; the extent of the domain is also linked to the intensity and actual distribution of the loads. One possibility, the most straightforward, is to assume that the stiffness of the roller and the foundation support is so much larger than that of the beam as having infinite value. In this case the domain of analysis can be specified as that shown in Figure 7.2 where the beam is considered to be encastré at a rigid boundary and simply supported at the point where the roller touches the beam. At this point it is appropriate to remind the reader of

[1] Of course, the actual analysis of the total car will not be simple as the analyst will soon discover that, among other complications, there is a very stiff engine block attached to a relatively flexible car body structure by elastic engine mounts and this introduces a number of interesting analysis problems.

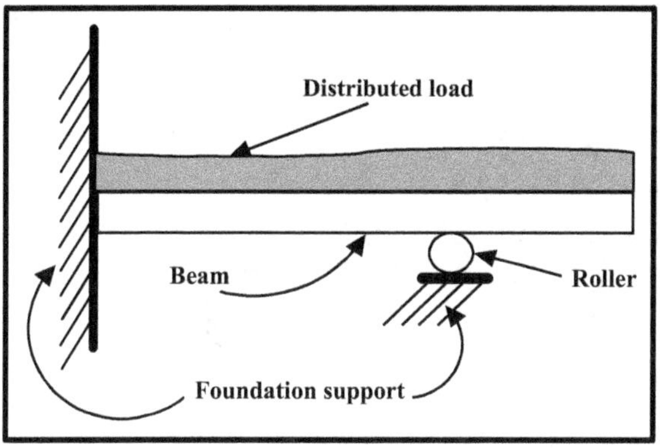

Figure 7.1 Elastic beam on elastic supports subjected to a distributed load.

the potential problems associated with the application of point constraints that are outlined in Chapter 6.

A second option would arise if the analyst felt that the left-hand support structure and the roller were not sufficiently stiff as to play no role in the analysis. In this case at least a portion of the support structure would need to be included within the analysis domain together with the roller as illustrated in Figure 7.3. In this case the analyst must decide on how much of the support structure to include, illustrated in the figure by the radius R. In an actual analysis the dimension R might be sufficiently large as to include other structural members and this greatly adds to the complexity of the analysis.

A further extension would be required if it was considered that the material providing the foundation support to the roller had to be included in the analysis. This is illustrated in Figure 7.4, and represents

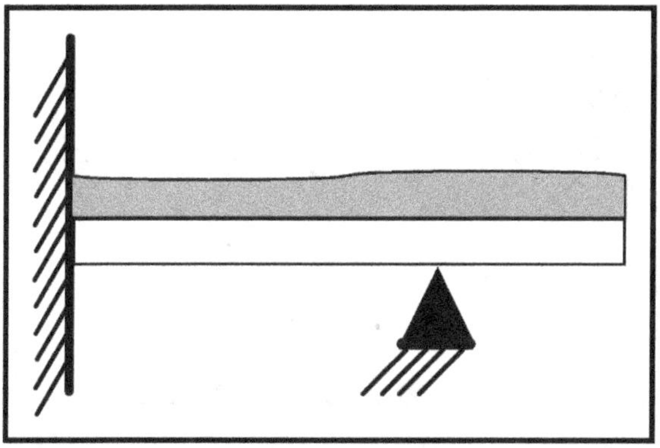

Figure 7.2 Simplest domain of analysis.

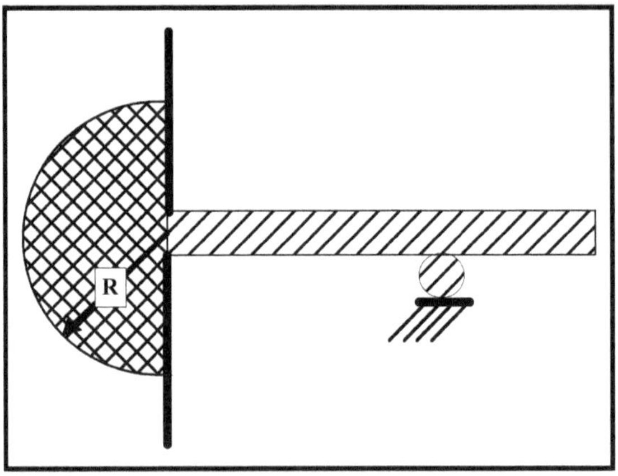

Figure 7.3 The first extended domain of analysis.

the final form of the potential domain of analysis for this problem. In addition to including a portion of the left-hand foundation support, a portion of the foundation support underpinning the roller has been included. As with the previous example the analyst must now decide on how much of the new support structure is to be taken into account, i.e. the values to be ascribed to the dimensions *a* and *b*.

Clearly if the domain of analysis is that shown in Figure 7.4, but a decision is made to use that illustrated in Figure 7.2, then serious errors can be injected into the analysis. This observation will apply whatever

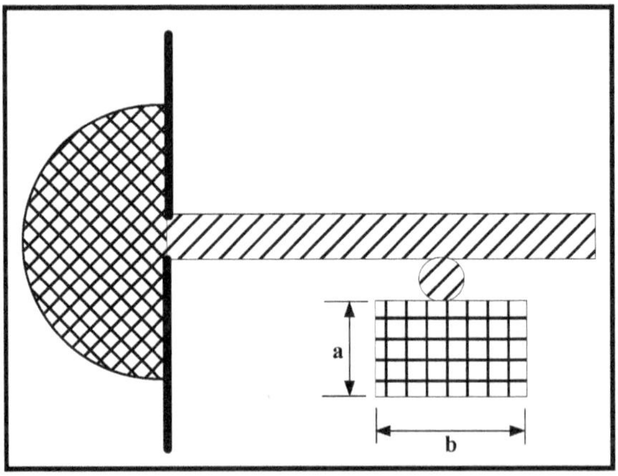

Figure 7.4 Final extended domain of analysis.

the nature of the analysis: static, dynamic, non-linear, etc. Indeed, selecting an incorrect domain of analysis is likely to have a much larger impact when a structure is subjected to dynamic loads or is responding to the loads by following a non-linear behaviour profile than is the case when a linear static analysis is undertaken. For these non-linear and dynamic cases the domain may not have a fixed extent. For a dynamically loaded structure the extent of the domain involving the foundation support may be a function of the frequencies as a larger region may participate in responding to some frequencies and not others. Take the simple case illustrated in Figure 7.3. The first extended domain of analysis in this observation would imply that the radius R varies with frequency and the extent of this variation may require separate analyses to quantify the impact of uncertainties. An example for a non-linear analysis would be the need to modify the domain when parts of a structure that are, initially, physically separate come into contact during the course of the deformation history.

This very simple discussion based on the analysis of the illustrative example in Figure 7.1 has shown that the question to be answered, under this heading, is 'How much error will be injected into an analysis for the uncertainties introduced through the process of specifying the domain of analysis?' The methods that allow the analyst, at least, to attempt to answer this question are given in Chapters 8 and 9. With this information to hand the analyst can make a decision on which domain of analysis to employ and attempt to quantify the extent of the uncertainties associated with the choice of domain.

7.3.2 Domain Reduction

The selection of the domain of analysis may represent the major domain reduction but there are a number of reasons why a domain may be further reduced. Some of these relate to the reduction of the RWS to the RRWS in sub-process 1 illustrated in Figure 5.1 and discussed in Chapter 5, Section 5.1, others may be due to an analyst's decision to work with a reduced model in order to explore rapidly the behaviour of a structure or a component. The model may also be reduced because the analysis is undertaken by several different groups using separate sub-structures or because the analyst exploits symmetry or anti-symmetry in the model. Potential errors from sub-structuring are dealt with in the section devoted to uncertainties associated with boundary conditions.

Clearly the creation of the RRWS may reduce the domain of analysis and has the potential to introduce into the analysis the type of uncertainty discussed in Section 7.3.1. This is particularly true if the reduction process eliminates part of the RWS. Eliminating a structural feature or component in a dynamic analysis can introduce very serious distortions in the predicted dynamic behaviour. For many dynamically loaded structures certain components or parts of the structure play a key role in defining the structural response and their removal or their reduced influence can radically distort the predicted behaviour. In the case of a structure exhibiting a non-linear response to the applied loads, domain reduction has to be undertaken with great care as the eventual role played by certain parts of the structure cannot be predicted before the analysis is undertaken.

Domain reduction is sometimes undertaken on the basis that it has been custom and practice within a company or industry but this hypothesis should always be challenged and subjected to the kind of rigorous examination discussed in the following chapters – particularly in cases where the company level of experience does not match the degree of complexity or novelty of the analysis problem.

In the case where an analyst decides to work with reduced models within a process where a finite element model is being built up, within a methodology that quantifies the impact of errors or uncertainties, then domain reduction is perfectly acceptable. Indeed, with such a control methodology in place, domain reduction can be a very effective tool for reducing the initial size of a problem, allowing the analyst the opportunity to start with a heavily reduced domain and then progressively increase the degree of complexity and detail as the total analysis proceeds.

Finally, domain reductions from the use of symmetry and anti-symmetry should not introduce any additional errors into a finite element analysis. However, employing symmetry and anti-symmetry in analyses with dynamically loaded structures or in the presence of significant non-linear behaviour is not straightforward and can give rise to user-induced error if the analyst lacks appropriate expertise in these areas.

7.4 LEVELS OF ABSTRACTION

One of the key aspects in understanding the way that the Finite Element Method solves structural analysis problems is the realisation that the method obtains solutions by employing one of the mathematical

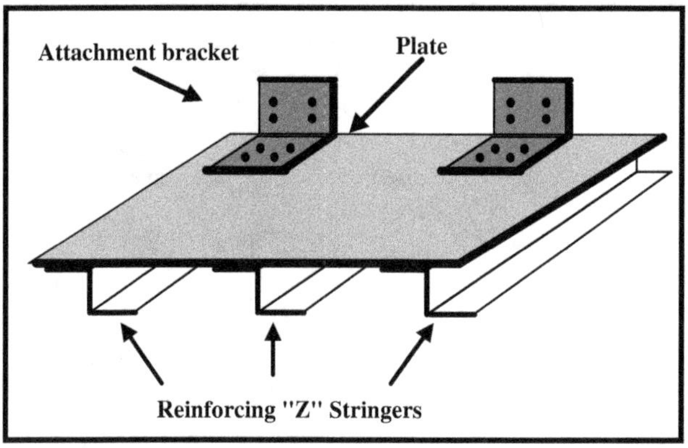

Figure 7.5 Z-stringer reinforced panel.

descriptions of structural behaviour. When the method models the behaviour of a thin shell, it will use elements constructed using thin shell theory as the mathematical model. A mathematical model is an abstract representation of physical reality and the analyst must choose that which is appropriate to the analysis problem requiring a solution. In a complex problem selecting an appropriate abstraction can be far from straightforward and may require many different models to match the behaviour of different types of component dependent upon their shape, loading configuration, anticipated response type, etc. This process of selecting a level of abstraction or mathematical model is one aspect of the process of hierarchical decomposition employed in Chapter 9 and detailed in references [1] and [2]. In order to give a simple illustration, consider the structure shown in Figure 7.5 comprising a panel with a set of Z-stringers attached to the lower surface and attached to a wall by means of two attachment brackets.

Consider first the panel and its reinforcing stringers. There are a number of possible abstractions that can be employed to model the behaviour of this combined structure. If we focus first on the stringers, there are several that can be used to model their behaviour and the appropriate model depends on a number of factors relating to the role being played by these structural members, the load acting on them and their interaction with the panel. Some of the possible models are shown in Figure 7.6 where three options are illustrated.

Each of the three models shown in Figure 7.6 represents the behaviour of the stringers in a different way and the resulting behaviour of the combined panel/Z-stringer will differ, dependent upon the model chosen.

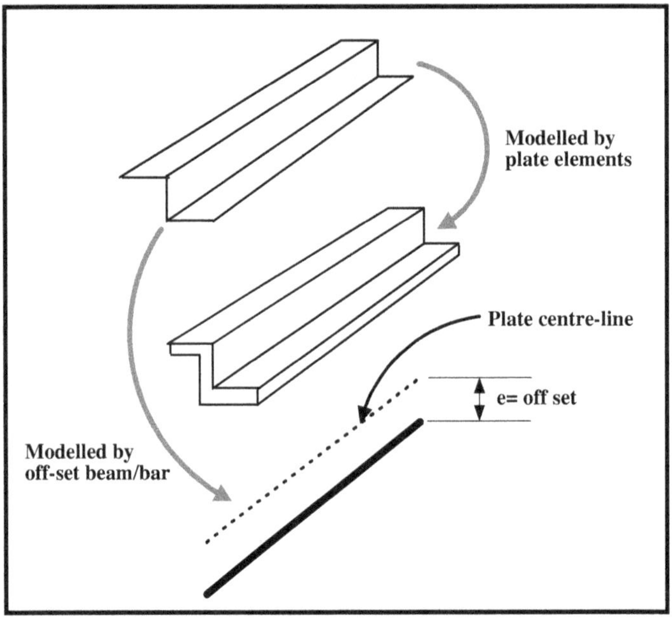

Figure 7.6 Stringer abstraction models.

Two of the models represent the stringer as an offset member, either as an offset bar or offset beam. The use of a bar element would be satisfactory if the stringer had very thin walls that could not accommodate bending and the bending support to the panel is achieved simply because the centre-line of the stringer is displaced from the panel centre-line. If the stringer has sufficient wall thickness as to possess intrinsic bending stiffness, then the offset beam representation would be appropriate. Finally, if the thickness might be such that the plate elements are required, then the third model shown in Figure 7.6 would need to be used.

This would also be an appropriate abstraction set if the structure were subject to a vibration-inducing load. These three alternatives clearly represent a different level of abstraction and it is not appropriate to consider one as more accurate than another. However, there is certainly an increase in the level of complexity as one moves from an offset bar to a full plate modelled structure or even to a representation using three-dimensional solid elements. It is sometimes assumed that the more complex abstraction level represents the top of a hierarchy of models from an accuracy viewpoint, but this is not necessarily the case. A structural component has a specific behaviour pattern dependent upon a number of factors, its three-dimensional shape, type of fixation, type of loading, etc., and the job of the analyst is to select that model which most closely represents the actual behaviour of the component and not the most complex. There will normally be differences between the results found by undertaking a finite element analysis using these levels; differences should not be thought of as errors but as discrepancies as they essentially represent different computed "measurements" of the structural behaviour of the stringers.

In essence, the above discussion has, by implication, focused on linear behaviour but the problem illustrated in Figure 7.5 could exhibit a significant non-linear behaviour that might not be immediately apparent. Consider the attachment brackets and assume that the 'wall' or 'foundation' to which the structure is fixed is very stiff so that the attachment bracket might exhibit the complex non-linear behaviour illustrated in Figure 7.7. In this figure the bolt is subjected to a direct tension T and a bending moment M that lift the surface of the attachment bracket from the wall surface (or this could also be the panel surface). The fact that surfaces part company and the bolt undergoes a large bending deformation turns what at first sight is a linear problem into a non-linear analysis. Combining this consideration with those discussed above concerning the behaviour of the stringers gives rise to a complex set of interacting mathematical models with plenty of opportunity for the generation of error.

In considering the level of abstraction, the approach presented in this section applies irrespective of the type of deformation or response. Thus all types of analysis problem – static, dynamic or non-linear – are treated in the same way. Of course, the complexity associated with addressing and assessing the level of abstraction may increase as one moves from static to dynamic or non-linear analysis problems.

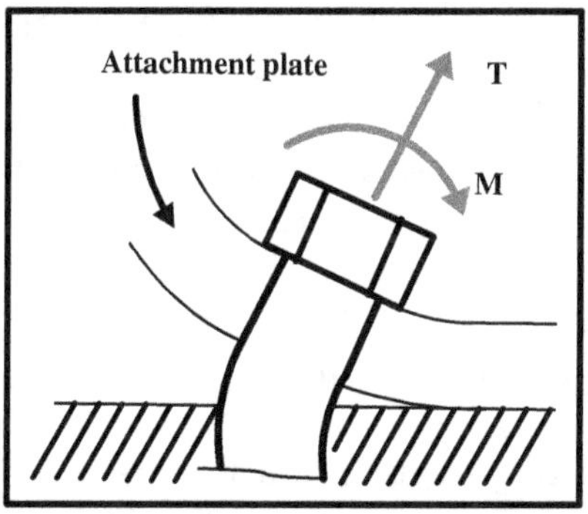

Figure 7.7 Bolt deflection.

As we shall see in the following chapters, the discrepancies between the various abstractions can be exploited to control the uncertainties within an analysis or, possibly, offer an opportunity to bound their impact on the target parameters for the analysis.

In order to illustrate some of the issues being raised in this chapter, consider the Z-stringer reinforced panel with the dimensions shown in Figure 7.5 where the panel and stringers are taken as aluminium with Young's modulus 60 GPa and the rest of the components as steel with Young's modulus 200 GPa. For convenience we are taking the domain of analysis as the panel–stringer–bracket combination and it is assumed that the support structure to which this is attached through the L-shaped attachment bracket is rigid; rigidity is not a realistic assumption but allows us to work with a simplified problem for illustrative purposes. In this example, the panel is subjected to a uniformly distributed pressure load on the top surface of 30 Pa.

An overview of the element layout is shown in Figure 7.9 and, for the first model where thin shell elements are employed, a detail of the model in the region of the bracket is shown in Figure 7.10. The bolts are modelled by cylindrical shell elements. The bracket and panel are assumed to be in contact at the fastener bolts. This represents one level of abstraction for which the maximum stress occurs on the bracket where the bolts make contact. The value for this stress is 52.7 MPa.

The second level of abstraction models the bracket, panel and stringer with solid elements and the bolts using the same element planform as that shown in Figure 7.9. The through-thickness element layout is shown in Figure 7.11. In this case the maximum stress occurs at the same point but the value for the stress is now 49.9 MPa. Clearly these two levels of abstraction predict different stress maxima and these differences are significant and would require resolution. This is covered in Chapter 10 where a walkthrough example is displayed.

As indicated in the Introduction, with modern finite element packages operating on very high-performance computers, it is natural to ask the question 'Why bother with levels of abstraction? Why not simply use three-dimensional solid elements and allow automatic mesh generators and adaption methods to do the work?' There are a number of reasons why such a 'black box' approach is not recommended unless there are very sound reasons for adopting such a policy:

1. In many cases the structure under analysis will have a form that lends itself to a particular structural model. Thin shells behave as thin shells except at attachment points or at points where highly

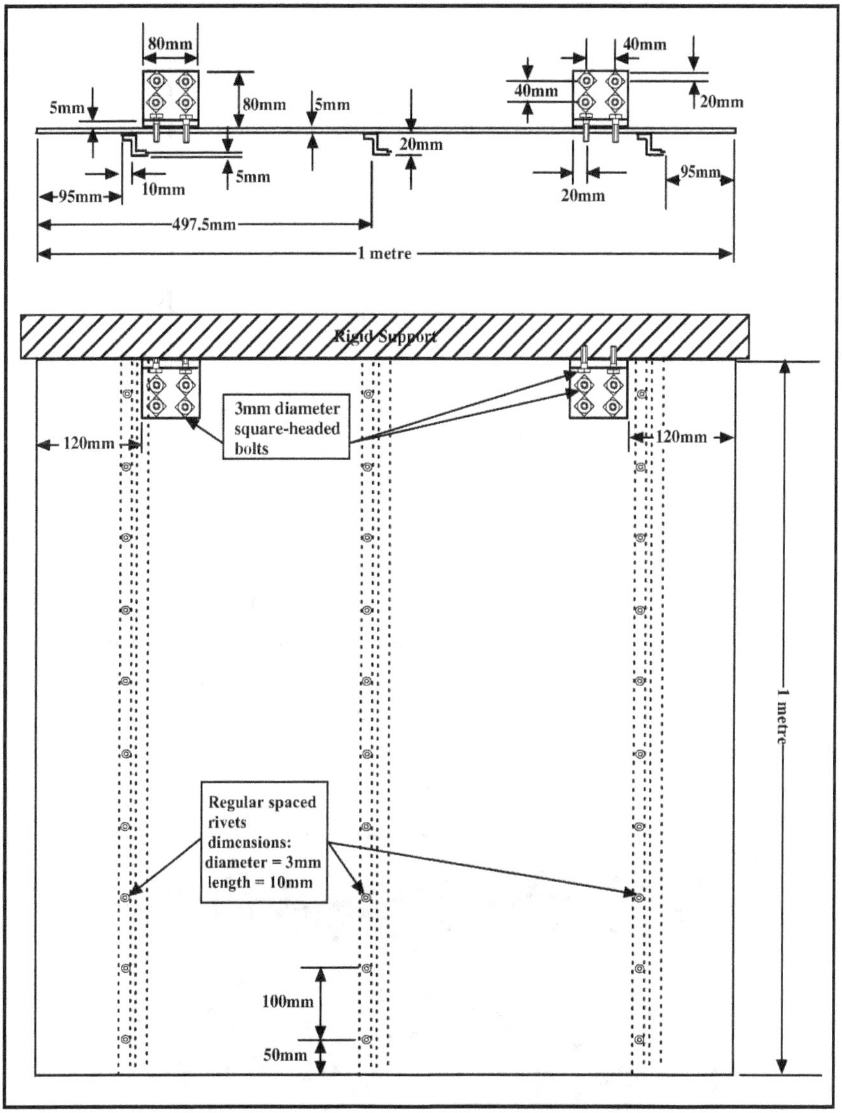

Figure 7.8 Dimensions for Z-stringer reinforced panel.

localised loads give rise to local regions of very high stress. Thus, thin shell elements are the appropriate elements for the major part of a thin shell structure. 'Horses for courses' is a good modelling strategy.

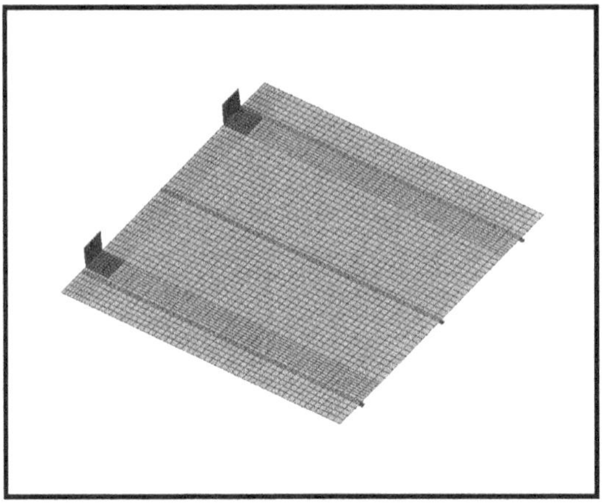

Figure 7.9 Element layout.

2. For large-scale structural analysis problems, exploiting the fact that a structure exhibits a specific structural behaviour normally significantly improves the computational efficiency. Mopping up all the available computational capacity is expensive and unpopular with the analysis budget holder and with the other analysts trying to solve their problems.

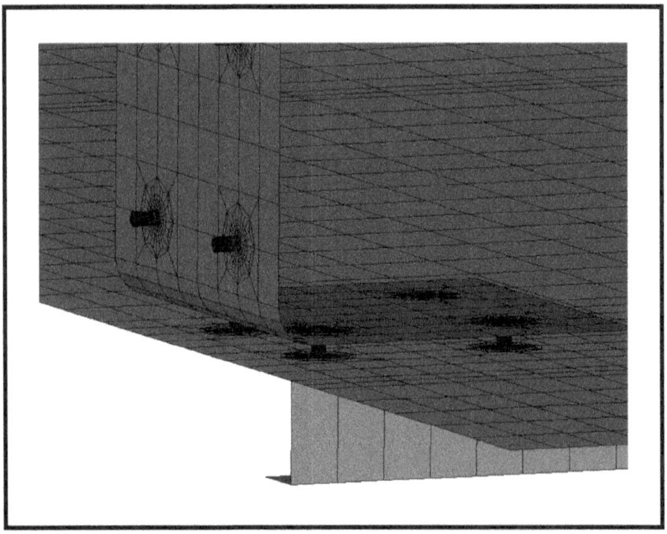

Figure 7.10 Detail of thin shell model.

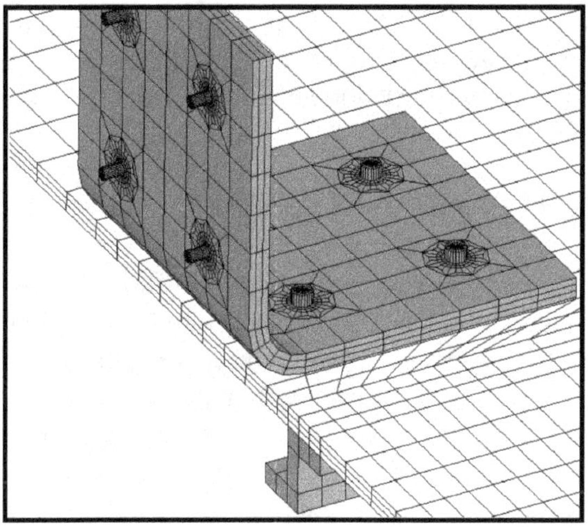

Figure 7.11 Detail of solid model.

3. Giving a Finite Element Method a large number of degrees of freedom allows it to generate spurious modes. In the case of structural dynamics there are as many natural frequencies and mode shapes as there are degrees of freedom. Not all of these frequencies and mode shapes are, necessarily, physically realisable.
4. As illustrated in Chapter 6, an unthinking use of automatic meshing or adaptive methods can give rise to problems with an analysis, inappropriate element shapes can be generated and convergence may not take place if singularities are introduced in the model.
5. Finally, thinking about and exploiting options for levels of abstraction means that the analyst builds up an appreciation of the way that the structure should behave when subjected to the applied loads. *In the engineering world faith is not a substitute for thinking so finite element models should always be selected on the basis of careful consideration of the type and nature of the structural analysis problem under review.* This type of information has saved many analysts from accepting results from a computed finite element analysis that are in error.

All these points illustrate that the use of three-dimensional solid elements does not remove the need to give serious attention to the decisions relating to modelling the structural behaviour but simply replaces one set of problems by another set. It is also worth noting that all the other error or uncertainty sources detailed in this chapter apply equally to structures modelled by solid elements as they do to structures modelled by lower dimensional elements.

7.5 BOUNDARY CONDITIONS

In this section we discuss errors that can be introduced into an analysis by the use of incorrect boundary conditions. Boundary conditions define the information that is transmitted across interfaces. Broadly there are two types: external boundary conditions that define the way the structure interfaces with the external world at the outer edge of the domain of analysis, and internal boundary conditions that occur at the interfaces between the structural features within the domain of analysis. The number of internal boundary conditions is directly linked to the depth of analysis discussed in Chapter 5, Section 5.4.2. The terms to be included in these boundary conditions are dictated by the mathematical models employed and thus relate to the levels of abstraction discussed in Section 7.4.

The boundary conditions and potential uncertainty sources at the points where the structure interacts with the external world at the outer extremity of the domain of analysis were fully covered in Section 7.4. The second set of boundary conditions are those internal to the structure and are concerned with the way that the various structural features interact and pass loads and displacements across feature boundaries. If the depth of analysis is very large, then there will be a large number of internal boundaries that have to be considered and these can have a large degree of uncertainty with associated error. In addition to the depth of analysis, the load and displacement transfer mechanism is a function of the particular abstraction model selected for the features interacting at an internal boundary. If we consider the very simple problem depicted in Figure 7.12, the representation of the boundary conditions at the interface between the two features is different if they are modelled by thin plate/shell theory elements, thick plate/shell theory elements or three-dimensional elements.

In order to discuss some of the uncertainties associated with modelling considerations that can lead to error we return to the simple panel and Z-stringer structure shown in Figure 7.5. This could be a structure in its own right so that the attachment brackets represent the edge of the domain of analysis, but we shall take the case that it is a component within a complex structural form. In order to examine the possible

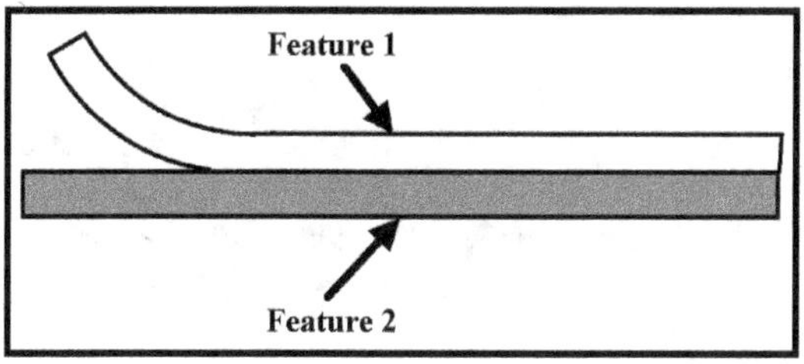

Figure 7.12 Simple double feature boundary.

sources of uncertainty in this structure due to internal boundaries, we need to break it down into its basic component features. In Figure 7.13 the structure is shown consisting of three features and has been subjected to a single level of decomposition; that is, the component has a single level with respect to the depth of analysis but the overall analysis has more depth since this is one component within a larger structure.

In defining the analysis problem for this structure as one that has three features, a certain degree of simplicity of analysis is being assumed. The basic assumption is that the loading and displacement fields are such that all three Z-stringers can be taken as acting in a very similar manner and that the two attachment brackets also have very similar behaviour patterns. If this were not the case, then it might be necessary to consider all of these as separate features with, potentially, different interface boundary conditions.

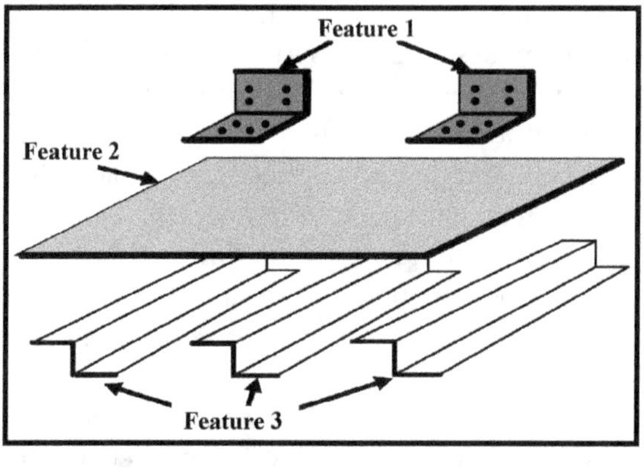

Figure 7.13 Single-level decomposition.

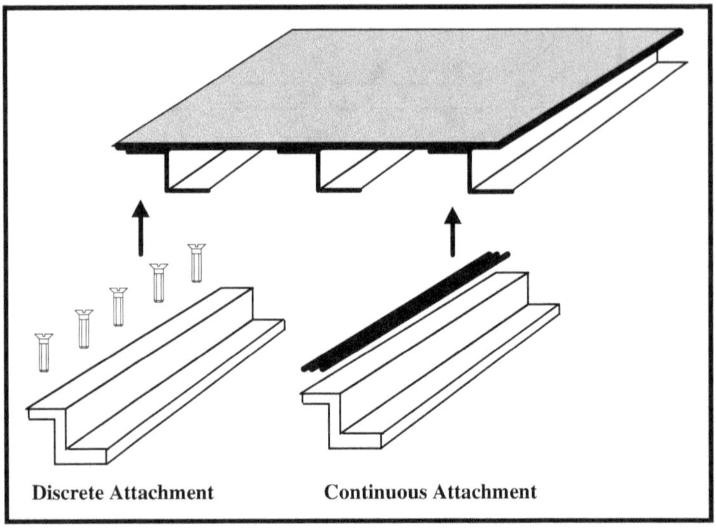

Figure 7.14 Panel/Z-stringer attachment possibilities.

Taking the interface between feature number 2 and feature number 3, then the method of attachment of the stringers to the panel is of critical importance. For illustrative purposes we consider only two of the possible attachment methods and these are shown in outline in Figure 7.14. In one case there is a discrete attachment method employing screws, but these could also be rivets or bolts; the other is a continuous attachment possibly involving an adhesive-bonded joint. While some load conditions and displacement responses give rise to interface behaviour patterns that are independent of the attachment method, there are others that will give rise to significant differences depending on the attachment method employed. It should be noted that the discrete attachment system allows movement to take place normal to the panel–stringer interface surface that cannot, in principle, take place with a continuous attachment system.

If the structure being analysed is subject to a dynamic load that gives rise to a free or forced vibration response, these internal boundary conditions have a major influence on the finite element results in terms of frequencies and mode shapes. Where a structure has a number of levels involving features that are attached to each other, there are three very broad attachment types that need to attract the analyst's attention:

- features may be connected to each other in such a manner that they are relatively free to vibrate independently;

- features may be connected in such a manner that they have a limited degree of dependence – for example, during a part of the vibration cycle they operate together and in harmony while in another part they may separate and behave independently;
- finally, the features are attached in such a manner that they behave as though they were an integral part of the main structure.

For the example shown in Figure 7.14 the discrete attachment system using screws or rivets could fall into the second of these categories by allowing the stringers to exhibit vibration modes that are partially independent of the panel modes as illustrated in the dynamic structural analysis example in Chapter 10. However, the continuous system shown in Figure 7.14, equivalent to a bonded joint, would be covered by the third attachment category. Even in a static analysis these two attachment systems can feed loads across the interface boundaries in a significantly different manner. This is particularly true in the case where such an analysis is being undertaken to calculate the buckling loads of the structure. The interaction between local and global buckling modes is critically dependent on the way that features interact.

In the case of the attachment bracket–panel interfaces, similar arguments can be made. In this case there is the added difficulty that there is now a two-sided attachment by means of which the panel/stringer structure is attached to the rest of the structure undergoing analysis. If the analyst considers that a situation displayed in Figure 7.7 might arise, then the bolt will have to be modelled in detail and the depth of the analysis increased as a set of more refined features will need to be employed to model bolt behaviour.

Clearly there is extensive scope for the introduction of errors into a finite element analysis due to an incomplete understanding of the behaviour of interfaces at internal boundaries. Consideration of interface behaviour is complicated by the fact that it is intimately connected to the levels of abstraction selected for the various features, the depth of analysis, the nature of the applied loads and the domain of analysis. In the case of a non-linear analysis, the problem can be further complicated by load paths moving from one part of the structure to another due to load shedding. This type of movement can radically change the way that forces are transmitted across internal (or even external) boundaries which may require the introduction of new joint models as the analysis progresses and previously separate components are forced into contact. The process of continual reassessment of the joint models is not exclusive to the non-linear problem because reassessment is a continuous process that takes place throughout the analysis of any major structure as the analysis process matures.

As a simple illustration, consider the reinforced panel shown in Figure 7.5 and, in particular, let us focus on the influence that the modelling of the boundary between the panel and the two brackets can have on the predictions of maximum displacement and maximum stress, i.e. the error associated with uncertainty concerning the mechanical process that transfers loads and displacements across the bracket–panel interface. The panel, stringers and brackets are modelled using 8-noded thin plate finite elements and, in order to keep things simple, the stringer–panel attachment bolts and rivets are not modelled and the Z-stringer is assumed to have a continuous attachment to the panel as shown in Figure 7.14. Two cases are now considered: one where the interface between the panel and brackets is assumed to be so tight that it can be modelled as continuous, the other where the brackets and panel are attached at the bolt points only. Clearly these two cases would bracket the actual performance of the bracket–panel interface. The results obtained for these two cases are displayed in Figure 7.15 and Figure 7.16 respectively.

As might be expected in the case where there is continuous contact at the interface surface for the bracket and panel, the maximum stress is located at the bracket corner with a value of 38.4 MPa and the maximum displacement occurs at the outer panel edge with a value of

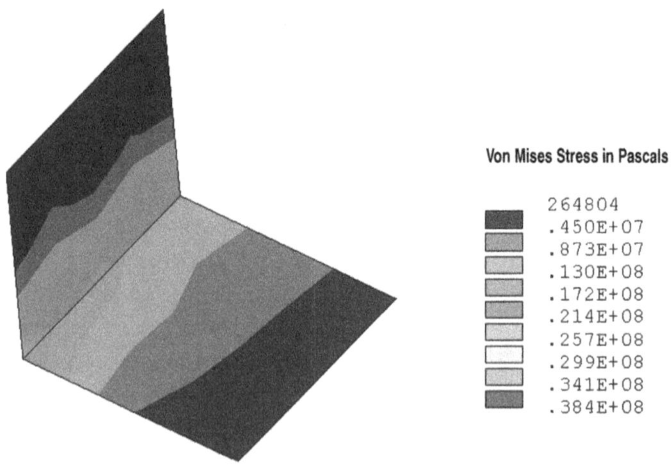

Figure 7.15 Bracket and panel assumed to be attached at all points on their interface surface. (See also Plate 2.)

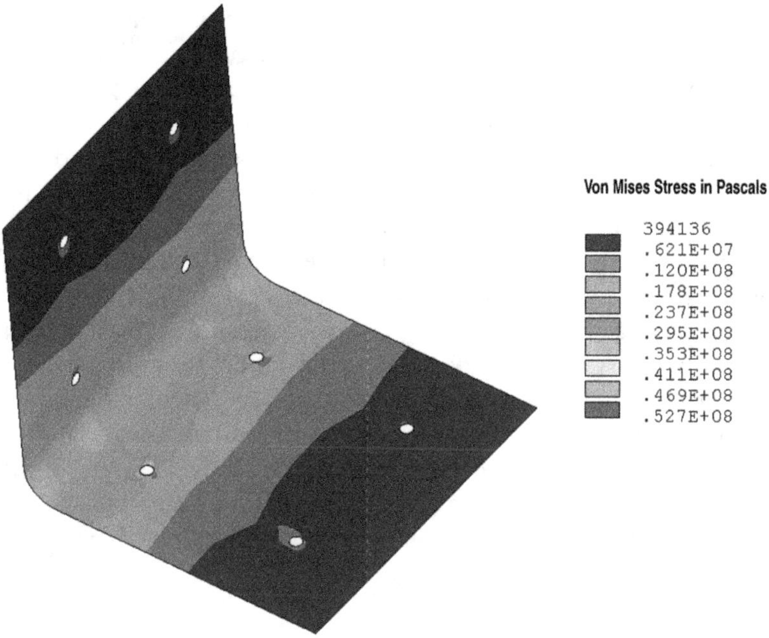

Figure 7.16 Bracket and panel assumed to be attached at bolt positions. (See also Plate 3.)

2.515 mm. If the interface boundary condition for the bracket–panel interface is changed to being one where contact occurs at the bolt points, the resulting stresses are as shown in Figure 7.16. In this case the maximum stress is now located at the points where the bolts are connected to the two components and the maximum stress has increased to 52.7 MPa. The location of the maximum displacement does not change but the value has increased to 3.654 mm. Whether or not these differences are important would depend on the qualification parameters and the level of accuracy that the analyst requires, but these results clearly do illustrate the existence of one source of uncertainty that falls into the category of an error source that is a direct consequence of the modelling assumptions made for a specific internal boundary condition.

Finally, there are a number of 'artificial' boundaries that can be introduced into the analysis by the analysis team or individual analyst. As explained in Chapters 2 and 3, a useful way to reduce the problem size is to exploit any symmetry or anti-symmetry that the design problem exhibits. This process should be error free as use of these two properties is a simple manipulation operation. But, human nature being what it is, there is no guarantee that these will be error free, particularly in those cases where symmetry/anti-symmetry properties exist in some structural features and not others. Condensation is another useful procedure for reducing the size of a problem and is frequently employed in dynamic analyses, and boundary conditions should not have any role to play in this process. It is also used when a part of a structure is being analysed with a much greater density of elements than the remaining structure. In this second case there is a boundary between the local, high-density analysis model and the remaining low-density model(s) associated with the rest of the structure. This interface does offer opportunities for error generation, particularly when different levels of abstraction meet across the interface. Condensation has a major role to play in creating sub-structures for use in analyses involving more than one analysis team. This environment offers great scope for the introduction of human error, particularly when a large number of teams are working on a common design problem but are separated by both distance and language!

7.6 MATERIAL PROPERTIES

Often an analyst will take for granted that the properties of the materials being used in a structure will have been catalogued and that the values supplied have very high accuracy. While this is the case for many conventional materials used in the construction of a structure that is subject to static loads, giving rise to relatively low stress levels, such a benign situation is not universal. If new and exotic materials are to be employed, or the analysis involves dynamic loads and responses, the accuracy of the supplied data may be very suspect. In these situations the analysis and the testing programme providing the material properties must be intimately linked so that appropriate information is used in the analysis. By linking analysis and test, the accuracy of the values ascribed to material properties can be deduced or inferred.

There are many potential sources of error or uncertainties associated with material property values and the discussion below is an attempt to thread a pathway through these sources. It is not intended to be comprehensive but more of a guide as to how one might approach identifying such sources, as these need to be catalogued by the analyst before trying to introduce the control methods discussed in Chapter 8. For an experienced analyst, this section might be considered as an aide-memoire and for an inexperienced analyst as the basis for constructing an error source catalogue. The discussion considers linear static, linear dynamic and non-linear analysis problems as separate entities but, in reality, an analysis might range over some or all of these areas.

- Linear static problems:
  - For isotropic materials most properties are obtained from suppliers' lists and these have, normally, small measurement errors associated with them and the analyst can take these errors as insignificant. However, if the material is one not in common use, then errors might not be insignificant if the tests have been undertaken by an independent laboratory!
  - For anisotropic materials the situation is often less clear as the properties listed for the material could have been derived from tests that the analyst cannot effectively trace. In addition, the properties will usually relate to tests performed on a coupon of material and errors of scale can be present if the analyst wishes to have a more refined model than the size of the coupon will allow.

- Linear dynamic problems:
  - The same reservations with respect to measured materials properties for the linear static case should be taken into account for linear dynamic problems.
  - In addition, the accuracy of the finite element analysis critically depends on the values used for the material damping properties. Damping factors are not 'static' quantities and for a given material will change as a function of temperature, external pressure, humidity, etc. For composite materials the damping factor can change significantly with relatively small variations in lay-up or adhesive properties. The influence of damping and the effects of non-linearities can often be confused.

- Non-linear problems:
  - Obtaining non-linear reliable material properties is often difficult even for low-speed phenomena as these properties can depend on the loading sequence, particularly if the sequence involves both loading and unloading.
  - For high-speed loading sequences, the material properties are a function of the loading and unloading speed.

- In the case of multi-phase phenomena where the material experiences changes that move it from one physical state to another, i.e. from a solid to a liquid phase, then reliable data is very difficult to obtain and major errors in the results can be anticipated.

7.7 LOADS AND MASSES

7.7.1 Loads

Modelling loads within a finite element analysis is often difficult as the actual loads seen by a structure are often not completely quantifiable or, in some circumstances, fully understood. In certain cases uncertainty exists with respect to the intensity of the in-service loads; in other cases the intensities may be relatively well known but the loading frequency has a high level of uncertainty, e.g. the buffeting example used in Chapter 5. Finally, the analyst may be confronted with a case where the full load spectrum may have high levels of uncertainty.

The analyst has to know how to deal with uncertainties in the loading environment. In certain cases uncertainty is removed or constrained by a requirement to use design codes that define the loading conditions or the requirement is to design a structure such that the as-built structure passes specified certification tests. However, we are unaware of any certification process involving design codes or tests that completely remove all uncertainties from the analysis loads. High levels of uncertainty often lead designers to impose large safety factors, but it is the analyst's job to find a way to reduce these factors so that the resulting design has an improved efficiency. Although reducing uncertainty in the applied loads or load spectrum is difficult, there are approaches that can help and these are discussed in Chapter 8. Uncertainty reduction in loads is one of the areas where analysts, test engineers and designers have to work together as a multi-disciplinary team.

Once the analyst has made an evaluation of the applied loads and load spectrum with respect to their intensity and distribution, these have to be applied to the element nodes. Broadly there are two methods for applying loads to elements. In the case where the load is distributed over several elements, the consistent method employed in Chapter 4 can be used. As shown in that chapter, the method assumes that the loads are distributed to nodes in the same way that the displacement field is represented by nodal displacements. This approach can give rise to errors in the solution if the nature of the distributed load is very complex so that the analyst needs to employ the approach advanced in Chapter 6 for selecting the size and types of the elements to be used in the final meshed structure. An alternative approach is to employ a lumped load methodology where the loads are distributed to the element nodes on the basis of a pre-analysis where the loads are applied using an assumed structural deformation field. This requires experience and a sound understanding of how the structure is going to behave when the loads are actually applied. For complex analyses this approach may give rise to a need to do a number of repeat analyses. A lumped approach is most readily applicable where the applied loads are highly localised, but local loads need to be treated with care as they can be a significant source of error.

7.7.2 Masses

For analysis problems involving dynamic loads and responses the analyst has to arrange for the mass of the structure to be 'seen' by the finite element analysis. As with loads there is ample opportunity for the appearance of uncertainties, particularly when the structure is to be manufactured using several different materials as in the case of carbon fibre reinforced composites. In this situation the analyst must liaise with those involved in creating the material to be used in the final manufactured structure.

When decisions have been made regarding the masses and mass distribution within the structure under analysis, the same problems occur with distributing the masses to the element nodes as those introduced in Section 7.7.1. However, in the case of mass distribution, the use of lumped masses, as opposed to consistent masses, is more common because the dynamic effect of the masses within the structure is often more accurately modelled employing a user-interpreted lumped distribution. Of course, the opportunity for error now greatly increases in inverse proportion to the experience of the analyst!

REFERENCES

1. Morris, A.J., *et al. Hierarchical Decomposition.* 1992: SAFESA Report 9034/TR/CIT/2002/0.0/15.11.92, Cranfield University.
2. Vignjvec, R. and Rahman, A., *Hierarchical Decomposition.* 1994: SAFESA Report 9034/TR/CU/2005/1.0/30.03.94, Cranfield University.
3. Morris, A.J. and Vignjvec, R., Consistent finite element analysis and error control. *Comput. Methods Appl. Mech. Eng.*, 1997. **140**: 87–108.

8

Error Control

8.1 INTRODUCTION

Throughout this book it is assumed that a finite element analysis is undertaken to support the design and entry into service of a product. As has been emphasised in earlier chapters, the problem is that this is a real-world structure often with many components assembled by welding, riveting, bonding, etc., and is very different from the relatively smooth, mathematically idealised world associated with a finite element analysis. In addition, the loads to which the real-world structure is subjected may also be different from the idealised loads used in the analysis.

The quest is to link the outputs from a finite element analysis to the actual behaviour of the in-service structure so that differences between the two are known and quantified. The first step is to decide on the qualification parameters that are going to be used to measure the performance of the real-world structure, such as maximum stresses at specified points, frequencies, etc. These are the primary focus of interest as the analyst must decide on the tolerance limits for these parameters defining acceptable values of error or uncertainty. Methods have then to be employed that ensure the finite element analysis results do not have variations that exceed the defined tolerances when compared to the values observed in the structure when it is in operational use.

This chapter falls within this broad analysis framework and introduces methods that can be used to treat the error and uncertainty sources identified in Chapter 7. It assumes that a structure is being analysed in order to be compliant with a set of qualification criteria that

A Practical Guide to Reliable Finite Element Modelling A. Morris

provide evidence it is fit for purpose. As commented earlier, in certain industries, such as aeronautics or nuclear power, the qualification criteria are generated by an external regulator; in others that are less safety critical, the criteria may be based on industry or on internal company quality requirements. This chapter and Chapter 9 make no distinction between these various types of qualification requirements but assume that a quality regime is in operation and that an analysis must comply with its demands. These chapters are linked: Chapter 9 introduces and explains an error or uncertainty control process lying within the overall qualification process, while this chapter discusses the control methods and technique.

8.2 APPROACH AND TECHNIQUES

Chapter 7 identified a range of potential error or uncertainty sources that can cause the predicted behaviour or performance of a structure from a finite element analysis to differ from that measured on the actual structure once it has been entered into service. This section describes techniques that can be used to obtain estimates for these variations between the finite element and the real world. The first step is to place these within an overall approach that is reformulated in Chapter 9 to form an error and uncertainty control process. The second step describes and illustrates the actual techniques employed and shows how these naturally form into a hierarchical set, as originally defined in reference [1].

8.2.1 Approach

The methods introduced in this chapter are being used to estimate or, at least assess, the impact that variations in structural properties, modelling assumptions, applied loads, built state, etc., have on the qualification parameters. These represent estimates of the impact that errors or uncertainties in the analysis model have on its ability to represent adequately the behaviour of the in-service structure. These errors and uncertainties are associated with all, or part, of the structure being analysed or even its surrounding domain. Some of the areas linked to a specific error or uncertainty may be highly localised, as would be the case where there is uncertainty with respect to the slippage of a single attachment bolt within a large structural component. Alternatively, the area could be large if there is uncertainty in a distributed property for a large component. An example would be uncertainty in the stiffness property of the aluminium alloy surfaces of an aircraft wing of a civil airliner where the associated area might include the upper and lower surfaces of the main wing.

Before applying the methods in Section 8.3.2 below, the analyst needs to go through the structure on an area-by-area basis assessing the likelihood that errors or uncertainties within these areas have an impact on the qualification parameters. At this stage the analyst is not being asked to quantify their effect but simply to identify that certain features have sources of errors or uncertainties that impact the parameters defined by the qualification process. The initial review may not capture all the sources and it is likely that this stage will need to be repeated as analysis or test evidence becomes available and throws up new causes for concern.

The process being introduced is one where the structure being analysed is broken down into individual features as dictated by the presence of errors or uncertainties. If the analysis problem is very simple, e.g. in the static linear analysis of a homogeneous beam, the only uncertainty involves the value to be given to the elastic stiffness property of the beam, for then the entire structure can be treated as a single feature. This implies that there is one term being bound in expression 5.3 and one layer to the depth of analysis. Most structures are more complex and have considerable depth giving rise to a set of features depending on the location of the sources of error or uncertainty and the analyst now has to deal with as many bounding terms M_i, from expression 5.3, as there are error or uncertainty sources.

In order to illustrate the process of linking error or uncertainty sources to the number and location of structural features, consider the static analysis of the reinforced plate used as an illustrative example in Chapter 7 and shown again in Figure 8.1. This figure shows the structure unattached to any support structure so that the domain of analysis is the panel, the attachment plates, the Z-stringers and the fasteners holding the structure together. In reality, the domain of analysis would include at least a portion of the supporting structure as discussed in the illustrative examples in Chapter 7, Section 7.3, but omitted here to simplify our discussion. It is assumed that the structure is loaded on the top surface of the panel by a uniform pressure and the qualification parameter is taken to be the tip deflection measured on the bottom surface of the middle stringer.

The analyst is now faced with a preliminary review of the structure that leads to an initial view of the number of features and error or

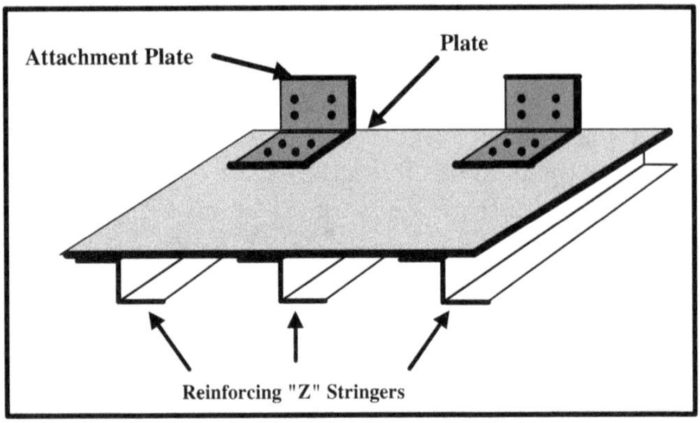

Figure 8.1 Reinforced panel.

uncertainty sources. The number depends on the behaviour of the components making up this structure and their properties. The process can be considered as hierarchical:

1. If the structure is made from a single material and the role played by the fasteners is insignificant, the only uncertainty influencing the values of the qualification parameter would be the stiffness of the material. In this situation it would be possible to argue that a single uncertainty factor is sufficient so that the entire structure can be taken as a *single feature.*
2. An assumption that a single feature is sufficient would ignore the interaction that is likely to take place between the three main components of this structure. While a single level of decomposition might be acceptable when all components have the same material properties, it would be inappropriate if the different components had different material properties. In this situation the uncertainties relating to the elastic stiffness of each of the major structural components would contribute to an overall uncertainty in the accuracy of the tip deflection predicted by the finite element analysis. Accommodating this requires decomposing the structure into *three features* as shown in Figure 8.2.
3. Finally, it might be considered that uncertainty in the stiffness properties of the bolts and screws could significantly impact the accuracy of the predicted value of the qualification parameter generated by the finite element analysis. In this case the number of features increases as the bolts and screws are now added to create a *five-feature* set as illustrated in Figure 8.3.

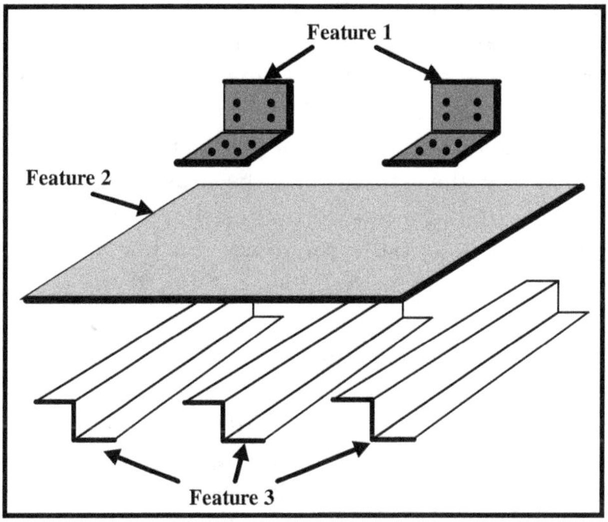

Figure 8.2 Three-feature decomposition.

8.2.2 Techniques

Thus far, discussion has focused on the broad principles underpinning the process of applying techniques or methods to control or quantify the effects of errors or uncertainties on the finite element predicted

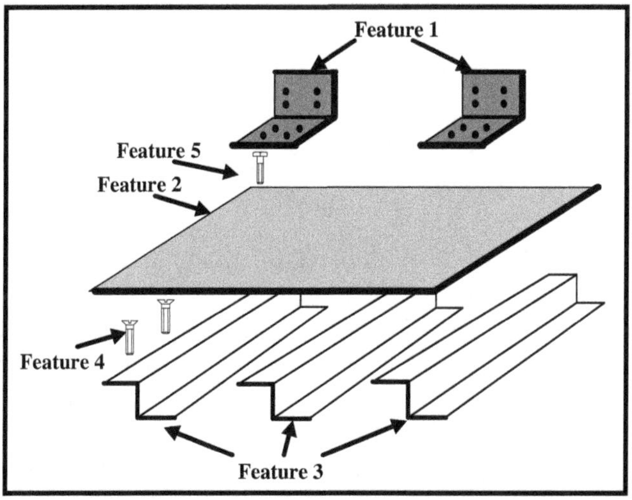

Figure 8.3 Augmented feature set.

values for the qualification parameters. The debate now moves forward to discuss the details of computational and other techniques for error and uncertainty control. It is assumed that those responsible for a finite element analysis have gone through the qualification assessment so that a set of qualification parameters has been established. It is further assumed that the structure has been assessed from a 'feature' viewpoint so that it has been decomposed into features associated with the location of potential error sources or uncertainties that may introduce deviations in these parameters from those seen in the in-service structure.

Before moving on to discuss control techniques, we need to take into account the fact that the qualification parameters are often functions of underlying structural properties. If the qualification parameter is, for example, maximum stress, the underlying structural properties that could impact the accuracy of its calculated value could be material properties, boundary conditions, etc. That is, the accuracy of the calculated value of this qualification parameter is directly influenced by uncertainties in the values ascribed to materials properties or in the behaviour of the structure at some boundary. One way of accounting for the impact of variabilities in the underlying structural properties is by means of sensitivity analyses.

Let p be such an underlying property that influences the ith qualification parameter q_i; then, within an error control framework, the qualification parameter becomes a function of this property, i.e. $q_i(p)$. If there is an uncertainty in the value of p, given by δp, then the resulting change in the qualification parameter is:

$$\delta q_i = \frac{\partial q_i}{\partial p}\delta p \tag{8.1}$$

on the basis that the value of δp is relatively small. This is a reasonable assumption as the presence of large uncertainties would render the analysis valueless! If δp is known or can be estimated and the term $\partial q_i/\partial p$ can be calculated or, in some other way, found (possibly from a test), then δq_i can be obtained from expression 8.1 and then inserted into expression 5.3. In reality, there would be more than one contributing uncertainty sources in expression 8.1 but this situation is addressed in Section 8.3.

8.2.2.1 Initial considerations and calculations

By working out the values of the bounds M_i the analyst is defining a tolerance on the acceptable level of deviation in a parameter from that experienced by the in-service structure. The first step is to try to calculate the values for these bounds from a reference structure using *simple engineering formulae*. These formulae can be found in many text books but particularly by using standard reference books such as *Roark's Formulas* [2] for a static analysis or Blevins [3] for a dynamic analysis or, in today's world, more directly by using the many Internet sites that provide this type of information on demand.

To illustrate the use of formulae, consider the case of a static analysis where a structure has irregularities in its form that arise from the presence of notches or corners and give rise to high localised stresses. The analyst can estimate the influence of these irregularities from the value of the *stress concentration factor*. This gives the ratio of the actual maximum stress to that computed using the net section ignoring the irregularity. As a specific example take the case where a regular structure, such as a bar, has a change in cross-section that involves a radius curve. If there is uncertainty in the radius of curvature of the curve, the influence of this uncertainty on the fatigue life can be assessed from tabulated values of stress concentration factors. This, in turn, requires that the analyst can obtain a value for the net section stress. Providing the structure has a relatively simple form, the net section stress, i.e. the reference value, can be found using standard mechanics formulae. As a second example consider the reinforced Z-stringer taken as a single feature in Section 8.3.1. Because the structure is to be taken in its entirety it may be remodelled so that the reference model is an orthotropic plate subject to a distributed load. In this instance an appropriate set of formulae would be applied to create the orthotropic model and then orthotropic plate formulae used to quantify the behaviour of the tip deflection and thereby compute the bound associated with uncertainty in the plate stiffness.

It might be thought that the use of simple engineering formulae for assessing the impact of errors or uncertainties would be applicable only in those cases where the analyst is encountering a structural analysis problem that lies within the experience base. In fact, these formulae should always be used as the first port of call in any analysis and have great value in the case where a problem has a high degree of novelty or complexity by shedding light on the likely behaviour of the structure. In addition to providing potential bounding information, these formulae assist in guiding the analyst in developing the analysis plan and analysis implementation. They can be used to assess the impact of error or uncertainty in any of the types and sources detailed in Chapter 7 across the entire range, including the domain, the loads, the boundary conditions, the level of abstraction, etc. The use of simple engineering formulae to represent reference values represents the initial or simplest set of assessment techniques in an attempt to create the error or uncertainty bounds M_i in expression 5.3 before more complex methods are employed. However, in the case of complex built-up structures this may not be possible due to the difficulty of assembling the components of the built-up domain into a form readily applicable by simple formulae.

In certain cases the influence of an irregularity on a qualification parameter may be negligible as its influence does not extend to the region where the parameter is located so that it can be ignored. At this initial stage the decision to ignore an irregularity can be based on applying a set of 'rules of thumb'. If a feature contains some form of reinforcement, a rule might say that this can be ignored if the reinforcement is not continuous and is less than a certain dimension or volume when compared to those of the feature as a whole. Before applying 'rules of thumb' the analyst needs to substantiate their validity. This may be done by reference to earlier analyses where the rule was justified, by reference to test data or the rule may have been imposed by an external qualification body.

Application

1. *Simple formulae can be used to provide an initial assessment of the influence of errors and uncertainties throughout the structure on the qualification parameters for all the sources listed in Chapter 7.*
2. *Simple formulae can assist the analyst in decomposing the structure into a complete set of features.*
3. *Simple formulae may be able to provide all the required information relating to the impact of errors and uncertainties on the qualification parameters for non-complex structural forms and layouts.*

8.2.2.2 Levels of abstraction

One potentially major source of uncertainty lies in the selection of the mathematical model(s), as explained in Chapter 7, Section 7.4.2. Following the approach of Section 8.2.2.1, decisions regarding the choice for the mathematical model can be made using 'rules of thumb' which, in this case, are usually derived from past experience. An example is the 'Taig' rules, reference [4], for deciding whether an aircraft structural component should employ reference models using shell/plate elements or solid elements as shown in Table 8.1. Such rules are clearly not comprehensive and may not be reliable, but can be a useful guide in assisting the analyst in selecting a specific level of abstraction for an analysis.

Rules of thumb can be a useful starting point in an attempt to select an appropriate set of mathematical models for a finite element analysis. However, these are rarely more than a guide and recourse to using a sequence of finite element models is normally required to complete the selection process. We now need to look a little more carefully at what is implied in employing sequential finite element models.

Different levels of abstraction can be considered as different measurements of what constitutes the real-world mathematical model that replicates the behaviour of the in-service structure. If a structure is

Table 8.1 Ian Taig's 'rule of thumb' for selecting a shell/plate or solid model in the stress analysis of an aircraft structure.

Shell/plate model	Solid model
The structure has a form characteristic of shell/plate or is built up from such members	Structure has the general characteristics of a general three-dimensional body
The thicknesses are small compared with other significant dimensions	Thicknesses are significant compared with other relevant dimensions
Homogeneous, isotropic material employed	Non-homogeneous or highly anisotropic material employed
Through-thickness or out-of-plane shear stresses unlikely to be significant	Strength through the thickness or out-of-plane shear is considered to be significant
Mainly interested in stresses at the extreme fibres or in displacements	Stress distribution through the thickness is important or there are stress raisers in the depth-wise direction
Geometric non-linearity only may arise	Progressive non-linearity may develop moving in from the surface

plate-like, it could be modelled as a membrane, a thin plate, a thick plate or a three-dimensional body depending on the physical dimensions, the applied loads or the boundary conditions. Each of these models would predict a value for the qualification parameter(s). It is argued that employing a sequence of such finite element models can generate a hierarchical sequence that converges towards the behaviour of the real-world structure. In other words, we should be able get increasingly accurate predictions of the real-world behaviour by using evermore sophisticated finite element models.

The process can be outlined by considering a simple linear static analysis problem where the loads on the structure are fully defined. For the real-world structure this implies that the behaviour can be represented in terms of a stiffness matrix K_R and load matrix P_R such that:

$$K_R u_R = P_R$$

where u_R represents the actual real-world displacements resulting from the application of the real-world applied loads. If the hierarchical modelling procedure is being used, then there is a sequence of user-generated finite element models and for the ith reference model we have:

$$K_{(i)} u_{(i)} = P_{(i)}$$

Similarly for the next reference model in the sequence:

$$K_{(i+1)} u_{(i+1)} = P_{(i+1)}$$

The two models at step i and step $i+1$ do not necessarily employ the same sample points, i.e. nodal positions, or type of representations for the displacements $u_{(i)}$ and $u_{(i+1)}$; this is clearly the case if one model is a thin plate representation whilst the next uses thick plate theory. The difference in models and connection quantities also explains the difference in the load representation in the two expressions above. The underlying assumption for the hierarchical modelling approach is based on the concept that it is always possible to link directly the stiffness properties of these finite element models to that of the real-world structure in the following manner:

$$K_R = K_{(i)} + e_{(i)} = K_{(i+1)} + e_{(i+1)}$$

where the errors $e_{()}$ relate to a specific subset of the total errors in the analysis process which are considered amenable to a hierarchical modelling process; these include the mathematical model, the boundary conditions and the loads applied to the structure. Inspection of this expression implies the existence of an error norm such that:

$$|u_R - u_{(i)}| \xrightarrow{i \to \infty} 0$$

on the assumption that errors which cannot be treated by model improvement are ignored. While, in principle, this is a correct argument, in practice it does not work as there is no concrete method to allow the analyst to know that in going from one model to the next, the analysis outputs are genuinely converging to the real-world values.

Clearly if the analyst can generate solutions for all the possible reference models relating to the structure, then the real-world behavioural model has to be one of them and the spread of values for M_i computed from these solutions represents a range within which the actual value must lie. This now brackets the actual value for M_i. Of course, this observation assumes that the reference models selected give a complete coverage of all the possible levels of abstraction. Unfortunately, nature is not that kind and it is usually not possible to guarantee completeness in the model selection. However, we can approach this problem in a logical manner that will provide the best possibility of achieving the objective of capturing the true behaviour of the real world – and it will be a useful defence in a court of law.

In order to explore how we might achieve our objective of capturing the true behaviour, consider once again the reinforced panel shown in Figure 8.1 which could be an entire structure or a component within a larger structure with all components manufactured from an isotropic homogeneous material. As before, we simplify the structure by taking the case where this is the entire structure and assuming it is attached via the attachment plates to the rigid support. If the analyst has concluded that the first stage involving one single feature is not appropriate, it will be necessary to move down the hierarchical chain and take the next level model involving the three features shown in Figure 8.2.

If the distributed load is 'moderate', then it will be reasonable for the analyst to consider that the attachment plates and the main panel behave as thin plates. The Z-stringers could contribute to the overall stiffness in several ways depending upon their relative thickness with respect to the plate and the method of attachment. Three modelling options are shown in Chapter 7, Section 7.4: in Figure 7.6 each model requires analysing, one for each of these different assumptions but using in all cases thin plate theory elements for the panel and attachment plates. Because it is important to compare like with like, these models should be run employing adaptive meshing programs so that each reaches the same level of internal accuracy. As we do not know which of the three results is nearest to the in-service value, we can take them as measurements of the potential uncertainty and use them as bounds on the potential error that can be injected into the finite element model if the wrong abstraction is used. This is assuming that the analyst has some grounds for accepting these as true measures of the potential uncertainty and this, in turn, assumes that this type of analysis has been undertaken before. In this relatively benign situation one would expect to observe a convergence process coming into play or even all the models giving very similar solutions. If there is no basis for accepting these three values as the basis for creating a measure of uncertainty, then the worst case that gives the largest reference values for the qualification parameter should be accepted or the analyst must have recourse to test. However, the test can now be precisely defined as it is intended to validate one of the models and provide values for the uncertainty bound related to model choice. In the case of a first-time analysis, a further set of reference analyses may be required in which the panel, the attachment plates and the Z-stringers employ a thick plate model using appropriate Reissner–Mindlin finite elements. Ideally, both the thin and thick plate finite element reference models should be used in conjunction with an adaptive meshing procedure in order to allow a comparison to be made on a level playing field. In a moderately loaded structure there should be little difference between the thin and thick plate models.

Moving to the case where the distributed load produces large forces but with responses still within the domain of linear elasticity theory, then the thickness of the various structural members will greatly increase and a reference model using thick plate theories would be required in addition to the thin plate model. If the difference between the results from these two analyses is significant, then a third and much more complex reference analysis will be required involving three-dimensional finite elements. In this case it is possible that the stresses in the panel are much less than those seen in the attachment plates. This would imply that the thickness of the two attachment plates is greater than the thickness of the panel. However, the analyst should not be tempted into believing that it should be possible to model the panel with a thin plate theory finite element and the attachment plates by a thick plate element. Because of the nature of these two theories the thick model has a through-thickness direct stress modelling capability which is not present in the thin plate model. The thin plate would not 'see' the thick plate that is attempting to transmit direct forces across the contact plane, thus negating the very reason for selecting a thick plate model. There is an important message here: if different levels of abstraction are used within a complex analysis, these have to be, or have to be made to be, compatible.

The final level of abstraction to be considered is when the loading environment is such that the presence of the fasteners has to be taken into account leading to a situation where there are five sets of features as illustrated in Figure 8.3. This is a far more complex modelling problem than the simple cases discussed above. The fasteners were omitted on the basis that their contribution to variations in the qualification parameters due to the presence of error or uncertainties is insignificant. Now the fastener features have to be brought into the error and uncertainty control programme. This can be done by including models for the fasteners and the local structure near the fasteners within an overall global reference model. Such an approach may be acceptable for the type of small-scale problem discussed here but would be impractical for a large-scale structure that included a large number of attached components. The alternative is to use a global–local analysis procedure. In the present case this would involve using the three-feature model for assessing the influence of errors or uncertainties in the bulk of the structure with additional local reference models for assessing the part played by the additional features represented by the fasteners.

Taking just one example to illustrate the additional complexity by introducing the fasteners, consider the region local to the attachment plate – the main plate interface shown diagrammatically in Figure 8.4. Clearly this is now a three-dimensional situation requiring the use of three-dimensional finite elements and it is not possible to increase

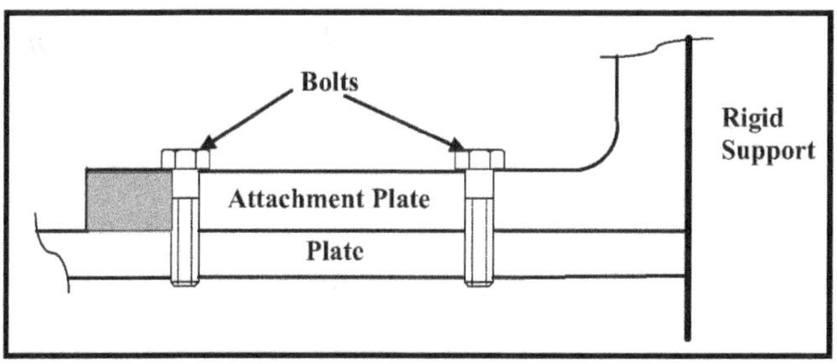

Figure 8.4 Bolt model.

further the level of abstraction. This is not a critical issue as the uncertainties will relate to such factors as the presence and action of friction forces, the elastic behaviour of the screw threads and the fact that the surfaces may separate particularly the region shown shaded in Figure 8.4 which can be assessed by other methods. In particular, the influence of some of these sources of uncertainty can be assessed using the sensitivity methods discussed in the next section. In addition to the more general type of analysis, an assessment of the influence of errors or uncertainties when there is a separation of surfaces originally in contact requires a non-linear analysis. Once again the creation of appropriate bounds M_i may not be possible from the application of formulae or finite element analysis alone and recourse to test will then be required.

Using adaptive methods, either p-adaption, h-adaption or h–p-adaption must be undertaken with some caution but, as shown above, these have important roles to play in selecting an appropriate level of abstraction for a given analysis problem and in generating bounds on the influence of errors or uncertainties. One role, already noted, is to ensure that all the models being compared in a model selection process are equally 'optimal' from a posterior error viewpoint. A second role is to ensure that once mathematical models have been selected, because they most closely match the real-world behaviour, the analyst gets the 'best' out of this model. An excellent example of using p-adaption to select and refine a model as proposed herein is given in the final chapter of the book by Szabó and Babuška, reference [5].

> *Application*
>
> 1. *Hierarchical modelling is used to select the mathematical models for all the features within the structure and the supporting structure(s) that most closely match the behaviour of the real world.*
> 2. *Model assessment can also be used to assist in deciding on the region that constitutes the domain of analysis.*

8.2.2.3 Finite-element-based sensitivities

When an appropriate level of abstraction has been established, the analyst(s) has a finite element model available. It is now possible to use this as a reference model to assess the impact on qualification parameters of uncertainties in the underlying structural properties based on the discussion in Section 8.2.2. The assessment process introduced here follows on directly from this discussion and uses sensitivities to quantify these variations. It seeks to find the rates of change of the qualification parameters with respect to rates of change of the material property, the boundary conditions, etc.

The sensitivities to be used in the quest to control errors and uncertainties associated with the analysis of a real-world structure can be obtained in two ways. One exploits our ability to extract rates of change for parameters such as displacements, stresses, frequencies and buckling loads from the finite element analysis in a direct manner using first-order derivatives. This requires relatively little re-analysis. The other is much more computationally expensive as it requires full structural or sub-structural re-analysis in order to perform a set of numerical experiments with varying values for the terms under examination. It comes into play in cases where the structure is complex or has high levels of interaction making the direct calculation of derivatives very difficult or impossible. Both approaches are described below.

In a major analysis the sensitivity methods can generate a large number of bounds relating to the impact on the qualification parameters of errors or uncertainties. Following such a sensitivity study all the bounds associated with significant errors or uncertainties have to be included in the global model and their influences assessed simultaneously so that their combined influence is measured. This can be very difficult, but in certain cases accumulating bounds into a single bound for a specific qualification parameter is achievable.

Direct evaluation of sensitivities Direct evaluation of sensitivities can be employed in an analysis where the linear theory of elasticity forms the basic assumption of the analysis and where there is no complex time-dependent history. It exploits an approach employed by structural optimisation algorithms, references [6] and [7], and allows the analyst to compute the effect of error or uncertainty in analyses involving static, dynamic and buckling problems. The approach is applicable in cases where the errors or uncertainties relate to a single finite element within the total system or to clusters of elements combined into superelements. It is also applicable in cases where substructures are being employed and one group involved in an analysis and responsible for a sub-structure or structural component is concerned about the quality of the models provided by other groups.

Static analysis In this section K_i is the stiffness matrix for the element i within a global system, u is the set of global nodal displacements, P the nodal loads associated with K, the global stiffness matrix, after the boundary conditions have been applied. Because the boundary conditions have been applied, this system is non-singular. It should be noted that this nomenclature differs slightly from that employed in Chapter 2 (where current matrix K is denoted by KR). The matrix equations for this system are given by:

$$P = Ku \tag{8.2}$$

Now consider the case where this represents an analysis where the qualification parameters are stresses and displacements. The question being asked is relatively simple in concept. The analyst has identified that uncertainties are present in the underlying structural properties in a certain part of the structure that give rise to changes in the global stiffness matrix K. The question to be addressed is 'Can the resulting changes in structural displacements and stresses be estimated?' This section shows how it can be done in the case of a structure subjected to static loads only and, because it has been assumed that the analysis is being undertaken using displacement finite elements, it is convenient to start the discussion by examining the process whereby the analyst can extract the sensitivities of the nodal displacements to changes in structural properties or parameters.

Displacement sensitivities Consider the case where an analysis has been performed and a global set of displacements u found by solving the matrix equation 8.2 for a set of applied loads P and this provides the reference model. Suppose the analyst wishes to quantify the impact that the uncertainties in a set of underlying properties have on the qualification parameters. Take the case where the uncertainties now appear or can be taken as a change in the stiffness properties of the structure that modify the terms in the global stiffness matrix by an amount δK. This change in the stiffness matrix properties gives rise to a modified global displacement vector δu and the analyst now requires a process for evaluating the terms in this vector.

If the loads applied to the structure do not change, the modifications in the underlying structural properties give rise to a variation in the global stiffness and displacement vector only so that equation 8.2 becomes:

$$P = (K + \delta K)(u + \delta u)$$

and after expansion:

$$P = Ku + \delta K u + K\delta u + \delta K \delta u \tag{8.3}$$

By taking note of equation 8.2 and neglecting the second-order terms $\delta K \delta u$, equation 8.3 can be written:

$$\delta u = -K^{-1}\delta K u \tag{8.4}$$

Normally the uncertainty or potential error will not apply to the total structure, nor will the analyst be interested in the rate of change of all the nodal displacements. In order to develop this concept assume that the qualification criterion has identified the specific nodal displacement j as being critical and it is considered that the uncertainty relates to changes in the stiffness of one element only, i.e. element i. In this situation the term δK becomes very sparse. Recalling from Chapter 2 how the global matrix K is assembled from the individual element stiffness matrices, then the matrix δK has zeros everywhere except where the individual element k_i was originally inserted into the global stiffness matrix, at which point the terms associated with the change in that element's stiffness matrix are now inserted. Thus the term δK is now replaced by the term $\{\delta k_i\}$ that has zeros everywhere except where the original element made its contribution to the global stiffness matrix. The term δu is a vector of the rate of change of all the nodal displacements with respect to changes in the stiffness of element i, whereas the requirement is to calculate the rate of change of one specific displacement u_j. In order to extract the required specific displacement rate of change δu_i, the vector δu needs to be pre-multiplied by a vector e^T given by $\{0\,0\,0 \ldots 1 \ldots 0\,0\}$ so that $e^T \delta u = \delta u_j$. Note that the unit term in e^T is positioned so that when multiplied by the vector u, all the displacement components are multiplied by zero except the term u_j. Equation 8.4 can now be manipulated using these terms to give:

$$\delta u_j = e^T \delta u = -e^T K^{-1}\{\delta k_i\} u \tag{8.5}$$

Note that $e^T K^{-1}$ is equivalent to solving the structural analysis problem with a unit load applied at the node where we wish to assess the impact on the nodal displacement of a change in stiffness of element i and gives rise to the vector v. The vector v is, therefore, a vector of displacements that would be generated in the structure by the application of a unit load at the nodal displacement (or degree of freedom) j in the absence of the actual applied loads. Noting all these aspects of the rate of change calculation, then equation 8.5 compresses down to:

$$\delta u_j = -v_i^T \delta k_i u_i \tag{8.6}$$

where:

- δu_j is the rate of change in the displacement j due to changes in the stiffness of the structure in the element i.
- v_i is the component of the vector of nodal displacements due to the application of a unit load at nodal displacement j that act on the nodes of element i.
- δk_i is the element stiffness matrix for the element i with the terms in that matrix replaced by a set of terms corresponding to the assumed or known values of the changes associated with errors or uncertainties. It is termed here the static *modification matrix*.
- u_i is the component of the vector of displacements due to the application of the real loads that act on the nodes of element i.

Note that expression 8.6 is now equivalent to expression 8.1 because displacements are the qualification parameters so that δu_j now represents the term δq_i.

In the case where stress is the qualification parameter, a modification matrix needs to be constructed that allows a variation of element stresses to be assessed against changes in the underlying structural properties. Assume, once again, that a change has occurred in an underlying structural property that changes the stiffness of element i but the qualification parameter is now the stress in element p. Because the stress in element p is a direct function of all the element nodal displacements, a set of computations is required that repeatedly applies equation 8.6 to generate rates of change for all the nodal displacements of this element. The modification matrix is then compiled by using the normal process for calculating element stresses from the values of the nodal displacements as demonstrated in Chapter 2, except that these nodal displacements are now rates of change terms equivalent to δu_j above.

Before commenting further on equation 8.6, it is useful to see how the process actually operates! Consider the static analysis problem first used

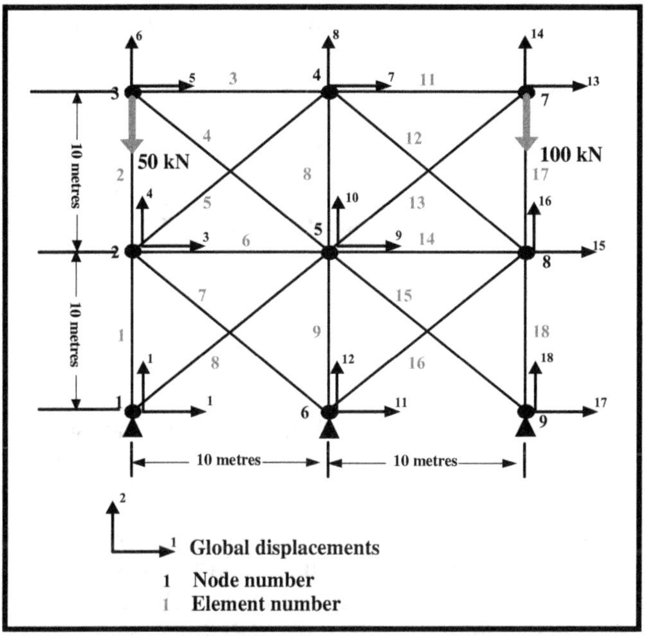

Figure 8.5 Multi-member bar frame subject to static loads.

in Chapter 2 and shown again in Figure 8.5 as the reference model and assess the impact of making a 10% change in the terms in the stiffness matrix for element 4 on displacement (degree of freedom) number 15. It has, therefore, been assumed that the uncertainties in the underlying structural properties create a 10% change in stiffness, i.e. $\delta p = 0.1$ (equation 8.1), and that the qualification parameter of interest q is that represented by displacement number 15. The structure is analysed using the actual loads shown and separately with a single unit load applied along displacement 15 which gives rise to the following displacement vectors v_4 and u_4 at the nodes of element 4 for the unit load case and the applied load case respectively:

$$v_4 := \begin{Bmatrix} 5.408 \times 10^{-10} \\ 1.133 \times 10^{-10} \\ 3.59 \times 10^{-10} \\ -3.44 \times 10^{-11} \end{Bmatrix} \qquad u_4 := \begin{Bmatrix} 1.36 \times 10^{-5} \\ -2.629 \times 10^{-5} \\ 4.147 \times 10^{-6} \\ -1.334 \times 10^{-5} \end{Bmatrix}$$

Recalling that the initial value of the elasticity modulus $E = 200\,\text{GPa}$, then replacing the terms in the stiffness matrix for element 4 with terms equal to 10% of this initial value gives the modification matrix δk_4:

$$\delta k_4 := \begin{Bmatrix} 7.071\times10^7 & -7.071\times10^7 & -7.071\times10^7 & 7.071\times10^7 \\ -7.071\times10^7 & 7.071\times10^7 & 7.071\times10^7 & -7.071\times10^7 \\ -7.071\times10^7 & 7.071\times10^7 & 7.071\times10^7 & -7.071\times10^7 \\ 7.071\times10^7 & -7.071\times10^7 & -7.071\times10^7 & 7.071\times10^7 \end{Bmatrix}$$

Putting these into equation 8.6 gives:

$$\delta u_{15} = -v_4^T \delta k_4 u_4 = -5.402 \times 10^{-8}\ \text{metres} \tag{8.7}$$

The results from the analysis of the unmodified reference structure, given in Section 2.3.1, Table 2.2, yields $q(\text{ref}) = u_{15} = 1.056 \times 10^{-5}$ metres, so the sensitivity analysis has indicated that a 10% increase in the stiffness of element 4 gives rise to a 0.5% (approximate) reduction in displacement 15. This implies that if there is 10% uncertainty in a structural property that can be represented by a 10% change in the element stiffness, then the bound on the corresponding uncertainty in the qualification parameter is $M_{15} = 0.5\%$. Because this structural problem is statically determinate, the element stresses are insensitive to changes in element stiffness as stress is directly related to element dimensions and not element stiffness.

Turning to an analysis involving a sub-structure, consider the simple beam theory problem example illustrated in Figure 8.6 which shows a beam 2 metres long encastré at the left-hand end and propped at the right-hand end. The beam has a depth of 200 mm, a breadth of 6 mm, an elastic modulus of 200 GPa and is subject to a distributed load of 12,000 N/m along half its length. The qualification parameter q for the problem is the maximum deflection of the beam.

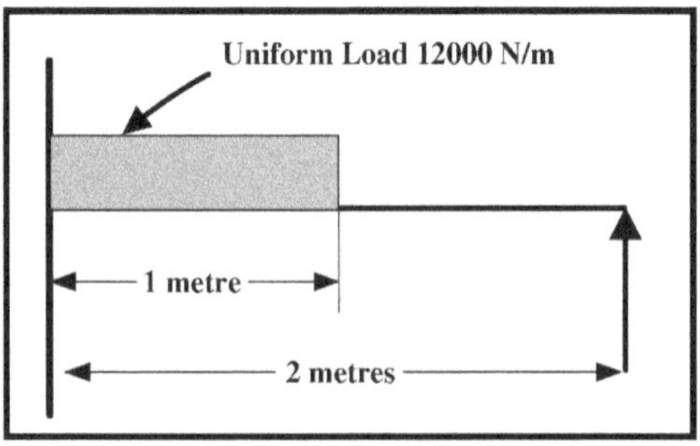

Figure 8.6 Encastré beam.

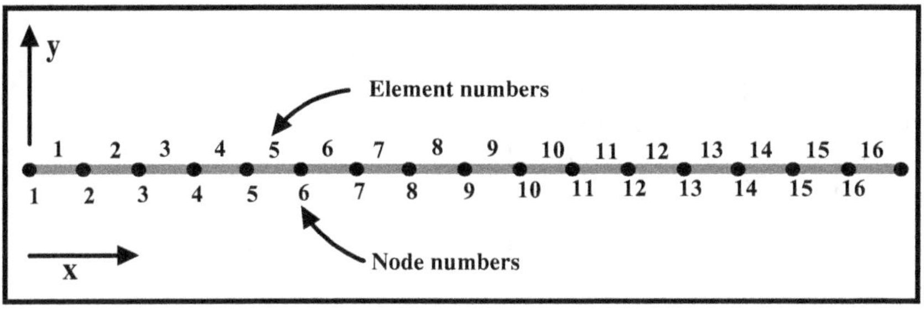

Figure 8.7 Element and node numbers.

The reference analysis of this problem is undertaken using 16 straight beam elements as illustrated in Figure 8.7. A plot of the vertical displacements along the beam is shown in Figure 8.8 which indicates that the analysis gives a maximum vertical deflection of -5.022×10^{-4} metres in the y-direction at node 10 so that this nodal deflection now becomes $q(\text{ref})$. Assuming that there is a potential uncertainty of 10% in the properties of the material in all the elements 1–4 which could increase the value of the elasticity of these elements by 10%, we now wish to estimate the influence this has on this qualification parameter.

Denoting u_{10} and v_{10} as the displacement of the nodes of the sub-structure consisting of the elements 1–4 due to the application of the real loads and a unit load applied at node 10 (i.e. dof 18), these are found to be given by:

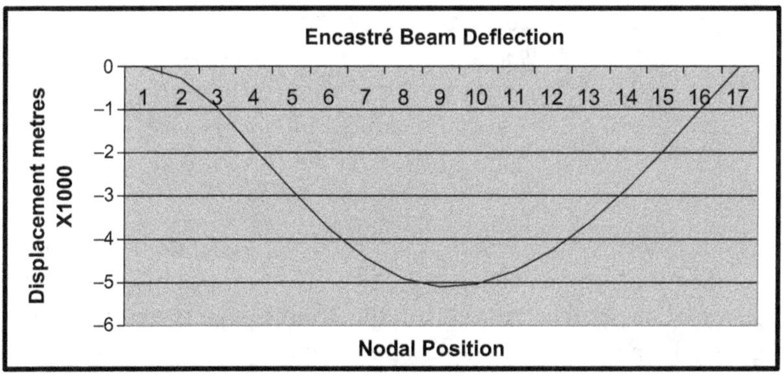

Figure 8.8 Vertical displacements along the beam.

$$u_{10} := \begin{pmatrix} 0 \\ 0 \\ -2.873 \times 10^{-5} \\ 4.274 \times 10^{-4} \\ -9.939 \times 10^{-5} \\ 6.757 \times 10^{-4} \\ -1.914 \times 10^{-4} \\ 7.743 \times 10^{-4} \\ -2.879 \times 10^{-4} \\ 7.523 \times 10^{-4} \end{pmatrix} \qquad v_{10} := \begin{pmatrix} 0 \\ 0 \\ -3.205 \times 10^{-9} \\ 4.928 \times 10^{-8} \\ -1.182 \times 10^{-8} \\ 8.655 \times 10^{-8} \\ -2.434 \times 10^{-8} \\ 1.118 \times 10^{-7} \\ -3.928 \times 10^{-8} \\ 1.251 \times 10^{-7} \end{pmatrix}$$

and the modification matrix by:

$$\delta k := \begin{pmatrix} 4.915 \times 10^{8} & -3.072 \times 10^{7} & -4.915 \times 10^{8} & -3.072 \times 10^{7} & 0 & 0 & 0 & 0 & 0 & 0 \\ -3.072 \times 10^{7} & 2.56 \times 10^{6} & 3.072 \times 10^{7} & 1.28 \times 10^{6} & 0 & 0 & 0 & 0 & 0 & 0 \\ -4.915 \times 10^{8} & 3.072 \times 10^{7} & 9.83 \times 10^{8} & 0 & -4.915 \times 10^{8} & -3.072 \times 10^{7} & 0 & 0 & 0 & 0 \\ -3.072 \times 10^{7} & 1.28 \times 10^{6} & 0 & 5.12 \times 10^{6} & 3.072 \times 10^{7} & 1.28 \times 10^{6} & 0 & 0 & 0 & 0 \\ 0 & 0 & -4.915 \times 10^{8} & 3.072 \times 10^{7} & 9.83 \times 10^{8} & 0 & -4.915 \times 10^{8} & -3.072 \times 10^{7} & 0 & 0 \\ 0 & 0 & -3072 \times 10^{7} & 1.28 \times 10^{6} & 0 & 5.12 \times 10^{6} & 3.072 \times 10^{7} & 1.28 \times 10^{6} & 0 & 0 \\ 0 & 0 & 0 & 0 & -4.915 \times 10^{8} & 3.072 \times 10^{7} & 9.83 \times 10^{8} & 0 & -4.915 \times 10^{8} & -3.072 \times 10^{7} \\ 0 & 0 & 0 & 0 & -3.072 \times 10^{7} & 1.28 \times 10^{6} & 0 & 5.12 \times 10^{6} & 3.072 \times 10^{7} & 1.28 \times 10^{6} \\ 0 & 0 & 0 & 0 & 0 & 0 & -4.915 \times 10^{8} & 3.072 \times 10^{7} & 4.915 \times 10^{8} & 3.072 \times 10^{7} \\ 0 & 0 & 0 & 0 & 0 & 0 & -3.072 \times 10^{7} & 1.28 \times 10^{6} & 3.072 \times 10^{7} & 2.56 \times 10^{6} \end{pmatrix}$$

Placing these terms into equation 8.6 predicts a change in the value of the qualification displacement of 2.113×10^{-5} metres. Thus a uniform change of 10% in the stiffness of the four elements creates a 4.2% change in the qualification displacement, and the corresponding bound on the uncertainty in the qualification parameter from expression 5.3 is $M_{18} = 4.2\%$.

The direct sensitivity process for static analysis The use of direct sensitivities has to fit into the total error and uncertainty control process. This implies that the qualification process has identified the critical parameters in the analysis so that the analyst is faced with the problem of estimating the impact of any errors or uncertainties on these target parameters. It is at this stage that the analyst can start to consider the use of direct sensitivities and apply it to the problem in hand. Clearly, the application of this method requires that the errors or uncertainties are connected with terms that directly influence the stiffness of the structure. It might be undertaken because there is uncertainty in material properties influencing the structural stiffness, or in a physical dimension, or in attachment properties if the element is at the boundary of an analysis region. If the method is applicable, then the process described below can be initiated.

The Process

1. The analyst identifies the regions where the sources of error or uncertainty are located and then creates the modification matrix δk_i. There may be a number of separate regions that have error or uncertainty sources and the analyst must ensure that all these are captured.

   1.1. Breakout points: Before proceeding, the analyst must be satisfied that a number of conditions have been met, otherwise initiate a breakout from the process at this point.

      1.1.1. That the terms in the modification matrix or matrices must be significantly smaller than the terms in the original element, sub-structure or superelement matrix or matrices because the process is employing first-order Taylor expansions which require $K \gg \delta K$. Breakout is recommended if the terms in the modification matrix are greater than 10% of the original value.

      1.1.2. That the size of the superelement or the number of individual elements is so large that it is not cost effective to proceed and the alternative indirect sensitivity process introduced on page 236 should then be employed.

2. The process for obtaining direct sensitivities begins with the analyst using best estimates for the element properties to create a finite element model for the reference structure. A full analysis is undertaken using the actual applied loads and unit loads placed at the selected degrees of freedom to generate the v_i matrices in expression 8.6. Thus far the discussion has focused on displacements only. If rates of *change of stresses are required*, then unit loads must be placed separately at all the nodal degrees of freedom of the element(s) where the variation in stress is required. In a normal commercial system, placing unit loads at specified points is not difficult as these can be applied using the multiple load facility.

   2.1. Breakout point: If there are a large number of elements for which stress sensitivities are required, the problem size could be too large and the analyst should consider breaking out of the direct sensitivity calculation process at this point.

3. The analyst has accumulated all the required information to allow sensitivities to be computed with an appropriate use of expressions 8.6 and 8.1.
4. The final stage is a results interpretation stage where the outputs from the sensitivity analyses are compared with the target error limits defined in the very early stage of the error and uncertainty control process. If the results show that the displacement or stress variations exceed the error or uncertainty limits, then the analyst has to take action as the design is now vulnerable to failure.

Application to Static Analyses

1. *Direct sensitivities can be used in an analysis of a structure subject to static loads to provide an assessment of the influence of errors and uncertainties throughout the structure on the qualification parameters for all the sources listed in Chapter 7.*

2. *Direct sensitivities provide values for estimating bounds on the impact of an error or uncertainty on a qualification parameter as defined by expression 5.3.*
3. *Direct sensitivities can only be derived when the input of error or uncertainty can be interpreted as a change in element stiffness.*
4. *Direct sensitivities should be used when the estimated change in element stiffness is less than or equal to 10% of the reference value.*

Dynamic analysis (free vibration) In the case of dynamics, the direct sensitivity method can be applied in this domain but is only robust in the case where the structure is in a free vibration state. As a result no attempt is made in the current text to include a discussion of direct sensitivities for forced responses. The results from a vibration analysis give a set of natural frequencies ω_i and modes U_i where $i = 1 \ldots n$ with n being the number of degrees of freedom of the finite element model. If the mode shapes are normalised with respect to the mass matrix M, then for each natural frequency we have:

$$\begin{aligned} \{K - \omega_i^2 M\}U_i &= 0 \\ U_i^T M U_i &= 1 \end{aligned} \tag{8.8}$$

Following the static case it is assumed that the error or uncertainties introduce changes in both the mass and stiffness matrices so that the first of equations 8.8 become:

$$\Big\{\{K + \delta K\} - (\omega_i + \delta\omega_i)^2\{M + \delta M\}\Big\}\{U_i + \delta U_i\} = 0$$

If the natural frequency term in the above equation is expanded:

$$\{\{K + \delta K\} - (\omega_i^2 + 2\omega_i\delta\omega_i + \delta\omega_i^2)\{M + \delta M\}\}\{U_i + \delta U_i\} = 0 \tag{8.9}$$

ignoring the second-order and higher terms, equation 8.9 reduces to:

$$\{K - \omega_i^2 M\}U_i + \{K - \omega_i^2 M\}\delta U_i + \{\delta K - \omega_i^2 \delta M - \delta\omega_i^2 M\}U_i = 0$$

Noting that the first term in this expression is identically zero leads to:

$$\{K - \omega_i^2 M\}\delta U_i + \{\delta K - \omega_i^2 \delta M - \delta\omega_i^2 M\}U_i = 0 \tag{8.10}$$

Pre-multiplying this equation by U_i^T and taking note of the second term in equation 8.8, i.e. the normalisation equations, allows 8.10 to be modified leading to a sensitivity expression that gives the rate of change of the *i*th natural frequency $\delta\omega_i^2$:

$$\delta\omega_i^2 = U_i^T\{\delta K - \omega_i^2\delta M\}U_i \tag{8.11}$$

In the case where the stiffness and mass changes are in a single element, i.e. the *j*th element, then 8.11 becomes:

$$\delta\omega_i^2 = u_{ji}^T\{\delta k_j - \omega_i^2\delta m_j\}u_{ji} \tag{8.12}$$

where:

- u_{ji} is that part of the *i*th mode shape vector operating on the nodes of element *j*.
- δk_j and δm_j are the element stiffness and mass matrices for the element *j* with the terms in these matrices being replaced by a set of terms corresponding to the assumed or known values of the changes associated with errors or uncertainties. These are termed here the *dynamic modification matrices*.

As with the static analysis, the modification matrices can apply to an individual element or to a superelement or combinations of elements.

Sensitivity equation 8.12 gives the changes in the natural frequencies for changes in the mass and stiffness matrices and can be interpreted as equivalent to expression 8.1. An analyst may be interested in how the mode shapes change in response to the inclusion of errors or uncertainties in the mass and stiffness matrices, particularly if dynamic stresses and strains are being evaluated. If mode shape sensitivities δU_i are required, then it is found that 8.11 is replaced by:

$$\begin{Bmatrix} K - \omega_i^2 M & -MU_i \\ U_i^T M & 0 \end{Bmatrix}\begin{Bmatrix} \delta U_i \\ \delta\omega_i^2 \end{Bmatrix} = \begin{Bmatrix} -\{\delta K - \omega_i^2\delta M\}U_i \\ -\frac{1}{2}\{U_i^T\delta M U_i\} \end{Bmatrix} \tag{8.13}$$

Some care has to be exercised in using equation 8.13 because the principal minor is singular and the analyst may prefer to employ a more indirect process for calculating mode shape sensitivities as discussed in later sections.

Because dynamics problems require more computational effort to achieve a solution than comparable problems subject to static loads, the use of equation 8.12 for the calculation of frequency changes due to changes in structural properties is illustrated using the multi-member frame shown in Figure 8.5. In the static case, the structure is shown with two applied loads of 50 kN and 100 kN, but, in the present case, these are not applied so that we are considering a free vibration problem. As with the static case, the cross-sectional area is 0.1 m^2, the initial modulus of elasticity is $E = 200$ GPa and the initial density of the material is taken to be 1000 kg/m^3. The qualification parameter for this problem is the lowest natural frequency of the structure. This type of qualification environment is one where this frequency is required to lie outside a specified range.

The dynamic analysis of this problem gave rise to 12 natural frequencies with 12 associated mode shapes. The lowest, or fundamental, frequency for this structure, shown in Figure 8.9, has the value $\omega_1 = 438.68$ radians/second. Let us now answer the question 'By how much does this frequency change if the stiffness and mass matrices for element 1 increase by 10%?' In order to apply equation 8.12 the modification matrices for element 1

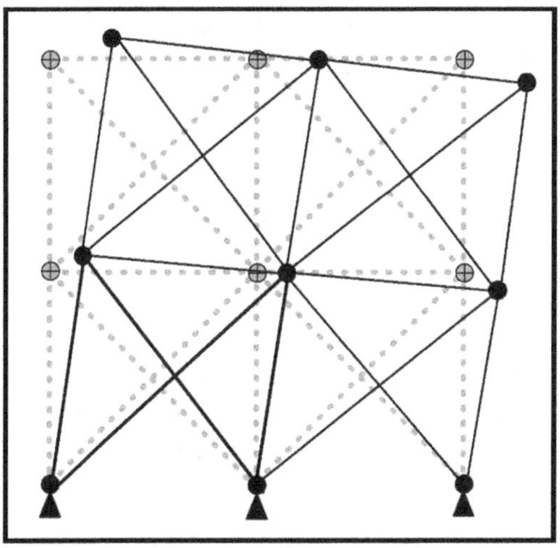

Figure 8.9 Fundamental mode shape for initial structure.

are required and these are given by:

$$\delta k_1 = \begin{Bmatrix} 0 & 0 & 0 & 0 \\ 0 & 2\times 10^9 & 0 & -2\times 10^9 \\ 0 & 0 & 0 & 0 \\ 0 & -2\times 10^9 & 0 & 2\times 10^9 \end{Bmatrix}$$

$$\delta m_1 = \begin{Bmatrix} 0 & 0 & 0 & 0 \\ 0 & 333.333 & 0 & 166.667 \\ 0 & 0 & 0 & 0 \\ 0 & 166.667 & 0 & 333.333 \end{Bmatrix}$$

The sensitivity equation also requires the component of the associated normalised mode shape, taken to be mode shape 1, that acts on element 1 and this is found to be:

$$u_{11} = \begin{Bmatrix} 0 \\ 0 \\ 7.886\times 10^{-3} \\ 3.856\times 10^{-3} \end{Bmatrix}$$

Placing these values into equation 8.12 gives the following estimate of the rate of change of the selected natural frequency:

$$\delta\omega_1^2 = \{0 \;\; 0 \;\; 7.886\times 10^{-3} \;\; 3.856\times 10^{-3}\}\left\{\begin{Bmatrix} 0 & 0 & 0 & 0 \\ 0 & 2E+9 & 0 & -2E+9 \\ 0 & 0 & 0 & 0 \\ 0 & -2E+9 & 0 & 2E+9 \end{Bmatrix}\right.$$

$$\left. -438.68^2\begin{Bmatrix} 0 & 0 & 0 & 0 \\ 0 & 333.333 & 0 & 166.667 \\ 0 & 0 & 0 & 0 \\ 0 & 166.667 & 0 & 333.333 \end{Bmatrix}\right\}\begin{Bmatrix} 0 \\ 0 \\ 7.886\times 10^{-3} \\ 3.856\times 10^{-3} \end{Bmatrix}$$

Evaluating this expression gives a predicted change in the square of the frequency $\delta q(\text{ref}) = (\delta\omega_1)^2 = 3.27$, thus the 'new' fundamental frequency ω_1(new) can now be computed from the expression:

$$(\omega_1(\text{new}))^2 = \omega_1^2 + \delta\omega_1^2$$

This predicts that the new value for the first natural frequency is $\omega_1(\text{new}) = 441.95$ radians/second. The implication is that a change in the properties of element 1 giving rise to 10% changes in the values of the stiffness and mass will alter this specific natural frequency from expression 5.3 by a mere 0.75%! Thus, for this reference problem the bound for qualification parameter ω_1 is $M_1 = 0.75\%$.

We can now compare the estimated change in frequency with the 'actual' change in frequency by putting the modified values for the mass and stiffness matrix into full global mass and stiffness matrices in the finite element model and re-computing. This yields a new value for the frequency $\omega_1(\text{actual}) = 441.76$ radians/second which is slightly below the estimated value. If a 30% change in the properties of element 1 is made, the corresponding results are $\omega_1(\text{new}) = 448.41$ radians/second, giving a corresponding qualification parameter uncertainty bound $M_1 = 2.219\%$; but the 'actual' value of the frequency due to the 30% change is $\omega_1(\text{actual}) = 446.90$ radians/second. This indicates that the estimation method, given by equation 8.12, is now beginning to provide values for the frequency that are drifting away from those obtained after changing the element property values and performing a full finite element analysis on an updated reference model.

The direct sensitivity process for dynamic analysis The use of the direct method to calculate the sensitivities for a structure subjected to dynamic loads requires a process that is very similar to the static case described on p. 225–226. As with the static case the qualification process identifies the targets for the dynamic analysis and the analyst has to estimate the impact that any error or uncertainty will have on the analysis targets. If these errors or uncertainties affect the properties of the elements used in the analysis, then a direct sensitivity analysis can be considered. Thus, uncertainty with respect to material properties that could influence the stiffness or mass properties or in physical dimensions having a similar impact can be handled directly. Uncertainty in the connection properties between structural components or with respect to the attachment stiffness or mass properties also falls within the scope of a direct sensitivity analysis.

The Process

1. The process begins with an analysis employing a full finite element model using the analyst's best estimates for the reference state element properties within the structural model in order to generate the frequencies and mode shapes for the full structure.
2. The analyst identifies the regions where the sources of error or uncertainty are located and then creates the modification stiffness and mass matrices δk_i and δm_i. There may be a number of separate regions that have error or uncertainty sources within them so that there may be several modification matrices requiring several applications of the sensitivity analysis. In certain cases the matrices will relate to an individual element or to a modified superelement if the region is modelled by a number of elements.

   2.1. Breakout points: Before proceeding, the analyst must be satisfied that a number of conditions have been met, otherwise break out of the process at this point.

      2.1.1. That the terms in the modification matrices are significantly smaller than the terms in the original element or superelement matrix or matrices. As with the application of the method to a static analysis case, the use of a first-order Taylor expansion requires $K \gg \delta K$ and $M \gg \delta M$. However, as shown elsewhere in this book, in the case of a frequency analysis the Finite Element Method converges more rapidly to a solution than is the case with a static analysis. This offers the possibility that modification matrices with values greater than 10% can be employed.
      2.1.2. That the size of the superelement or the number of individual elements is so large that it is not cost effective to proceed and the alternative indirect sensitivity process introduced on page 236 should then be employed.

3. The analyst is now required to extract the appropriate element mode shape vectors u_{ji} for the frequencies to be examined. In the case of frequency sensitivities the extracted information can be fed into equation 8.12 to obtain frequency changes for

specified variations in the structural properties or structural modelling assumptions. If rates of change of dynamic stresses/ strains are required, then equation 8.13 must be employed or the direct process abandoned and an indirect approach adopted as discussed on p. 236.

4. At this point the analyst has accumulated all the required information to allow sensitivities to be computed with an appropriate use of 8.12 if frequency change only is required, otherwise 8.13 can be employed, providing the singularity of the principal minor can be handled.
5. The final stage is a results interpretation stage where the outputs from the sensitivity analyses are compared with the target error limits defined in the very early stage of the error and uncertainty control process. If the results show that the frequency or stress variations exceed the required qualification limits, then the analyst has to take action as the design is now vulnerable to the presence of the defined errors or uncertainties.

Application to Dynamic Analyses

1. *Direct sensitivities can be used in an analysis of a structure undergoing free vibrations to provide an assessment of the influence of errors and uncertainties throughout the structure on the qualification parameters (frequencies) for many of the sources listed in Chapter 7.*
2. *Direct sensitivities provide values for estimating bounds on the impact of an error or uncertainty on a qualification parameter (frequency) as defined by expression 5.3.*
3. *Direct sensitivities can only be derived when the input of error or uncertainty can be interpreted as a change in element stiffness and/or mass properties.*
4. *Direct sensitivities should be used when the estimated change in element stiffness and/or mass is less than or equal to 10% of the reference value.*
5. *Direct sensitivities can be used to compute mode shape changes but care is required as numerical problems can be encountered.*

Buckling analysis Buckling analysis is associated with structures consisting of slender bars or thin-walled components and involves an interaction between membrane and bending forces. This interaction is modelled in a finite element analysis by the introduction of a geometric stiffness matrix K_G which is independent of the elastic properties of the structural material and is a function of both element geometry and membrane forces. The linear buckling stability of a structure is determined by solving the eigenvalue problem:

$$(K + \mu K_G)\eta = 0 \tag{8.14}$$

where μ_i are a set of eigenvalues, with $i = 1 \ldots n$ for an 'n' dof analysis, and are given by:

$$\mu_i = \frac{-\eta_i^T K \eta_i}{\eta_i^T K_G \eta_i}$$

where η_i is ith buckling mode. The buckling load of a structure is given by $P_{cr} = \mu P$ where μ is the lowest eigenvalue and P the set of applied loads.

In order to calculate the sensitivity of the buckling load, the elements of the global stiffness and geometric matrices we start by taking the lowest eigenvalue μ and its associated buckling mode η and noting that:

$$\eta^T(K + \mu K_G)\eta = 0$$

and assuming that the presence of uncertainties and errors causes changes in both stiffness and geometric stiffness matrices, the above expression becomes:

$$2\delta\eta^T(K + \mu K_G)\eta + \eta^T \delta K \eta + \delta\mu \eta^T K_G \eta + \mu \eta^T \delta K_G \eta = 0$$

where ΔK_G is the *buckling modification matrix*. Taking note of 8.14 and performing some manipulation, the above reduces to:

$$\delta\mu = \frac{-\eta^T \delta K \eta}{\eta^T K_G \eta} - \frac{\mu \eta^T \delta K_G \eta}{\eta^T K_G \eta} \tag{8.15}$$

If the buckling mode is normalised with respect to the geometric stiffness matrix, i.e. $\eta^T K_G \eta = 1$, then 8.15 further reduces to:

$$\delta\mu = -\eta^T \delta K \eta - \mu \eta^T \delta K_G \eta \tag{8.16}$$

The sensitivity expression for the buckling load is $\delta P_{cr} = \delta\mu P$ with $\delta\mu$ given by 8.16. The buckling sensitivity expression has two parts. One reflects the influence of changes in the elements making up the global stiffness matrix and these relate to uncertainties in the values given for material properties, structural stiffness, boundary stiffness, etc. The second reflects the influence on the buckling load of changes in the geometric stiffness matrix and this may be due to uncertainties in the values of any geometric initial imperfections or changes in load paths caused by uncertainties or errors in other parts of the structure under analysis.

Although examining the influence of uncertainties or errors on the structural buckling load is important, the analyst should always bear in mind that exceeding the buckling load can give rise to *catastrophic failure*. Attention should always be paid to the post-buckling behaviour of the structure, particularly if the design is being optimised to achieve the highest levels of structural and economic efficiency.

Application to Buckling Analyses

1. *Direct sensitivities can be used in a buckling analysis of a structure to provide an assessment of the influence of errors and uncertainties throughout the structure on the qualification parameter (buckling load) for the sources listed in Chapter 7. The main source of uncertainty in this type of analysis is likely to be initial imperfections in structural shape or form.*
2. *Direct sensitivities provide values for estimating bounds on the impact of an error or uncertainty on a qualification parameter (buckling) as defined by expression 5.3.*
3. *Direct sensitivities can only be derived when the input of error or uncertainty can be interpreted as a change in element stiffness and/or geometric stiffness properties.*
4. *Direct sensitivities should be used when the estimated change in element stiffness and/or geometric stiffness is less than or equal to 10% of the reference value.*
5. *Direct sensitivities do not provide any estimate of post-buckled behaviour.*

Creation of modification matrices There are a number of methods for creating the various modification matrices used in the sensitivity expressions 8.6, 8.12, 8.13 and 8.16. Many commercial finite element systems have built-in structural optimisers employing algorithms requiring gradient information that can be used in the generation of the modification matrices. Certain systems allow the user direct access to gradient information to assist analysts in undertaking sensitivity studies which can be directly used to construct modification matrices. In the case where gradient information is generated but hidden within the structural optimisation algorithms, some ingenuity may be required on the part of the analyst to get hold of the information necessary for the creation of the modification matrices.

An alternative is to generate the matrices through the creation of a separate model or models involving those elements or components that play a role in the uncertainty or error sensitivity study. This requires that the finite element system allows the user to extract element matrices, superelement matrices or, in the case of buckling, an entire global matrix. These extracted matrices will have been generated using appropriate percentage changes in their properties so that they accurately replicate the modification matrices. If the finite element system is an in-house program it should not be too difficult to arrange that the relevant matrices are made available. In the case of bought-in systems, life may be a little more problematical as suppliers are often reluctant to let users get close to information normally internal to the system.

Indirect evaluation of sensitivities The direct method assumes the behaviour of a structure can be represented by a linear system so that the variation of a qualification parameter such as displacements or stresses with respect to a specified structural stiffness or material property can be represented by a straight line function. In reality most structural behavioural functions follow a curved path when the underlying structural parameters are changed. For example, a plot of the stresses in the members of a statically indeterminate structure against changes in structural dimensions is not a straight line. If the analyst is confronted with uncertainties that are greater than 10% of the reference value of the qualification parameter, then the direct method for evaluating sensitivities may be inappropriate. In addition, if there are many regions with error or uncertainty sources or the analysis involves a complex built-up structure with a large number of joints, then the direct method may become too clumsy and impractical. In these situations recourse must made to the indirect method for evaluating the sensitivity of the qualification parameters to the presence of errors or uncertainties.

The indirect evaluation of sensitivities simply involves a full re-analysis of the structure or of major independent components or even small components if the analyst employs a combination of implicit and explicit modelling that requires independent models for specific parts of the structure. The process requires that an initial analysis is performed using the reference state values for the element properties. Appropriate changes are then made to the properties of the elements where the sources of error or uncertainty are located and the structure re-analysed. The method requires an analysis for each identified error or uncertainty source. Comparing the values for the qualification parameters from these analyses indicates by how much they change for the assumed presence of an error or uncertainty with a specific value. Providing the assumed values used to represent the changes in element properties due to the presence of error or uncertainty bracket those likely to occur when the structure is in service operation, then the bounds derived from 8.1 can be employed.

As an illustration consider the analysis problem displayed in Figure 8.6 and Figure 8.7. When the direct sensitivity method was employed it was shown that the bound on the qualification parameter is $M_{18} = 4.2\%$. The indirect method requires that we increase the stiffness of the four elements identified as having an uncertainty of 10% by that amount in order to evaluate the resulting change in u_{18} which constitutes the qualification parameter. The analysis yields $u_{18} = 4.821 \times 10^{-4}$ resulting in a value for the error and uncertainty bound $M_{18} = 4.001\%$. The indirect method gives a slightly lower bound when compared with the direct method as it is not based on a linear approximation.

The process for bounding uncertainties in the applied loads can be considered as an application of direct sensitivities. In the case of a static analysis the evaluation of the impact of load uncertainty is normally undertaken by applying additional right-hand sides to the reference loads. If this involves extending the maximum load states, then consideration has to be given to the possibility that catastrophic failure is in the offing. In the case of a linear analysis this may require moving into the non-linear regime involving material or geometric non-linearities or both! Once the non-linear regime has been entered, life becomes complex as the loading and the structural response may be time-dependent. This same caveat applies when the analysis involves a dynamically loaded structure where the response is a function of both the load history and material properties such as damping.

Applications

1. *Indirect sensitivities can be used for any structure subjected to any type of load to provide an assessment of the influence of errors and uncertainties throughout the structure on the qualification parameter for the sources listed in Chapter 7.*
2. *Indirect sensitivities are more expensive to obtain than direct sensitivities but are more accurate.*
3. *For dynamic and non-linear problems the evaluation of sensitivities can involve time-dependent analyses that require significant computational power.*

Artificial sensitivities The calculation of sensitivities thus far has assumed that all the sources of uncertainty or error relate to structural entities and properties that can be directly related to the properties of the finite elements used in the structural model. But in certain situations this may not be appropriate and an alternative is to have recourse to 'experimental' elements that can adequately represent a structural behaviour or property without attempting to be a true mathematical model. With these elements in place, sensitivity studies can be undertaken using either a direct or indirect methodology, but the sensitivities obtained are artificial as they have not been found using elements that purport to represent the true mathematical model of the component under examination. For example, the analyst may have decided on the region that constitutes the domain of analysis but may feel that there is still some uncertainty in the selected outer limit and does not wish to continue adding additional complex finite elements to re-analyse an extended domain. In this case multi-dimensional springs could be used as appropriate 'experimental' elements and distributed along the edge of the domain of analysis to permit a rapid assessment of the likely uncertainties associated with the selected domain. This approach is useful when the analysis involves structural components, joints, supports, etc., where the properties are not known with exactitude. The impact on the qualification parameters can be placed within upper and lower assessment bounds using appropriate 'experimental' elements and rapid re-analyses. Of course, the results emerging from this process might lead to the conclusion that the information on these properties must be known with greater exactness and that this requires additional tests.

In order to illustrate this approach consider the simple problem shown in Figure 8.5 and take the situation where the supports at nodes 1, 6 and 9 are known to have some flexibility in the vertical direction and the analyst wishes to explore the influence of this flexibility on the qualification parameter which is taken to be the displacement of node 8 in the horizontal direction, i.e. dof 15. This exploration can be undertaken by adding springs so that the reference structure is now attached to 'ground' in a flexible manner as shown in Figure 8.10. An additional constraint must be added at nodes 1, 6 and 9 in the horizontal direction to remove rigid body movement. Let us now assume that from simple calculations or past experience the stiffness of the support is known to be somewhere between 5×10^9 newtons/metre and 25×10^5 newtons/metre. In this situation the best approach is to use the indirect method for assessing the sensitivity of the qualification parameter to variation in the stiffness characteristics of the support. This requires re-analysing the structure using a range of stiffness values that bracket the range of possible stiffness values.

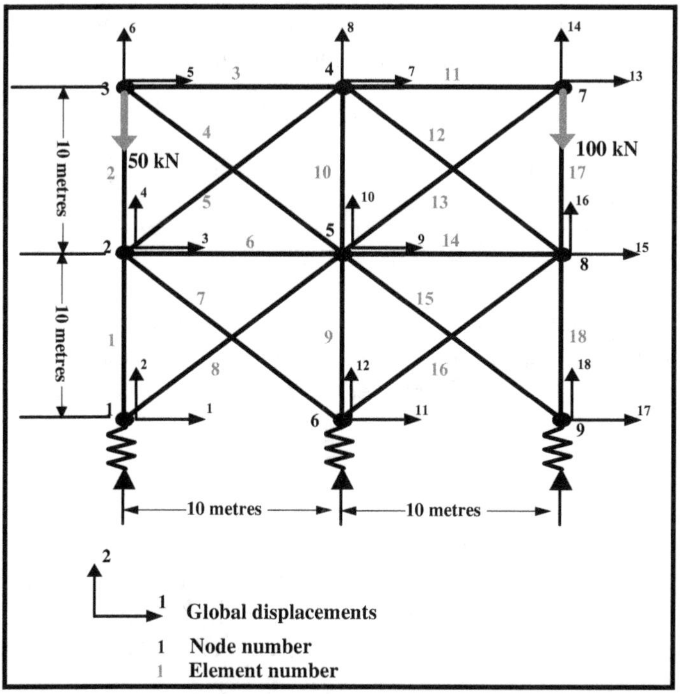

Figure 8.10 Structure with additional springs.

Table 8.2 Variation of qualification parameter with spring stiffness.

Spring stiffness value (N/m)	Displacement horizontally at node 8 dof 15 (m)
5×10^9	1.555×10^{-5}
15×10^9	1.222×10^{-5}
25×10^9	1.156×10^{-5}

Any number of different values for the spring stiffness can be used but three are sufficient to illustrate the point being made and the results from the analyses are shown in Table 8.2. In order to estimate the value for the bound on the uncertainty in the qualification parameter u_{15}, a reference value is required. Two possibilities present themselves: one is to take the mean value for the spring stiffness, i.e. 15×10^9 Pa, for which $u_{15} = 1.222 \times 10^{-5}$ m, the other is to take the mean value of the horizontal displacement which is $u_{15} = 1.355 \times 10^{-5}$. These are denoted by $u_{15}(\text{ref1})$ and $u_{15}(\text{ref2})$ respectively.

If the mean spring stiffness is used to generate the fractional uncertainty, there is a slight complication due to there being different upper and lower values for the bound so that we need to use $\pm\delta u_{15}/|u_{15}(\text{ref1})|$, in which case the bound on the reference value is:

$$-5.4\% \leq \delta u_{15}/|u_{15}(\text{ref1})| \leq 25\%$$

If the second reference value is taken, then expression 5.3 applies and the bound is given by:

$$M_{15} = 15\%$$

Normally an analyst would wish to use the single symmetric bound provided by the second reference value rather than the asymmetric version provided by the first reference value.

Application

1. *Artificial sensitivities can be employed to assess the impact of errors or uncertainties on qualification parameters in situations where the actual material properties of a structure or structural support are not available or when a very rapid assessment is required.*
2. *Artificial sensitivities can be employed in an identical manner to the direct and indirect sensitivity methods.*

8.3 ACCUMULATION OF ERRORS AND UNCERTAINTIES

As explained in Chapter 7, Section 7.2, a finite element analysis can be considered as a measurement of the actual behaviour of a real-world structure and this measurement contains uncertainties that lead to a difference between the behaviour of the finite element model and the real-world structure. In Section 8.2.2 the concept of bounding specific uncertainties is introduced and the remaining part of this chapter is devoted to providing methods of 'measuring' the likely influence of a given error or uncertainty source. Most finite element analyses are sufficiently large that an analyst will have to take into account a number of error sources that can be independently handled by the methods introduced in this chapter. The question now arises of how to combine these individual estimates into a single figure that provides a total bound on the impact that all the errors and uncertainties have on a qualification parameter.

Although combining individual errors into a single error estimate can be complex and the complete process is beyond the scope of this book, there are relatively simple methods that can be used to provide a total error estimate. Let us assume that a particular qualification parameter is influenced by four error sources e_1–e_4, each relating to a different underlying structural property. Denoting the change in a specific qualification parameter q_i with respect to a change in one of the error terms as $\delta q_i(e_j)$ for $j = 1$, or 2, or 3, or 4, using 8.1 to compute the variation in the values of the qualification parameter and then taking a simple-minded approach, the combined error can be written as:

$$\frac{\delta q_i}{|q_i(\text{ref})|} = \frac{\delta q_i(e_1)}{|q_i(\text{ref})|} + \frac{\delta q_i(e_2)}{|q_i(\text{ref})|} + \frac{\delta q_i(e_3)}{|q_i(\text{ref})|} + \frac{\delta q_i(e_4)}{|\delta q_i(\text{ref})|} \tag{8.17}$$

This combined error can be fed into expression 5.3 and a bound now sorts on the combined or total error for the qualification parameter.

As an example consider, once again, the analysis problem shown in Figure 8.5 and now take the case that the stiffnesses of elements 1, 4, 12 and 16 all increase by 10%. Recalling that the qualification parameter is the displacement at node 8 in the horizontal direction, i.e. degree of freedom 15, we denote the change in this parameter with respect to the changes in the element stiffness matrices as $\delta u_{15}(e_1)$, $\delta u_{15}(e_4)$, $\delta u_{15}(e_{12})$ and $\delta u_{15}(e_{16})$ and using equation 8.6 to obtain the individual variations in the qualification parameter gives:

$$\delta u_{15}(e_1) = 8.606 \times 10^{-8}$$
$$\delta u_{15}(e_4) = 5.402 \times 10^{-8}$$
$$\delta u_{15}(e_{12}) = 2.415 \times 10^{-8}$$
$$\delta u_{15}(e_{16}) = 7.223 \times 10^{-7}$$

Recalling that the original value for this displacement is $q(\text{ref}) = u_{15} = 1.056 \times 10^{-5}$, then equation 8.17 yields:

$$\frac{\delta q_i}{q(\text{ref})} = 8.4\%$$

and the value found by adding in all the stiffness changes and re-computing gives:

$$\frac{\delta q_i}{q(\text{ref})} = 7.4\%$$

As illustrated earlier, using equation 8.17 overestimates the impact of a combination of factors on the qualification parameter. The combined influence of changes in structural or other properties on a qualification parameter depends on a variety of factors including the qualification parameter itself and the terms thought to be influencing its accuracy. For example, if the qualification parameters in the example displayed in Figure 8.5 are element stress, then changes in element stiffness have no impact on these terms because the structure is statically determinate and the influence of element cross-sectional areas only influences the stress in the element where the change takes place. Trying to combine estimates and creating combined bounds is more complex for a dynamical analysis and even more difficult if the analysis has a non-linear behaviour pattern.

Nevertheless equation 8.17 can be usefully employed in a variety of analysis situations, but if more sophisticated methods are required, then the reader should refer to books on statistical methods. A useful starting point is the book on error analysis by Taylor, reference [8].

8.4 THE ROLE OF TESTING

The approach and methods being advanced in this book are attempting to make testing subservient to analysis. In essence, tests are now made necessary and defined by the need for additional information by the analyst or the structural designer. In the case of the analyst, tests are required to remove ambiguities and to create additional information on the value to be ascribed to uncertainties when the information obtained from analyses or direct calculation is inadequate. The structural designer or the certificating body may require validation tests where the predicted performance of a real-world structure from a finite element reference model is compared with experimental data.

Before considering how tests might be used to support an analysis in the control of error or uncertainty, it is worth recalling that tests are also subject to uncertainty. Experimental data is measured data and all measurements, however carefully made, are subject to uncertainty. The simple-minded concept that a measured value of a quantity is a true or correct value should be rejected. The analyst, in calling for experimental data to support an error control methodology or in responding to data supplied as part of a validation process, must take into account the presence of uncertainty in the test environment. There are two factors to be considered: first, the uncertainties in the actual physical quantities measured, and, second, the fact that these uncertainties propagate through the calculation processes used to produce a set of final 'measured' values. A simple example is that stresses throughout a test structure might be obtained from strain gauge values taken at specific and discrete points in the structure. Uncertainties in the strain measurements propagate through to uncertainties in the derived stress values. The analyst must, at all times, be alert to the fact that experimental uncertainties need to be taken into account if test results are to have any meaning. If the analyst is calling for a test programme, then it must be made clear to the test engineer that uncertainty measures are required. If the analyst is responding to test results within a validation process, such results must be accompanied by bounds on the experimental uncertainty.

Although it is the analyst's responsibility to ensure that test data is supplied with appropriate sources of uncertainties defined and bounds on their value supplied, it is not the analyst's responsibility to evaluate or calculate experimental uncertainty bounds. The challenges of estimating uncertainties and reducing them to a level that allows a proper conclusion to be drawn on their influence can be a very complex process requiring specialist expertise. Nevertheless, a professional test organisation or test engineer should be able to supply this information and provide a list of potential uncertainty sources, both systematic and random, together with bounds on these uncertainties and confidence levels on the bounds. If this information is not made available or cannot be made available, then refuse to accept the test results.

8.4.1 Analyst-Requested Tests

Although an analyst may request experimental data for a variety of reasons, there are three main ones that normally give rise to a request for test results. These relate to a need to remove ambiguities, to provide additional information on uncertainties in the analysis results or to explore the behaviour beyond the defined design range.

8.4.1.1 Ambiguities

As described in earlier sections, selecting the appropriate level of abstraction or mathematical model is a major decision in the control of uncertainties. Often this decision can be made using a hierarchical sequence of models, particularly in the case of linear analysis with homogeneous and isotropic materials. But, for analyses in which anisotropic materials are involved, employing a sequence of analyses using different mathematical models may not unambiguously reveal which model is correct for the analysis and reflects the behaviour of the real-world structure. In this situation tests are required to provide modelling information to allow an uncertainty-controlled analysis to be undertaken. These must be defined by the analyst in a precise manner that makes clear what physical or material properties require measuring. A trivial example would be a decision concerning the appropriate mathematical model for a plate structure to distinguish between its behaving as a thin or thick plate. These two models differ in that plane sections do not remain plane in the thick plate and thick plates also experience through thickness direct strains; these are measurable quantities and with this information the analyst can select which model to employ – so these are the quantities defined by the analyst for measurement by the test engineer.

8.4.1.2 Sensitivities

The methods introduced in this chapter are intended to create bounds on the likely level of uncertainty in the analysis variables and quantities.

There are two possible problems that can arise in this bounding process. One is that the analysis-based methods do not yield values that lie within the required bounds necessary for the construction and entry into service of a safe structure. The other is that occasions arise when it is not possible to differentiate contributions from individual materials or components within a complex anisotropic material or built-up structure. In this area the required tests need to focus on creating values for the sensitivities that can then be fed into the error and uncertainty control process. The definition of the data required has to be done with precision and care so that the test data allows the sensitivity of the qualification parameters to variations in the specific structural quantity to be known to within the required uncertainty bounds. Joints in a built-up structure often fall into this category and the variability in stiffness can only be found by undertaking tests on typical jointed structures. Of course this information may be available in situations where an analysis is being undertaken that lies within the analyst's or the analyst's company's experience base.

8.4.1.3 Testing the unknown

As has been emphasised in many of the above sections, there are dangers in using computed values for the sensitivities of the uncertainties. If the sensitivity of a buckling load is computed, this cannot be accepted without taking into account the post-buckled behaviour. While this behaviour can also be analysed using numerical methods, for complex structures in a safety-critical environment, testing may be the only way to give assurance that a catastrophic failure mode does not lie within the uncertainty bounds obtained from sensitivity calculations. Similarly, if a carbon fibre reinforced plastic material is to be employed having very complex material lay-ups, then testing for de-bonding may be required as part of the process of providing adequate confidence limits. In essence, we are drawing attention to the fact that some analyses are undertaken where aspects of the structural behaviour are unknown and can only be adequately revealed through appropriate, analyst-defined, tests.

8.4.2 Validation Test

Although validation tests are often imposed on the analyst by an external body or person, it is very important that both the analyst and analysis team are involved with the testing process throughout all its stages. There are two important reasons for this statement. The first is that the analyst must have a clear understanding of the structure to be tested, its fixation, loads and loading sequence to ensure that experimental data obtained for the test relates exactly to data obtained for the analysis. It is not sufficient simply to transmit data back and forth between the analysis and test teams. These two teams must work as a single joint team with the analysts spending considerable time physically examining the test rig, specimens, loading configuration. Misunderstandings based on ignorance are the chief cause of incompatibility between analysis and test, particularly if the analysis has been undertaken using the error and uncertainty control methods displayed in this book.

The second reason for working together is to ensure that uncertainties associated with the data output from the tests can be either incorporated within the finite element analyses or compared with the uncertainty measures constructed by the analyst. Although it is not expected that an analyst would have a sufficient level of expertise on the theory of experimental uncertainty, certain basic questions require answers:

1. What sources of uncertainty exist within the experimental set-up and measurement process – both systematic and random?
2. How are these uncertainties assessed and measured?
3. How are propagation effects measured or evaluated?
4. What are the values of the bounds on these uncertainties?
5. What are the confidence limits on all of this uncertainty data?

Some of this data will find its way into analysis uncertainties and can be treated using the methods of this chapter. Other data, which cannot be directly incorporated into the analysis uncertainty assessment process, will allow the analyst to assess the validity of the test data. By following this process there is some hope that uncertainty domains for the analysis and for the tests will have large areas of overlap and that experimental and analytical results will lie within this overlapping domain. If this situation occurs, then it can be concluded that test and analysis agree to within a boundable uncertainty set. If not, there is a problem that has to be resolved! When the test and analytical results do lie outside the bounded uncertainty domains, the analyst might like to consider the following truism: 'When test and analysis results disagree, the only person to believe the analysis results is the person who did the analysis; the only person who disbelieves the test results is the person who did the test.'

REFERENCES

1. Vignjvec, R. and Rahman, A., *Hierarchical Decomposition*. 1994: SAFESA Report 9034/TR/CU/2005/1.0/30.03.94, Cranfield University.
2. Warren, C.Y. and Budynas, R.G., *Roark's Formulas for Stress and Strain*. 2001: McGraw-Hill Professional, ISBN 007072542X.
3. Blevins, R.D., *Formulas for Natural Frequency and Mode Shapes*. 1995: Krieger, ISBN 1575241846.
4. Taig, I.C., Modelling for finite element method. In A.J. Morris (ed.) *Practical Application of Finite Element Analysis to Aircraft Structural Design*. 1986: AGARD (NATO): AGARD-LS-147, pp. 3–1 to 3–18.
5. Szabó, B. and Babuška, I., *Finite Element Analysis*. 1991: John Wiley & Sons, Ltd, ISBN 0471502731.
6. Morris, A.J., *Foundations of Structural Optimisation: A Unified Approach*. 1982: John Wiley & Sons, Ltd, ISBN 0471102008.
7. Haftka, R.T., Gurdal, Z. and Kamat, M.P., *Elements of Structural Optimization*. 1990: Kluwer Academic, ISBN 0792306082.
8. Taylor, J.R., *An Introduction to Error Analysis, 2nd edn*. 1997: University Science Books, ISBN 093570275X.

9

Error-Controlled Analyses

9.1 INTRODUCTION

This chapter links all of the discussions and methods of the preceding chapters into a single method that can be used as the basis for a coherent approach to the analysis of a structural design. The method operates on the understanding that the structure is in the design stage and, therefore, cannot be tested as a completed real-world entity before the analysis is complete. The method endeavours to place the emphasis for the creation of a safe design in the hands of the analyst with testing playing an important, but, in some ways, a secondary role. In reality, the major part of the method is directed at controlling errors which mainly consist of uncertainties that arise in the idealisation process. However, it relies on using finite element analyses to provide information on the impact of errors on the qualification parameters following the procedures described in Chapter 8. In performing these analyses the analyst needs to ensure that the correct finite elements are employed and that the mesh layout conforms to the principles illustrated in Chapter 6. The results from such analyses have to be examined to ensure that an appropriate level of convergence has been attained; again the contents of Chapter 6 and Chapter 4 provide guidance that aids the analyst in checking if a model has a sufficient level of internal accuracy.

At first sight the method seems to be complex and time-consuming. This may well be the case when applied first time and getting used to any new method takes time. With experience the time required to apply any method reduces and in the present case this is accentuated by the fact that the process, as it is applied to each new problem, must be fully

A Practical Guide to Reliable Finite Element Modelling A. Morris

catalogued. This cataloguing process ensures that any similarity between a new analysis problem and one already completed can be exploited. In addition, although the method is presented as a fixed sequence of actions, it should not be regarded as an inflexible template. All analyses are idiosyncratic and dictate their own sequence of steps depending on the location and relative importance of error sources. Also, all experienced analysts have their own way of approaching an analysis problem so that it is not possible to develop a methodology that suits all tastes!

The final question is 'Why bother with developing an error control process that requires time to employ and may be expensive to implement?' The method described was developed with the following aims:

- to provide a method to reduce the influence of errors in the design of structures that might be operating in a safety-critical environment;
- in a company that undertakes finite element analysis, to provide a route to improve the quality of the results obtained and the opportunity to create company-wide consistent and regulated analyses;
- based on the above point, to provide a company undertaking a finite element analysis of a complex structural component with an improved legal position should in-service failure actually occur;
- to provide a basis for an analysis methodology that can lay claim to being 'best practice';
- to reduce cost by reducing overall design cycle time.

This chapter starts by considering how to select an appropriate finite element system before any analysis is undertaken. It then moves on in Section 9.3 to consider how the method requires the construction of a catalogue of all the actions taken by an analyst or analysis team in controlling the impact errors or uncertainties within an analysis. This is called the Quality Report and it is this report that demonstrates how errors, in a particular analysis, have been controlled. It justifies all the decisions and assumptions made by an analyst or analysis team in creating a finite element model that claims to represent adequately the behaviour of the real-world structure as it operates in the in-service environment. It is this report that forms the basis for claiming that an analysis is satisfactory so that the resulting design may be declared fit for purpose – at least from a structural viewpoint. It then becomes part of the experience base of the analyst, analysis team or the company within which the analysis has been undertaken.

9.2 IS THE FINITE ELEMENT SYSTEM FIT FOR PURPOSE?

Before an analyst can undertake any work, some thought has to be given to ensuring the finite element system selected for solving an analysis problem is itself fit for purpose. That is, the system has the capability to match all the requirements to successfully undertake the analysis task and that it has passed appropriate verification and validation tests. There is no point in following the processes and procedures laid down in this and other chapters in this book if, fundamentally, the finite element system does not have an appropriate finite element library, meshing capability or solution processes that will allow it to solve the analysis problem being posed by the structural design. Clearly it is the responsibility of the finite element analysis expert or experts undertaking an analysis to establish that the finite element system has the required capabilities and has sufficient basic integrity for the problem in hand. Although it is unwise to try and establish that the system has the required capability and integrity on a problem-by-problem basis, it is usually the case that a user will build up a level of confidence with increasing use of a finite element system. There are two main questions to be answered:

1. What type of analysis problem does the finite element system solve, i.e. what does it do?
2. How does the analyst know that it can solve the problems it claims to solve, i.e. how do we know it does it?

9.2.1 What Does It Do?

Some finite element systems have been created with a particular industrial sector in mind so the first task for an analyst is to consult the Theoretical (or Theory) Manual. This manual should provide a full technical background of all the finite elements available within the system, the meshing techniques that can be employed, the solution methods, etc. The description of the elements within the element library should clearly explain the underpinning mathematical theory so that the analyst can exploit this information when an appropriate level of abstraction is being selected using the methods given in Chapter 8. The manual should also fully explain the various solution processes that can be undertaken relating to static, dynamic and non-linear responses. It should also include a full list of journal papers and books that have been used to provide the basic information that guided the developer when creating the system; if the analyst is confused by the information in the Theoretical Manual, these are vital sources. In addition to providing information on the basic capabilities of a finite element system, the Theoretical Manual should allow an analyst to assess whether or not the system has sufficient functionality and capability to allow an implementation of the error and uncertainty control methods explained within this book. In addition to using the Theoretical Manual an analyst should note that much of the information at the element performance level presented in this manual can be deduced from the methods for assessing element capabilities described in Chapter 6.

The Theoretical Manual, together with its source documents, should provide the information that allows a user to decide if a given finite element system is fit for purpose with respect to the analysis requirements relating to a specific design problem. Essentially it displays the inner workings of a finite element system but does not explain how to use it; this is the function of the User Manual. This second manual should be comprehensive and explain how all the capabilities of a finite element system can be called upon by the analyst in going through the full process of creating appropriate finite element models, applying loads and constraints and solving problems. From the specific reference point of this book, the User Manual will allow the analyst to evaluate the finite element system with respect to providing the data required to implement the techniques discussed in earlier chapters, particularly Chapter 8.

The last information source that allows the analyst to learn about the system's capabilities and functionality is the Tutorial. This describes how to use a finite element system by way of a number, usually a large number, of worked examples. The Tutorial can be a separate manual but is more often a set of executable data files that the user can run directly or in modified form in order to enhance understanding of the system.

9.2.2 How Do We Know It Does It?

The manuals described in Section 9.2.1 tell the analyst what a finite element system should be able to analyse and how to use it; they do not provide any assurance that the system can actually perform as stated.

In order to answer this question two processes must be undertaken, called verification and validation. These can be defined as follows:

- *Verification* is the process that ensures a finite element model within a given finite element system and the solutions derived from it accurately represent the developer's description of the model and the solution obtained from that model.
- *Validation* is the process that ensures models created within a specific finite element system are able to represent the real-world behaviour of a structure in a specific design environment.

In principle, verification is a very straightforward process and involves employing a test programme that demonstrates the finite element system actually does what it is supposed to do! If the manual says that an element is a thin plate element, the verification process should demonstrate that solutions derived from this element are mathematically equivalent to those obtained from classical thin plate theory. This should be demonstrated for all the element types within the system and for all the solution processes. In the dynamic case, for example, that an idealised finite element model predicted the same natural frequencies and mode shapes as those derived from an equivalent direct mathematical solution. The verification process relies on having, as far as possible, a comprehensive range of benchmark tests that cover the full spectrum of the system's capabilities. The range would include single element tests, patch tests, all the way up to complex models. Complex benchmarks would not normally result from closed-form solutions but from calculations that have not relied on finite element models but on an alternative simulation methodology such as finite differences. All the reputable suppliers of finite element systems will provide an extensive list of benchmark problems in the form of executable data files. When using a finite element system, the analyst should ensure that the finite elements, the solution methods, etc., to be used in an analysis are covered by the verification process with appropriate benchmark problems.

The process of validating a finite element system is much more complex than the verification process and is intimately connected to the error and uncertainty control processes discussed in this book. It relies on establishing a set of real-world problems that were analysed by the system and the results correlated with the measured performance of the in-service structure. As with verification this is a benchmarking activity and in order to establish that the system had been validated for a given design problem, the analyst would need to find a similar problem within the validation benchmark set. There are a number of difficulties in undertaking this activity. The first is that there is no clear definition of similarity to link one structural analysis problem to another, other than their being identical – in which case the benchmark results can be taken without further ado. Second, the analyst would need to ensure that an error and uncertainty control process had been employed before the in-service structural responses had been measured. This is to make sure that the analysed results had the same or similar error sources to the problem about to be analysed.

9.2.3 Is Size Important?

The power of modern computers increases daily and many users tend not to give any consideration to the size of the finite element model that is being presented for solution. In view of this custom it is relevant to ask 'Does the size of the finite element model matter?' To answer this question we must consider the way memory space and computing time increase with the size of the problem and it is convenient to use the constant stress element in order to quantify the effect of size. Consider the stylised 2-D and 3-D models shown in Figure 9.1 where m indicates the number of elements used along the sides of the two models. Table 9.1 shows the variation of memory requirements and time requirements as the number of elements used increases for a static analysis problem; note that these are not exact values but reasonable estimates.

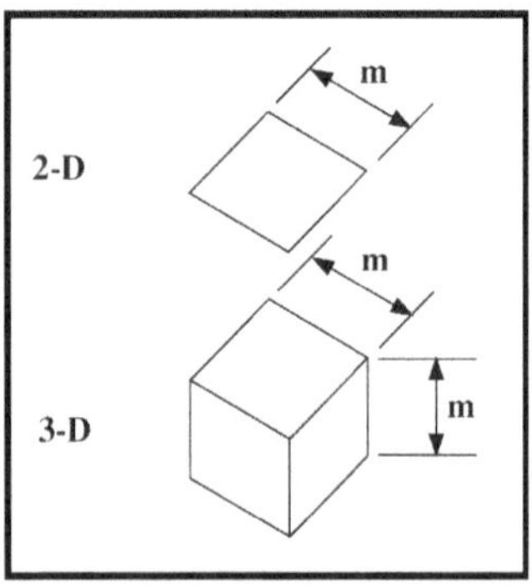

Figure 9.1 2-D and 3-D stylised models.

Table 9.1 Space and time requirements with size.

Value n	2-D memory	2-D time	3-D memory	3-D time
n	1	1	1	1
$2n$	4	16	16	128
$4n$	16	256	256	16×10^3
$8n$	64	4×10^3	4×10^3	2×10^6

Table 9.1 clearly demonstrates how the demands placed on the performance of the computer grow with the number of elements used. In the 3-D case, if the base element layout involving n elements takes 1 minute to solve, then a problem requiring $8n$ elements along a side, needs 4 years – this growth is exacerbated if the problem involves frequency calculations, non-linear responses, etc. Although the values of memory and time requirements are indicative and constructed using the constant stress elements, the message is clear – work out your problem's upper limit requirements before launching a major analysis. Then match this requirement against the measured performance of your computing system even if you have the world's largest supercomputer or largest parallel processing machine. Remember that over time problems increase in size until the computer cannot cope. If it is clear that the problem size is too large for the current computing capabilities available or budget, it may be possible to reduce the problem size using symmetry, condensation or sub-structuring and these features are described in Chapters 2 and 3.

The key point of this sub-section is that it is important to have measured the performance capability of the computing system operating with your selected finite element system. While this may not be important when a well-established finite element system has been used to solve problems on a given computer for some while, it is very important when either a new computer or a new finite element system is being selected. In order to make appropriate measurements the analysis must have available a set of finite element models essentially mimicking the $n \times n/n \times n \times n$ squares and cubes problems illustrated in Figure 9.1 in order to have representative numbers for solution times. Note that these problems are calibration tests and separate from any validation test problems. However, once the calibration exercise has been completed, the analyst is then in a position to make sensible assessments about the time required for the solution of any problem presented for finite element analysis.

9.3 QUALITY REPORT

The target for this book is a methodology for ensuring a finite element analysis is undertaken in such a manner that any potential errors or uncertainties are identified and their influence on the accuracy of the finite element results measured and bounded. In order to be effective the methodology must be formulated into a process involving a number of stages and steps which are presented in the next section. Although the process itself is important, the activities, decisions and consequences associated with applying the steps within the process have to be reported. This report is called the Quality Report (QR) and has four main purposes:

1. To provide a record of the actions undertaken by the analyst.
2. To provide a document that allows a third party to evaluate the quality of the analysis.
3. To be part of the 'experience' base of an individual analyst or analysis team or for a company that regularly undertakes commercial-level analyses.
4. To provide a document that could form part of a legal defence in the event of any claim or claims for negligence.

This report is the source document for justifying the analysis and demonstrating to third parties that all the potential errors and uncertainties that could be present in the model and the analysis have been adequately treated and quantified. If the analysis is an integral part of a certification or qualification process that is administered by a regulatory authority, the report would then be one of the fundamental documents used to support the case that the structure can be passed by this authority as fit for purpose. It could also form a significant component in a legal defence should there be a structural failure followed by a claim for damages.

In addition to being used to justify a particular structural analysis, Quality Reports represent codified knowledge for an analysis team or for a company that has regularly to undertake finite element analysis. If an analysis is to be undertaken that is significantly similar to an existing analysis for which a full report is available, then the starting point for the new analysis is the existing report. The judicious use of existing Quality Reports can greatly reduce the amount of time, effort and expense required in undertaking a new analysis.

The QR is compiled and written as the analysis process passes through the stages and steps given in the next section and grows as progress is made through this set of tasks and activities. In essence, the QR is a detailed written report of tasks and activities undertaken in each step within the various process stages together with any results and conclusions relating to errors and uncertainties.

Although separate from the AVP, see Chapter 5, Section 5.7, the QR records the point during the analysis history at which the first issue of the AVP is created and the points at which revisions may be made. The AVP should always be an appendix to the final QR document. Similarly, the major finite element data input and output files, i.e. CAD displays of meshed models, contour plots, data in numerical form, etc., are not part of the QR but must be directly linked to it. This could be done by simply listing where this finite element data is located, but with modern IT systems it is anticipated that the QR will not be a paper document, thereby allowing this data to be called directly from the QR. However, the QR must contain all the results from the error and uncertainty quantification process within the main document.

The detailed structure of the QR is, in some sense, evolutionary as it follows the analyst or analysis team in tracking through the error and uncertainty control methodology described in Section 9.4. Nevertheless, the overall structure is fairly simple with three major components: (1) a set-up information section; (2) a stage reports section involving as many reports as there are stages; and (3) a conclusions section.

Set-up Information

This section provides background information on who is responsible for the QR and assessing its quality together with information on the broad experience of those involved in undertaking the analysis and includes:

- Individual(s) responsible for compiling the QR named.
- Individual(s) responsible for peer reviewing the QR named.
- Qualifications of all individuals involved in the analysis – both formal, professional qualifications and informal, experienced-based 'qualifications'.
- Details of past major analyses undertaken by all companies involved with the analysis.

Stage Reports

This section records all the work done as the process goes through each of the steps in each stage. Each step within a specific stage has a number of actions that the analyst must undertake and these are recorded together with the consequences of enacting these actions. In this process the sources of error or uncertainty must be identified and action taken to generate bounds. If, at a given stage, some of these cannot be bound, then flags are raised indicating that further action is required in one or more of the remaining stages. Finally, this information is reviewed by engineers, designers and analysts at a review meeting where the consequences of the output from each stage are considered in depth and consequential action planned and implemented. It is worth noting that this may result in a looping back to earlier stages or even a decision to terminate the analysis process. The steps in this part of the process include:

- Stage introduction
  - Purpose of the stage
  - Inputs to stage.
- Actions undertaken in each step within the stage
  - Step Introduction
    - Purpose of step
    - Inputs to step
  - Details of processes followed in the step
  - Details of sources of errors or uncertainties
  - Details of methods employed to bound uncertainties
  - Details of actions taken to bound, control or eliminate errors
  - Outputs from step.

Conclusions

At the end of the process either all the potential errors or uncertainties have been captured and bounded or some have been found that cannot be bounded by analytic or numerical methods. For the bounded errors and uncertainties the conclusions list them together with values for the bounds. For those which are unbounded, the conclusions define sets of experimental tests than can provide appropriate information. The conclusions include:

- A list of all potential errors and uncertainties in the qualification parameters with bounds on their impact on the predicted behaviour of the in-service structure provided by the finite element analysis or analyses.
- In many cases errors (as opposed to uncertainties) will have been eliminated by the careful selection of appropriate elements, meshes, etc., following the instruction contained in Chapter 6. A rationale for believing that specific errors have been removed in the analysis build-up is mandatory.
- A review of those uncertainties that could not be bounded.
- Requests for specific tests to provide values for unbounded uncertainties.

When a stage in the process has been completed and the QR written for that stage in cases where the structural analysis problem requires the analysis being undertaken by a large team or a collection of distributed teams, a review meeting will be required. This meeting will use the information in the QR to assess the current position of the analysis process and make appropriate decisions with regard to modifying the validation plan and/or looping back to earlier stages.

9.4 THE ERROR AND UNCERTAINTY CONTROL METHOD

9.4.1 Introduction

All the background information and bounding techniques necessary for setting up an effective error and uncertainty control methodology whereby these entities can be assessed and quantified are now available. All the earlier work detailed in this book together with the associated methodologies can now be used to create a coherent process that can be applied to any structural analysis problem employing the Finite Element Method. The process consists of a number of separate stages starting with an assessment of the real-world structural design/analysis problem and concludes with a finite element model that has all uncertainties and error sources accounted for and assessed for their impact on the predicted values of the qualification parameters.

Although the process is presented as a sequence of stages with defined activities, it should not be taken as a rigid format. Rather it should be considered as a template or guide that can be adapted to personal, company- or problem-driven imperatives. Neither the sequence in which the stages in the method are presented nor their contents should be regarded as sacrosanct. The key principle is that the process should start by examining the design objectives and then accumulate information as the analyst goes deeper into the finite element analysis solution process. It is always useful to give a methodology a title so that it can be easily referred to or referenced; thus we have elected to call the method presented below the Finite Element Method Error Control or FEMEC.

9.4.2 FEMEC

It is assumed that this process, which eventually leads to the creation of a finite element model able to predict the behaviour of the in-service structure to within defined error and uncertainty bounds, begins with a design 'on the table' and, for most industrial or commercial problems, an initial team of experts, designers and test engineers. The membership of this team may change during the course of implementing the FEMEC process. Before the process can start, the first part of the QR has to be completed and this covers the details given in the 'set-up information' section of the document.

FEMEC has a number of scoping activities before moving to tasks that require implementing either full-scale or partial finite element analyses of the structure. Although the various stages in the process are different, they all have the same basic structure where the tasks and activities within the stage are to be described, the sources of error or uncertainty identified and then a process for bounding or controlling their impact defined and, in certain cases, implemented.

The first task for the team is to create a FEMEC implementation plan that outlines the stages and steps that will be undertaken in progressing towards a satisfactory analysis of a real-world structure. This plan must cover the points in the overview box, but the content details are not prescriptive. The golden rule is that any version of the implementation plan must be coherent and comprehensive.

FEMEC Implementation Plan Overview

- The plan sets out to create a finite element analysis model or a set of models that predict the behaviour of an operational in-service structure to within specified accuracy levels.
- The plan is laid out in a sequence of stages, each of which deals with a specific task or activity within the total analysis process.
- In each stage uncertainty sources are noted and their contribution to creating deviations in the final finite element analysis values for the qualification parameter values from those that occur in the in-service structure is measured and bounded.
- In a first phase of the FEMEC process an attempt is made to generate bounds using the experience base available to those undertaking the analysis of simple formulae or 'back of the envelope' calculations.
- If the first-phase activity does not quantify and bound all the uncertainty sources, further stages are required that use the methods described in Chapter 8.

9.4.3 FEMEC Implementation Process

Having outlined the purpose and processes associated with the creation of a FEMEC implementation plan, we can now go into the details of our version. The process is explained chronologically as a multi-stage process and within each stage there are one or more steps. Each step within a stage envisages an input from the QR document; it should be noted that the FEMEC method is, essentially, the compilation of the QR which is built up as the method progresses through the 10 stages which constitute the total process. Thus the 'input' statement at the beginning of each step often simply implies that, before commencing the step, full note is taken of the relevant information that previous stages and steps have implanted in this document.

Stage 1 Scoping the real-world problem

The aim of this stage is to link the real-world design problem with the finite element analysis problem; it also links the team members given the task of analysing the structure with those who are responsible for the design and eventual entry into service of the final product. Many of the steps in this stage require the active participation of designers, analysts, test engineers and manufacture/assembly experts. At the end of the stage the problem mainly becomes the responsibility of the analyst. However, before the process can be started, there is preliminary activity required of the analysis team to complete the set-up statements in an initial QR document.

Step 1.1 Analysis problem overview

Inputs:

- Initial QR document.
- Design requirements.
- Construction/manufacturing/assembly details relevant to the analysis.
- Qualification information – either from an appropriate code or from an internal company code, or from one derived by the analyst for this analysis task.

The process

This step quantifies the ovrall analysis requirements as they relate to the design requirements and consists of:

- A description of the design objectives encompassing the key aspects of the design:
  - In addition to the main aspects of the design, this description should include any special objectives, for example:
    - that the design must be a minimum weight structure – this implies that at least some of the structural dimensions are unknown at this start point;
    - that the design must have a specified functionality, e.g. that disabled persons are able to get easily into or out of a vehicle;
    - that the design must be innovative in order to attract a specific set of customers.
- A description of the design constraints that might include, for example (with the first two bullet points being obligatory):
  - That the design must comply with a specific qualification or certification code following the approach described in Chapter 5, Section 5.6.

- That the design must be able to carry safely a set of specific loads within a specific operational environment; it needs to list external load types, i.e. static, dynamic, etc., and magnitudes. This activity incorporates into the load set any specific loading types or configurations imposed by the certification codes.
- That the design must avoid a given frequency range.

- A description of limitations imposed on analysis by the above design requirements and constraints or by more general considerations including:
  - A high-level definition of qualification or certification criterion indicating the type of qualification parameters that the structure will be subjected to, but not the details. In the case of an external qualification code, quoting relevant paragraphs in the code together with the parameter type, e.g. 'the structure is subject to displacement limitations'.
  - Accuracy limits or margins set following the guidelines in Chapter 5, Section 5.6.2.
  - The physical limitations of the structure or the design environment that define an absolute outer limit to the physical domain of the structural analysis.
  - The impact of the analysis being done by several analysis teams – this could result either because the problem is too extensive in scope for a single expert team to cope with, or because the design environment has resulted in a multi-company design and manufacture policy.

- An uncertainty review:
  - The above activities must now be reviewed using the information contained in Chapter 7 to identify any initial sources of uncertainty.

Outputs:

- Revised QR containing the descriptions from this step, the limitations on the analysis and a list of any uncertainty sources identified.

Step 1.2 Initial analysis assessment

Inputs:

- QR document containing Step 1.1 additions.

The process

This step draws on the basic design information to form an initial view of the type of analysis that will be required. At this point there should be no attempt made to form an opinion on the details of the actual finite element analyses that will eventually be implemented. It draws on the information generated in step 1.1 and any other information relating to the manufacture/construction process, materials selected, etc., that have resulted from the design scoping which takes place before the FEMEC process is initiated. The step has two components as conceptualised here:

- Basic structural specification:
    - type construction
    - manufacturing or assembly processes to be employed
    - likely build quality
    - materials to be employed in the structure
    - bill of materials (if available)
    - CAD models.
- Analysis type: based on the information available at this point from step 1.1 and the above component, an assessment is made of the type of analysis or analyses that are going to be required to meet the design objectives. Hence this section indicates if the analysis task requires linear static, dynamic, geometric non-linear analyses, etc., or combinations.
- Uncertainty review:
    - The above activities must now be reviewed using the information contained in Chapter 7 to identify any initial sources of uncertainty.

Outputs:

- Revised QR containing the descriptions from this step and a list of any uncertainty sources identified.

Step 1.3 Review of available information

Inputs:

- QR document with step 1.2 revisions.
- Relevant company documentation relating to earlier analyses.

The process

This step draws together and reviews all the relevant information that might assist the team in forming an opinion on the value of any background information. It has two main tasks:

- Compilation of available information:
  - previous QRs for similar analyses (essentially exploiting data models, see Chapter 5, Section 5.4.3)
  - data sheets
  - existing test data
  - available reference books and handbooks, e.g. *Roark's Formulas for Stress and Strain.*

- Exploitation of available information leading to:
  - the creation of an experience database;
  - an identification of any uncertainty sources from the database information;
  - an identification of any information in the database that would provide bounds on uncertainties identified in the earlier steps or that might be useful in bounding errors later in the processes.

Output:

- Revised QR containing pointers to database information and listing uncertainties identified in step 1.3 together with any bounds on specific uncertainties.

Step 1.4 Uncertainty list

Inputs:

- QR document with step 1.3 revisions.

The process

This step examines the material in the QR, draws together all the compiled information on potential sources of uncertainties and starts the process of either bounding them or identifying their existence for future consideration.

- The step provides a catalogue of uncertainties that have been identified in earlier steps or are now considered here for the first time related to the qualification parameters introduced in step 1.2. The sources of these uncertainties to be derived from:
  - The physical limitations of the analysis from step 1.1:
    - actual limitations on size and extent imposed by either the analyst or the design;
    - artificial limitations due to the structure being analysed by several distributed teams.
  - Potential variations in build quality from information in step 1.2.
  - Information from past experience database – step 1.3.
- The identified uncertainties are now reviewed and bounded where possible using:
  - information from material compiled at step 1.3;
  - 'back of the envelope' calculations.
- The step includes a list of uncertainties identified that cannot be bounded at this stage and flags raised to indicate their presence to future stages.
- The step does an initial review to identify any requirement for specific tests to enable bounds to be generated.

Output:

- The listed uncertainties, any generated bounds on uncertainties, flags for currently unbounded uncertainties and any required tests written into the QR.

Step 1.5 Novelty, complexity and experience assessment

Inputs:

- QR document updated with inputs from step 1.4.

The process

This step makes critical decisions on the future progress of the analysis. It could result in the analysis being abandoned or radically changed.

Although presented here as a single task within the analysis process similar to the other steps in this stage, in reality it is likely that it would be undertaken by the full team involved with the analysis process.

- It assesses the novelty and complexity of the analysis problem, following the principles laid down in Chapter 5, Section 5.4, using:
    - outputs from steps 1.1 and 1.2 that define the nature of the design/analysis problem;
    - the uncertainty information from Step 1.4 that defines the overall nature of the uncertainty sources.
- It compares the novelty and complexity assessment with the experience base of the analysis team encapsulated in step 1.3 using the principles described in Chapter 5, Section 5.4.3.
- It makes an assessment of the gap between the experience base and the novelty and complexity.
- Based on the extent of this gap, recommendations are made for future progress that come within one of three categories:
    - The gap is too great and the project should be abandoned.
    - The gap is large but could be reduced if additional expertise is brought to bear – this requires deciding on the exact nature of the required experience and ensuring that appropriate expertise is available, i.e. back-up identified from expert resources that are not already fully committed to other projects.
    - The gap is manageable with the current experts working on the project.

Outputs:

- Updated QR with recommendations from step 1.5.

Step 1.6 Review

Inputs:

The current QR containing reports on all the steps in this stage together with all decisions and recommendations made.

The process

This is the point where decisions are made relating to the progress of the analysis taking into account all the information accumulated in the stage:

- Review recommendations from step 1.5 and decide on a go or no-go plan of action.
- Review the uncertainty material in step 1.4 and assess whether there is a need to include a feedback loop that requires revisiting any of the steps.
- Based on the above reviews, make firm decision on next steps.
- If necessary, revise the original FEMEC plan.

Outputs:

- Go/no-go decisions.
- An updated QR based on the decisions made.

Actions

- 'Go' decision – continue the process and move to Stage 2 after creation of initial AVP (for this latter point see below).
- Non-terminal 'no-go' decision on condition that the problem is modified to reduce complexity or novelty – make return loop back to step 1.1 and/or step 1.2 with acceptable revised design requirements or acceptable reduced qualification criterion.
- Terminal 'no-go' decision; the analysis cannot be undertaken by the existing team nor by a team augmented or replaced by available experts either in-house or bought in – analysis abandoned.

At this point a 'go' decision to progress with the FEMEC progress will require that the first issue of the AVP is written. It is recognised that, at this early stage, the plan cannot have great detail but it does begin the process of focusing the analysis on the targets set by the design and qualification requirements. An outline diagram of the processes undertaken in progressing through Stage 1 is given in Figure 9.2.

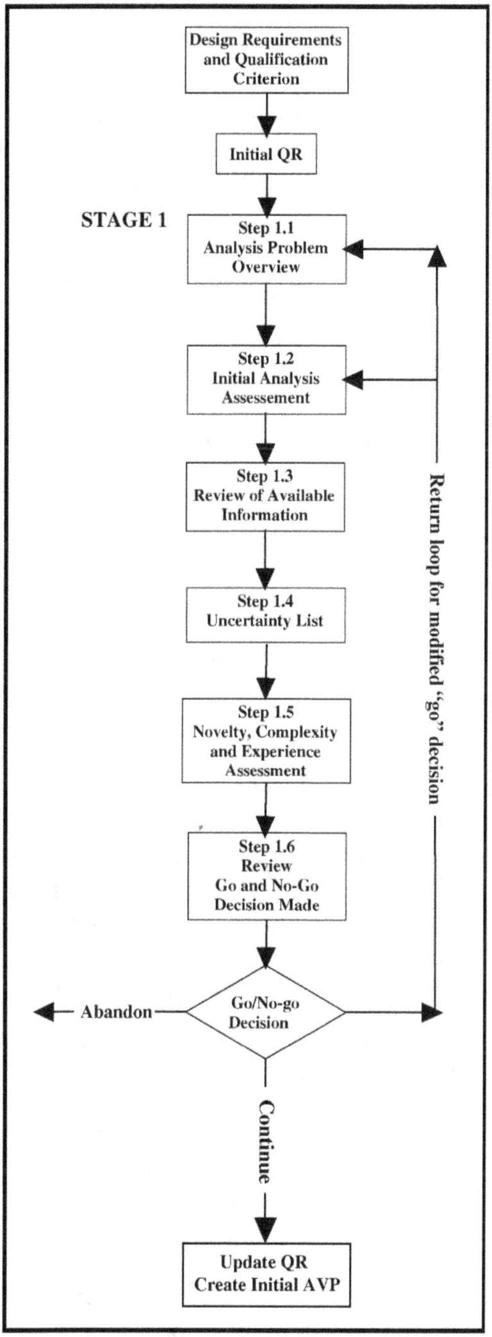

Figure 9.2 Outline flow chart for the steps in Stage 1.

Stage 2 Initial assessment

The aim of this stage is to take the outputs from Stage 1 and develop the information so that assessments can be made of sources of uncertainties from a higher level viewpoint where bounds are sought without the direct application of supporting finite element models. It also sets up the analysis process so that an analysis plan can be constructed before moving on to a detailed assessment of uncertainty sources. Because the AVP is not modified during the individual steps within this stage, it is not considered as a direct step input or output. However, it should be considered as a guidance document that is to be continually consulted.

Step 2.1 Definition of domain of analysis

Inputs:

- QR document from Stage 1, relevant section being the information input by step 1.1.
- CAD files/diagrams of structural layout.

The process

The step finalises the physical limits of the analysis by defining the domain of analysis (DoA) first introduced as a concept in Chapter 7. It takes preliminary information limiting the scope of the analysis domain from step 1.1 as initial data. The main work of the step is to implement the process described in Chapter 7, Section 7.3, and from this capture the sources of uncertainty. The process has three main tasks:

- Defining the DoA following the principles in Chapter 7, Section 7.3.
- From this activity list the uncertainties that have an impact or influence on the qualification parameters of boundary conditions associated with the process of defining the DoA (this to include any uncertainties that are associated with any additional domain reduction activity separate from the definition of the DoA).
- Measure and quantify impact on the qualification parameters:
    - Material from step 1.3 is employed in an attempt to generate bounds on the impact that identified uncertainties have on the qualification parameters.
    - Additional calculations undertaken by the analyst that directly employ structural analysis theory are also employed.

- Flag unbounded uncertainties for future consideration (note that these flags will be lowered as more sophisticated methods using finite element models are employed in Stage 4 to bound the impact of uncertainties).

Outputs:

- Revised QR containing:
  - If the QR is digital, a link to the CAD files that display the DoA; if not, a set of appropriate diagrams in the paper version.
  - A list of uncertainty sources.
  - Quantified bounds on the impact of these sources on specific qualification parameters.
  - Flagged unbounded uncertainties.

Step 2.2 Loading actions

Inputs:

- QR document, relevant section being the information step 1.1 on load cases;
- CAD files/diagrams of structural layout.

The process

This step traces the relevant loading actions, experienced during operational service, through the various parts of the structure. It should provide the analyst with sufficient information to define and list the features that play a central part in the uncertainty control procedure. The definition of what constitutes a feature and the role features play is described in Chapter 7, Section 7.2. The step takes the information from the QR document relating to the load definitions in step 1.1 and:

- assesses load paths for static analyses;
- assesses dynamic load transfer for analyses involving dynamic responses;
- assesses load path movement for non-linear analyses.

The sources of these actions could be from external or internal loads; an example of internal loads would be those occurring due to the presence of inertial body forces. In the case of internal loads a complicating factor is that structural stiffness may be an important factor in the way inertia loads are generated and consequentially influence the behaviour of structural components. In the dynamic situations the influence of non-structural masses may be very significant.

From this information the analyst must trace the passage of these various loading actions through the body of the structure noting how these pass through the components of the structure. This information is best recorded as part of the CAD records of the structure if a direct digital link to the QR is available but can be recorded as separate tables in the QR. Note that this information may change if the process resorts to looping back though the stages and steps of the FEMEC process.

Outputs:

- Revised QR with loading action information; if QR digital, through links to an appropriate CAD file, or through tables if a paper version.

Actions

The results of examining the loading actions in detail at step 2.2 may require revisiting step 2.1 if it is felt that the DoA was constructed without due note being made of the effect of one or more of the loading actions.

Step 2.3 Decomposition of structure

Inputs:

- QR document.
- CAD files/diagrams of structural layout.

The process

This step decomposes the structure into a set of features, as defined in Chapter 7, Section 7.2, with associated uncertainty sources identified and characterised. There are several components to this step which are enacted in chronological sequence:

- The first is to assess the influence that the loading actions have on the behaviour of the structure and structural components. For example, the presence of a bending loading action on a thin-plate-like component clearly indicates that this cannot be considered, in a structural sense, as a membrane.
- The second is to identify the sources of uncertainty using the source types and explanations given in Chapter 7, Sections 7.4–7.6 (this covers uncertainties with respect to joints, discontinuities, levels of abstraction, etc.).
- Finally, to record the identified features and uncertainty sources. If the QR is in an appropriate digital form, this record can be held in CAD data form with a link to the QR document. If not, then a paper record is required that lists the features and their potential uncertainty sources.

Outputs:

- QR document containing the information that identifies and locates the structural features and uncertainty sources.

Step 2.4 Initial uncertainty bounds

Inputs:

- Information in the QR from step 2.3.

The process

Step 2.3 has decomposed the structure into features and has identified the uncertainty sources associated with each feature. This step attempts to provide bounds on the impact these uncertainties have on the qualification parameters:

- Review uncertainty sources and assess the possibility of creating bounds on the qualification parameters using information from step 1.3 and other simple calculation procedures.
- For those uncertainties for which the simple methods or existing knowledge are applicable, create appropriate qualification bounds.
- Flag the remaining uncertainty sources for future consideration by more complex procedures.

Outputs:

- Bounds obtained for specific qualification parameters are added to the list of bounds in the QR.
- Flags identifying those uncertainties for which there are no bounds on the effect these have on the qualification parameters are added to the QR.

Actions

As a result of decomposing the structure into features with associated uncertainty sources at steps 2.3 and 2.4, it may be necessary to revisit step 2.1 if it is felt that the definition of the DoA zone may have to be modified due to the influence of a newly identified uncertainty source from step 2.3.

When this process emerges from Stage 2, a review is required that allows a comparison to be done with the QR document, assembled during the passage through Stage 2, and the AVP. If nothing has emerged during Stage 2 that jeopardises the validation process detailed in the AVP, then the FEMEC process proceeds to Stage 3. If a conflict has arisen with the validation process, the AVP must be revised by those responsible for its contents.

An outline flow chart of the activities involved in Stage 2 of the FEMEC process is shown in Figure 9.3.

Analysis approach

FEMEC has two distinct phases: the first which do consists of Stages 1 and 2 does not use any finite element analysis to acquire information that assists the analyst in assessing sources of uncertainty. The second relies heavily on finite element reference analyses but builds on the information from the first phase. Now that the process has reached the point where the domain of analysis has been defined, the features selected and uncertainty sources identified, it is possible to enter the second phase of the analysis process where finite element models are to be employed.

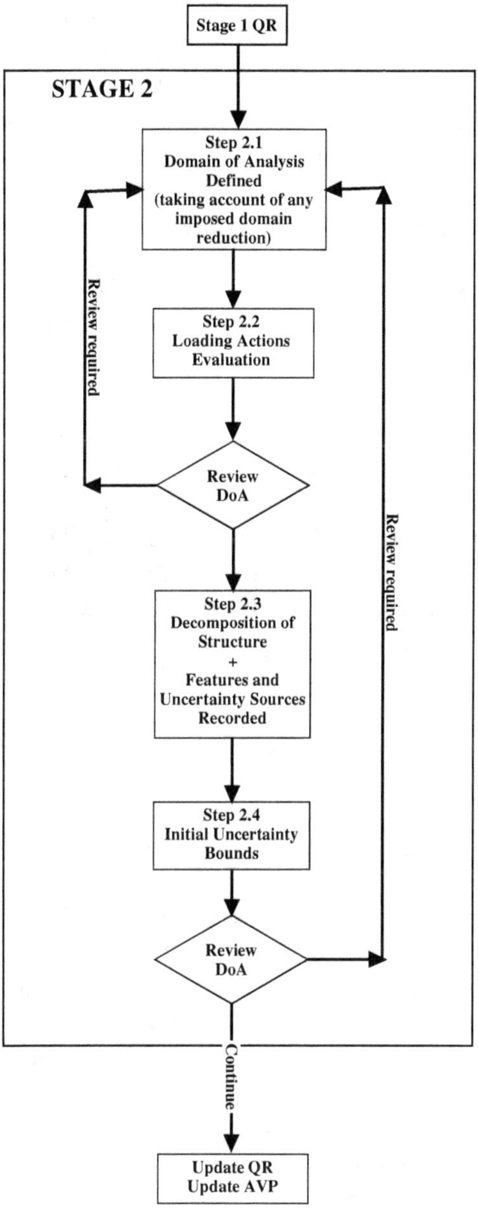

Figure 9.3 Outline flow chart for the steps in Stage 2.

It may be argued that the point has been reached where the finite element reference analyses that are going to be used in the FEMEC process have to be planned. However, FEMEC is dynamic in nature and in preference to writing a formal analysis plan it creates specific finite element reference models within an overall approach where decisions are dictated by the situation facing the analyst at each of the ensuing analysis stages. Some analysts may prefer to develop a formal analysis plan that would be undertaken earlier in the process. This is perfectly acceptable, but this type of analysis plan would play a different role in the total analysis process to that played by FEMEC where a gradual assessment of analysis requirements is envisaged as the process unfolds.

Within FEMEC the analyses undertaken have to accomplish two important tasks. One is the conventional requirement that the finite element reference analyses provide information that fully characterises the behaviour of the structure and shows that the structure adequately fulfils the design requirements. In most cases this devolves down to the analysis results clearly demonstrating that the qualification requirements are met. The second is to provide information that allows the impact of uncertainty sources on the qualification parameters to be quantified and bounded. The two are clearly intertwined as any analysis undertaken under one of these headings will normally be of value under the other heading. The analyses undertaken from this point onward, in the FEMEC process, must satisfy demands under both these headings.

Although FEMEC is moving away from the preliminary first phase, the process being enacted in this phase still needs to take account of the fact that certain parts of the structure may already have had an appropriate finite element model associated with them from the results obtained in step 2.4 or earlier. For example, the level of abstraction for a specific component may have already been fixed and the associated uncertainty bounds calculated. This situation might arise where a component has dimensions and loading actions that lead to the conclusion that it behaves as a membrane (say) and can, therefore, be modelled by membrane elements. This information would need to be drawn into any analyses at appropriate points in the second phase of the FEMEC process.

The major difference between the activities within the steps of the first phase and those of the second is the need to set up and exploit the results from finite element analyses. The stages in the second phase focus on setting up finite element reference models and then exploiting these to generate bounds on the impact that sources of uncertainty have on the qualification parameters using the methods described in Chapter 8. In addition to providing bounds these models are also available for providing design-based information.

Although we have indicated that the FEMEC method does not require the creation of a formal analysis plan as the details of the finite element models to be employed are generated in the global and detailed assessment stages described below, it is advisable that an initial review of the possible analyses likely to be required is undertaken. The main purpose of the review is to ensure that the analyst or analysis team understand the level of commitment in time and expertise that will be required in order to complete the FEMEC process. This way unexpected and expensive surprises can be avoided! It may be possible to ignore this particular step if the problem is very similar to existing analyses captured within the experience base of the organisation and for which a well-documented QR is available.

Stage 3 Analysis review and decisions

This stage defines the analysis requirements in order to create a set of finite element reference models that can be used to bound the impact of uncertainties on the qualification parameters. It is quite possible that the set identified at this stage may need to be augmented with additional models that result from the initial set of models, indicating that all of the potential uncertainty sources have not been detected by stages 1 and 2.

Step 3.1 Review

Inputs:

- QR document; the inputs from step 1.1 should mean that the design requirements are available in this document; if this is not the case, then QR input needs to be supplemented by an adequate description of the design requirements.
- AVP document.

The process

This stage takes the information from the QR, suitably augmented by design requirements if necessary, together with requirements from the

AVP and scopes the analyses required both to quantify the uncertainties flagged by earlier stages as unbounded and to design the structure. The dynamic nature of the analysis process means that the analyses envisaged at this stage may not represent the complete reference set that will eventually be required to complete the analysis process. These analyses will normally take place at the global and feature levels but will lead to a final reference analysis from which predictions will be made of the in-service structural behaviour of the design. By operating at both the global and feature level, FEMEC is able to follow the normal analysis procedure for attacking new designs by starting with analyses having limited scope that is increased as the analysis process evolves. The maxim is 'start small, grow big!'

The stage covers the following tasks:

- Global assessments that define the analysis requirements for a finite element reference model(s) with the least number of elements sufficient for the assessment task, called for convenience the 'reduced finite element model':

    - to allow an assessment of the boundary conditions related to the DoA and any other uncertainty sources that can be taken at the global level;
    - to size the structural members (if not pre-assigned) and track the internal loading actions so that features interactions can be assessed.

- Detailed assessment that defines the analysis requirements for finite element reference models for those features documented in the QR from step 2.3 with unbounded uncertainty sources sufficient:

    - To allow an assessment to be made of the level of abstraction required and to be able to undertake appropriate sensitivity studies using the techniques and methods from Chapter 8. Some of the assessments can be undertaken at the feature or component level but some may require global analyses employing more extensive models than the simple reduced models discussed above.
    - To allow an assessment of the use of feature models in an implicit–explicit analysis framework. An explicit–explicit analysis process employing sub-structures or superelements is often employed to optimise the finite element data generation process. In this case it is used when a detailed explicit analysis of a feature is required and subject to a loading environment derived from a

global or larger scale analysis in which the feature is crudely modelled.

- Using the information from the results of actioning the above bullet points, assess the level of novelty and complexity and compare with experience base.

Outputs:

- QR updated with finite element reference analysis requirements.
- QR updated with novelty/complexity and experience comparison.

Step 3.2 Decisions

Inputs:

- Revised QR from step 3.1.

The process

Using the information from step 3.1 a number of decisions have to be taken:

- Has the review indicated that Stages 1 and 2 have not been sufficiently comprehensive and that a loop back to one or more of the steps in these two stages is required?
- Based on the novelty/complexity/experience comparison, decide if the analysis is to be continued or not, i.e. make appropriate go/no-go decision.
- If no, then abandon analysis with appropriate explanation.
- If a qualified go decision is made, assess the nature of the qualification level of additional expertise required to overcome, and also identify available source of expertise.
- If the decision is to continue with the analysis process, assess the time, computing power and human resources to allow the process to meet the analysis objectives.

Outputs:

- QR updated with the decisions made and the reasons.

- QR updated with any qualifications and additional requirements that may be necessary to continue with the analysis.
- It may be necessary to modify the AVP to keep it in line with the decisions made.

> **Actions**
>
> - If 'no-go' decision, an adequate explanation is required.
> - If 'go' decision, this must be accompanied by a solid estimate of the required time and resources required.
> - If 'go' decision but additional expertise required due to lack of experience having been highlighted, appropriate 'back-up' must be identified and assurances obtained that this will be available as required.
>
> NB: the decisions relating to continuing with the analysis will, for most commercial analyses, require the approval of those responsible for the allocation of resources as well as the design team.

An outline of this stage with abbreviated descriptions is shown in Figure 9.4.

Stage 4 Global assessment

The aim of this stage is to undertake a finite element analysis covering the complete structural analysis problem using a reduced finite element reference model. The targets for the analysis are to generate appropriate bounds for global-level uncertainties flagged as unbounded by earlier stages and to provide basic information on the behaviour of the structure as a steer to the analyst(s) in generating more comprehensive finite element reference models.

Step 4.1 Initial calculations

Inputs:

- QR document with main focus on inputs from steps 1.2, 1.3, 2.1, 2.2 and 3.1.
- AVP document.

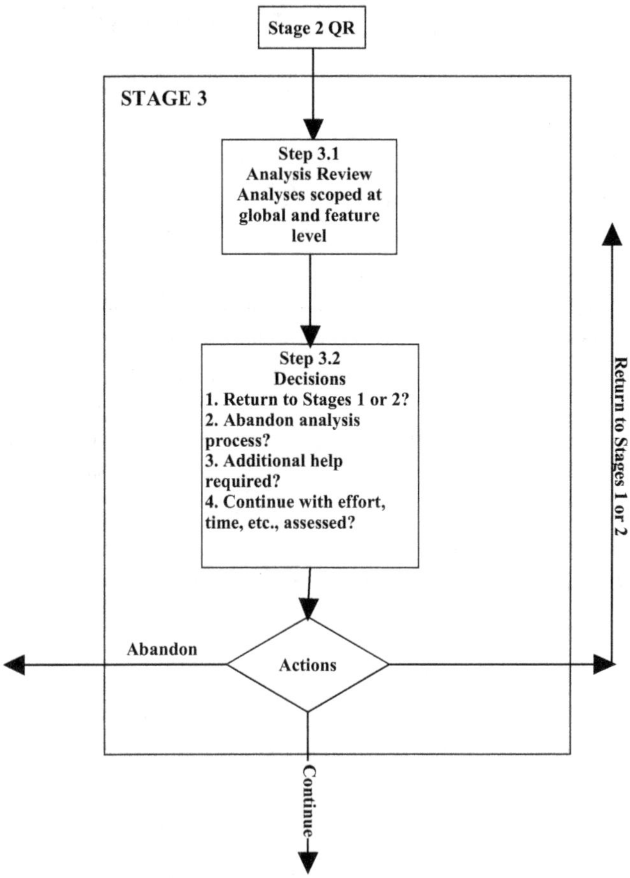

Figure 9.4 Flow diagram of Stage 3.

The process

This step attempts to generate an initial view of the likely behaviour of the structure. This is done to provide scoping information so that the global reduced reference finite element model used in step 4.2 is adequate for the task in hand. It also allows the analyst to obtain a broad understanding of the probable behaviour of the structure which can then be used as a comparator for the results. It supports the axiom that 'an analyst should never embark on a finite element analysis without knowing the answer'. Quantify structural responses to loading environment using:

- information from step 1.3 contained in QR and associated documents;
- simple calculations based on structural theory and simple analysis programs.

Outputs:

- Estimates of the basic characteristic behaviour of structure under the loading actions in terms of specific performance characteristics such as stress levels, displacement, frequencies, etc., incorporated in the QR or separate document linked to the QR. (It is accepted that these estimates will not give a comprehensive picture of structural behaviour and may only provide limited insight to the structural performance for complex analysis problems. Nevertheless, this step is an essential part of the analysis process even if the FEMEC process itself is not being undertaken.)

Step 4.2 Initial global assessments

Inputs:

- QR document focusing on information from steps 1.2, 1.3, 2.1, 2.2, 3.1 and 4.1.
- AVP.

The process

The aim of this step is to set up the reduced finite element analysis reference model for the structure lying within the DoA. The step also requires that the model is run to provide basic data on the behaviour of the structure and to generate bounds on uncertainties where possible. Detailed attention to bounding procedures is undertaken in the next step. The main tasks in this step are:

- Based on the input information, create an appropriate global reduced finite element reference model; the creation of this model is informed by the techniques and knowledge codified in Chapter 6. Apply loads and generate solutions.
- Review results against the outputs from step 4.1. *Note that if there is apparent conflict between the results obtained here and the output from step 4.1, it may be necessary to repeat the analysis with a refined finite element reference model.*

- The results from the finite element analysis are now drawn together to provide the basic structural behavioural information to support the process of building up structural performance knowledge.
- The results are also available for use in providing either firm bounds on the impact of uncertainties or estimates of such bounds.

Some aspects of these tasks are discussed in the simple illustrative example below:

- The results used to reassess the load paths and loading actions; this may require modifying the decisions on loading actions made earlier in the process.
- Reassess the features; modify and augment as appropriate.

Outputs:

- QR document updated with the results and decisions made in this step. Results may be held in CAD files that can be linked with the QR document, but even if such a link is available the QR should contain a synopsis of critical information.[1]

Illustrative example

In order to illustrate some of the points raised in step 4.2, take the case of the connecting rod problem illustrated in Figure 9.5 that is to be used in an automobile engine. The analysis of such a component requires a complex analysis as the structure is subject to a range of dynamic loads through the big and little ends, thermal loads, sliding friction at the two rotating surfaces, contact forces at the big-end split-joints, bolt forces and a number of stress raisers in the form of filets, etc. A possible lower limit finite element layout is shown, in section only, in Figure 9.6. It is assumed that a single brick element is

[1] Note that the information obtained from the finite element analyses that support the design of the structure rather than generating bounds on uncertainties is treated separately from the uncertainty-related information. Nevertheless the results focused on the design itself may have major consequences for the analyst. For example, in the illustrative example, the finite element results obtained together with Roark's formulae could indicate that the filet radii have to be greatly increased with consequential knock-on effects. This could lead to a major redesign and a requirement for new analyses on a modified structure. Clearly this represents a significant component in the analysis process but we do not include this type of analysis modification in the FEMEC process. FEMEC, while noting the need to take account of results from the designer's viewpoint, focuses on error and uncertainty bounding and control.

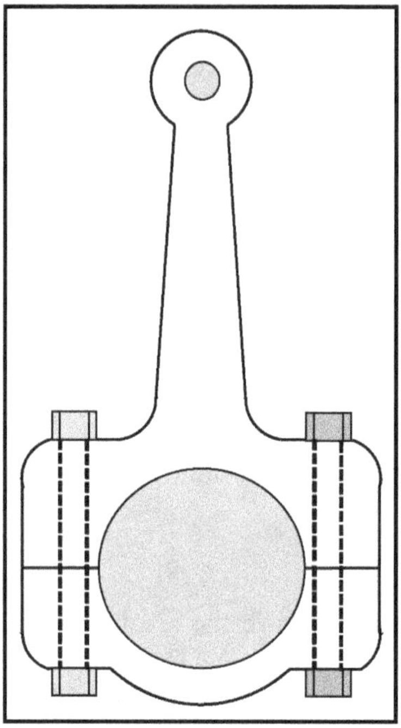

Figure 9.5 Illustrative connecting rod.

sufficient for the dimension normal to the page. The elements are taken to be either 20-noded hexahedral or 16-noded tetrahedral elements as illustrated at several points in Figure 9.6.

There has to be a reason why this particular lower limit element layout has been selected. This is shown in Figure 9.7 where it is seen that a set of representative loads is being used to evaluate the main dimensions of the design. The thickness and taper of the connecting rod itself, the diameter of the little end and the other main dimensions can be found from using this simple model. To achieve this, the loads have also been simplified so that the forces at the rotating surfaces are represented by a periodic function line distribution as shown in black in Figure 9.7 and the bolt forces are each represented by two equal and opposite forces depicted in grey. Other forces, such as thermal and inertia loads, would also need to be included but again be represented in an initial and simplified form.

Many of the main design driver features have been omitted, such as the filets, and there is no need in this model to represent the contact surfaces. Provided this model is able to generate the main loads and

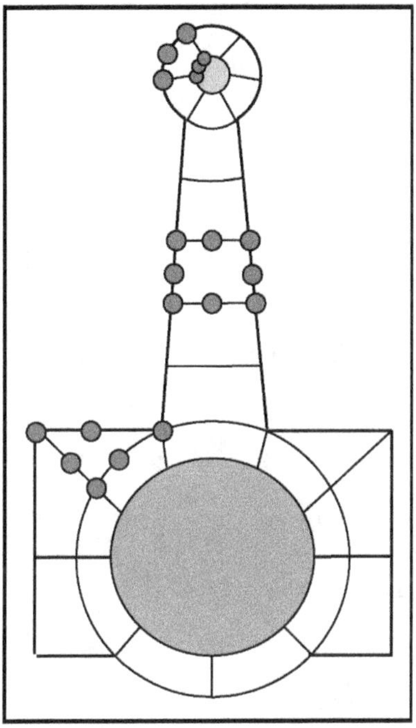

Figure 9.6 Possible lower limit element layout.

load paths within the structure, the ambient stress levels will be predicted with adequate accuracy to permit an initial sizing of the component and initial values of the maximum stresses from the main stress raisers such as the filets. This would be done by using the stress output from the elements in conjunction with the type of formulae found in such engineering books as *Roark's Formulas for Stress and Strain.*

Step 4.3 Uncertainty assessments and bounding

Inputs:

- Uncertainty flags listed in the current QR.
- Available global reduced finite element model.

The process

This step employs the methods and techniques introduced in Chapter 8 for bounding the impact of uncertainty sources on the qualification

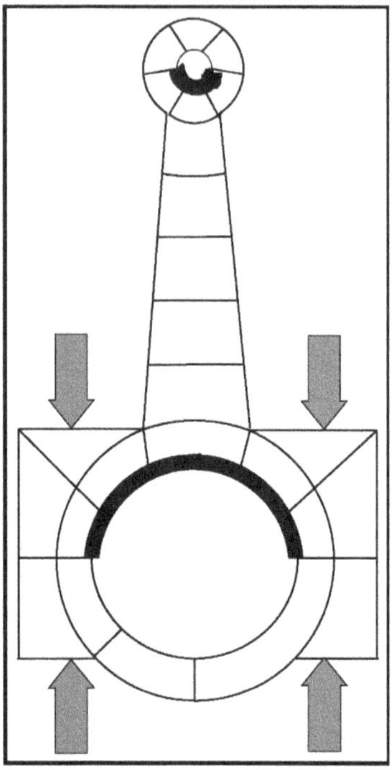

Figure 9.7 Simplified loads layout.

parameters. It uses the reduced global finite element reference model and the main targets are uncertainty sources connected with the DoA boundary conditions. However, it is likely that some of the uncertainties associated with specific features in the structure can be treated through the application of the techniques in Chapter 8 at the global level. The main task in this step is to perform analyses, using the reduced finite element reference model, in order to provide appropriate bounds in the following areas:

- The DoA:
  - The flagged uncertainty sources relating to the boundary conditions are handled and their impact on the qualification parameters bounded using the sensitivity methods from Chapter 8.
  - Any remaining uncertainty with respect to the extent of the DoA is considered and bounded using extended analyses. (Note that this implies the reduced finite element model is appropriately extended.)

- Feature-level uncertainties.

Although the primary focus for this assessment step is focused on global issues normally related to the DoA, it is possible that certain of the uncertainty sources related to specific features can also be handled at this stage using the reduced finite element model. Thus:

  - Some of the uncertainty with respect to the level of abstraction appropriate to a major feature may be resolvable at this level and may be achieved by applying the techniques discussed in Chapter 8, Section 8.2. (Note that this may require additional elements and additional solution runs in order to explore the level of abstraction possibilities but should not involve any changes to the reduced model layout.)
  - The impact of some of the other feature-level uncertainty sources flagged as having an unmeasured and unbounded impact on the qualification parameters may now be subjected to sensitivity analyses as described in Chapter 8, Section 8.2. (Note that this will certainly require additional analyses with new unit loads applied or analyses with modified element matrix properties but should not involve any changes to the reduced model layout.)

Outputs:

- QR updated with DoA uncertainty bounds.
- QR updated with uncertainty bounds on features handled at the global level.
- QR updated with information indicating that the DoA must be changed or that feature information critically affects the definition or number of features defined at the global level.
- QR updated with flags for those sources of uncertainty considered at this global level that have not been bounded – at this stage the flagged uncertainties should relate to the properties of features within the analysis.

Step 4.4 Review

Inputs:

- QR document.
- AVP document.

The process

The results from steps 4.2 and 4.3 are reviewed against the following criteria:

1. Have the steps shown that the initial decisions used to define the DoA require major modification?
2. Have the steps introduced new features or uncertainty sources?

If either of these bullet points is positive, then the analyst must consider how these new sets of problems are to be accommodated. Although it would not be good practice to loop back at this point, it may be necessary to return to earlier steps if their impact is such that further progress cannot be made without some rectification being undertaken. Normally the issues raised can be simply flagged and considered at a later stage in the process.

Outputs:

- Decisions made on the need to return to earlier stages in the process and activated.
- QR document updated with decisions and values for the bounds on the impact that the uncertainties considered at this stage have on the qualification parameters.
- Consequential changes to the AVP noted and included in the document.

Stage 5 Detailed assessments of uncertainty sources at the feature level

The aim of this stage is to undertake a series of finite element reference analyses that allow the impact on the qualification parameter of the uncertainty sources associated with each feature to be assessed and bounded. This is the most complex part of the FEMEC process and the tasks required to be undertaken depend on the depth of the analysis (see Chapter 5, Section 5.4.2), the type of analysis (static, dynamic, etc.) and the complexity of the loading actions. As a result, the assessment of an uncertainty may have to be undertaken using a major sub-structure reference model if the behaviour of a feature is strongly influenced by surrounding features or structural components or using reference models involving single features in isolation.

Before proceeding with a description of the steps and tasks within this stage there are two important caveats:

1. What follows is not intended to be a set of concrete rules that must be employed without any deviation, but guidelines to assist the analyst in approaching this part of the analysis process. For complex analyses the process could be very dynamic were the results from one set of analyses to radically change the analyst's understanding of what is required to achieve the required level of bounding on the sources of uncertainty.
2. At the other extreme, in the case of relatively simple or well-understood analyses it may be appropriate to merge Stages 4 and 5.

Step 5.1 Feature analysis set-up

Inputs:

- QR document with focus on the inputs from steps 1.2, 1.3, 2.1, 2.3, but with particular attention paid to the output from Stage 3 and any modifications to the conclusion reached in that stage by the output from Stage 4.
- Current AVP document.

The process

The QR information informs the analyst about outstanding uncertainty sources currently flagged and their associated features. Stage 3 has laid out the foundations for a set of analyses that are required but these may have been either increased in number or scope or decreased depending on the information from Stage 4. The net result of this combination of actions is that the type of analysis required for each type of uncertainty has been defined. In all but the simplest analysis problems, this will involve defining a hierarchical set of reference models, following the process described in Chapter 8, Section 8.2.2.2, to bound uncertainties relating to the level of abstraction. An illustrative example is displayed in Chapter 7 and repeated here for convenience as Figure 9.8. Additional sensitivity studies will be required to bound specific uncertainties, on (say) material properties, using the techniques described in Chapter 8, Section 8.2.2.3.

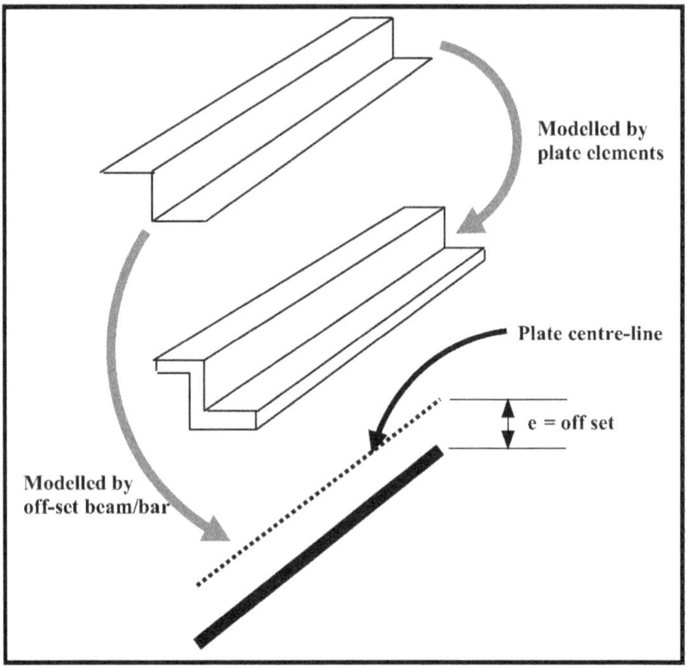

Figure 9.8 Abstraction levels for panel problem from Chapter 7.

It is worth noting that global assessment may well have left certain uncertainty sources only partially dealt with. For example, take the case shown in the illustrative study displayed in the discussion of the global assessment, Figure 9.5. In this case an initial assessment of the stress raising caused by the presence of one of the filets was discussed where it was indicated that an element layout, similar to that shown in Figure 9.6, could be used in conjunction with *Roark's Formulas for Stress and Strain* to obtain a value for the Stress Intensity Factor associated with a filet of specified radius. However, the use of a reduced reference model might not provide the level of accuracy required by the design team and the impact of the filet radius would have to be left as a flagged uncertainty source. This deficiency would be rectified in this stage by employing an explicit–implicit approach. The broad internal load paths are established by the reduced global model and then applied to an explicit model with a full description of the filet. A possible explicit reference model is shown in Figure 9.9 where 20-node hexahedral elements are employed which offer the advantages described in Chapter 6, Section 6.3.2, and some element mesh grading has been used following the advice given in Chapter 6, Section 6.2.3. The load

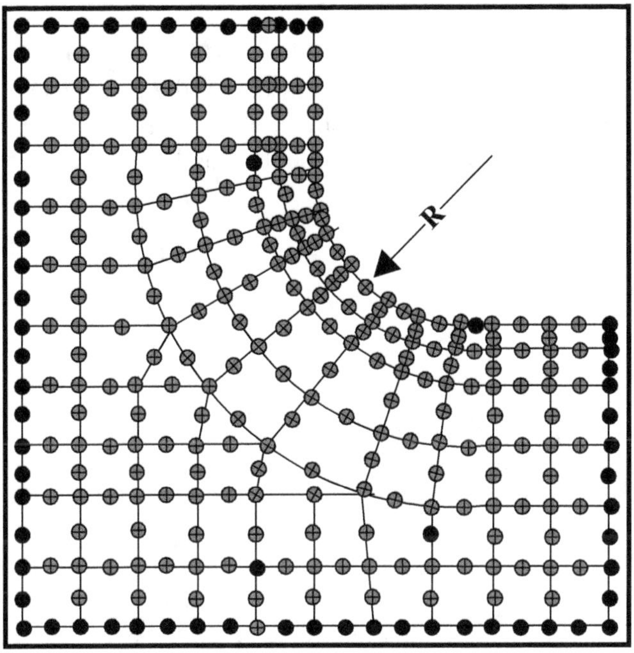

Figure 9.9 Detailed model for filet radius.

paths created from the results of the reduced model can now be applied to the exterior nodes of the explicit model and a more accurate stress profile generated. The impact of uncertainties in the value of the filet radius R on the Stress Intensity Factor can be assessed and bounded using the artificial sensitivity methods of Chapter 8, Section 8.2.2.3.

The tasks in this step amount to the following:

- Fully define the models, loading configurations and boundary conditions for models that are to be employed to assess the level of abstraction.
- Fully define the reference models, loading configurations and boundary conditions for the required sensitivity studies:
  - For direct sensitivities the machinery for generating the modification matrices (Chapter 8) has to be put in place.
  - For indirect sensitivities incrementally modified reference models have to be defined (Chapter 8).

Outputs:

- CAD and finite element input files for all the cases defined above.
- QR updated either with a direct link to the CAD models or with a description of the models, load cases and boundary conditions.

Step 5.2 Uncertainty assessments and bounding

Inputs:

- The definition of the finite element models from step 5.1 within the current QR.
- AVP document.

The process

The finite element reference analyses defined in step 5.1 are now run, the levels of abstraction considered and the feature-level uncertainties 'measured' and bounded using the methods described in Chapter 8. The tasks in this step, though complex in nature and often requiring large numbers of different finite element solution runs, can be simply defined as:

- a sequence of pre-defined analyses to establish the appropriate levels of abstraction for the feature reference models with bounds on any uncertainties;
- a sequence of pre-defined reference analyses to create modification matrices and using the output to compute the associated bounds on feature-based sources of uncertainty;
- a sequence of sensitivity analyses using the artificial sensitivity approach.

Outputs:

- QR updated with feature-level uncertainty bounds.
- QR updated with flags for those sources of uncertainty considered at this level that have not been bounded.

A flow diagram that combines Stages 4 and 5 and also shows the link to Stage 6 is shown in Figure 9.10.

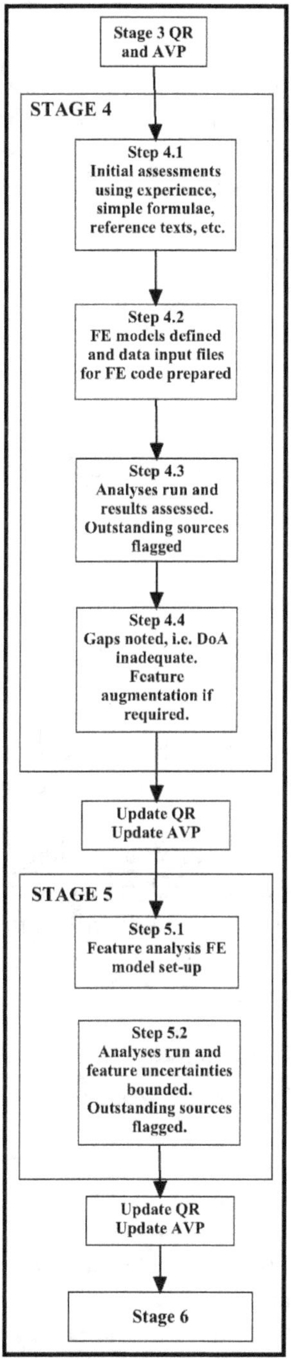

Figure 9.10 Flow diagram for Stages 4 and 5.

Stage 6 Overall assessment

This is one of the most important stages as it attempts to draw together all the results on uncertainty sources into a coherent picture and make decisions on what to do next. This could involve further analyses, defining tests or imposing maintenance standards.

Step 6.1 Assessments

Inputs:

- QR information on all current uncertainty bounds and all uncertainty sources flagged as currently unbounded.
- AVP document.

The process

This stage reviews the results from the previous stages and makes decisions on future stages. These could involve combining bounds on uncertainty sources, looping back to earlier stages, defining tests to supply information from which bounds can be derived or imposing maintenance requirements. These are covered by the following tasks:

- All the bounds on the impact of uncertainty sources on the qualification criteria are reviewed and where appropriate combined into single bounds using the method described in Chapter 8, Section 8.3.
- A critical evaluation of the flagged unbounded uncertainties and the following concluded:
    - That the global-level analysis be repeated with a more refined finite element reference model so that more accurate information can be generated.
    - That, for certain uncertainty sources, further analyses would not generate accurate bounding information and that experimental tests are required. This situation could arise if the analyst discovers, from sensitivity or other analyses, that variations in the qualification parameters are strongly dependent on the elastic properties at the boundary of the DoA and that these properties are not sufficiently well known or understood. Tests would now be called for, by the analyst, and defined in such a way that

specific information was supplied that allowed the required bounds to be generated. Note that the test programme is now the servant of the analysis and has to supply clearly defined and specified data.

- That certain bounds are very tight and that deviations in properties have to be controlled with maintenance implications. In order to illustrate the point consider the plate reinforced by Z-stringers shown originally in Figure 7.5. In this problem two possible methods are advanced for attaching the stringers to the main plate displayed in Figure 7.14 and again in Figure 9.11 for convenience. Taking the case where the design team has selected a continuous attachment resulting in a bonded joint, then the possibility of de-bonding must be examined by the analyst. If the analytic bounding methods show that the qualification parameters can move outside their accepted accuracy bounds for small amounts of de-bonding, then maintenance becomes an issue. The analyst has now to make clear that the analysis has been performed on the assumption that strictly limited amounts of de-bonding are present, so if inspection reveals that this limit has been reached, the structure must be withdrawn from service and the joint remade.

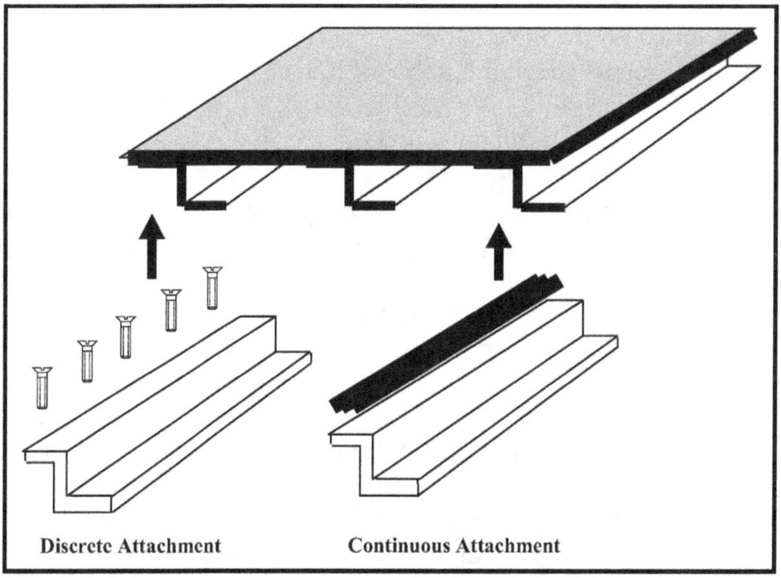

Figure 9.11 Attachment options.

Outputs:

- QR updated with bounds on the impact that uncertainty sources have on the qualification parameters.
- Remaining flags for those uncertainty sources that cannot currently be bounded.
- Maintenance requirements to limit variations in structural behaviour related to qualification parameters.
- Test requirements to supply information from which bounds can be generated for currently flagged uncertainties.

Step 6.2 Decisions and actions

Inputs:

- Updated QR document.
- AVP document.

The process

Essentially this step concludes the first pass through the FEMEC process and the next stages depend on the information that has been amassed up to this point. With the information in the current edition of the QR, the process now goes through a number of branch points:

1. Have all the uncertainty sources associated with the idealisation process been captured by the analytic approaches of Chapter 8 or have appropriate tests been defined that will supply the required bounds?
2. If the answer to 1 is yes, the next question to be asked is 'Have all the analyses used in the bounding and assessment process together with the test results provided sufficient information so that the designer can be given values for the qualification parameters with values for the bounds?' If the answer to this question is also yes, then the analysis is concluded. If not, then the analyst now moves to Stage 9.
3. If the answer to 1 is no, this implies that stages are repeated with more sophisticated models; normally the global model selected in order to undertake a repeat of Stage 4 could be the final finite element reference model discussed in Stage 9.

Outputs:

The outputs from this step are simply the decisions made in response to the three bullet points above and the consequential actions. These must also be recorded in an updated version of the QR.

Stage 7 Intermediate-level global assessments

This stage is essentially a repeat of Stage 5 so we will not repeat the details but simply outline the key features. It is brought into play if there remain flagged uncertainties connected to the global environment of the analysis or feature-level sources that can only be handled at the global level.

The need for a further review of uncertainties at the global level implies that the initial reference model is not able to capture all of the relevant uncertainty sources. This initial model needs to be replaced by one that the analyst believes is sufficiently detailed to undertake the required tasks but, while being more detailed than the initial model, carries less detail than the final analysis reference model. It may be considered as an 'intermediate-level reference model' with a mixture of model fidelities – high where the previous model was unable to model satisfactorily the uncertainty sources and low where the information already captured is adequate. Of course, this intermediate reference model must be constructed using the guidelines of Chapter 6 and Chapter 7.

The outputs from the stage are a new set of bounds on uncertainties and, possibly, fresh information on uncertainty sources associated with detailed features. It may also require further tests and further maintenance requirements. The QR and AVP documents are to be updated as at Stage 4.

Note: In practice it is likely that the need for an intermediate model will have become clear during Stage 4. As a result, this stage will have become a component of stage 4 making it redundant at this point in the process.

Stage 8 Additional feature-level assessments

This stage is a repeat of the operations detailed at Stage 5 and is called into play if Stage 6 indicates that feature-level uncertainty sources remain flagged due to the inadequacy of the original reference models at Stage 5 or because Stage 7 has indicated the presence of additional uncertainty sources. It may require more complex feature-level reference models than those deployed in Stage 5. Again, these models should be drawn up using the guidelines contained in Chapter 6 and Chapter 7.

The outputs from the stage are a new set of bounds on uncertainties on feature-based uncertainty sources. It may also require further tests and further maintenance requirements. The QR and AVP documents are to be updated as at Stage 5.

Note: In practice it is likely that the need for additional feature-level assessments will have become clear during the course of Stage 5. As a result, this stage will have become a component of Stage 5 making it redundant at this point in the process.

Stage 9 Final assessment and supporting test specification

Inputs:

- Current QR.
- Current AVP.

The process

At this point all of the analyses designed to compute bounds on the impact that uncertainty sources have on the qualification parameters have been completed. Nevertheless, for a complex analysis, there still remain a number of sources that are flagged as unbounded. It may be clear that some of these can be resolved analytically when the final analysis reference model has been built and run. The residual set of flagged sources is not amenable to computational methods and will have to be handled by means of a test programme. A description of the reasons why tests may be required, together with a discussion on the type of test, are given in Chapter 8, Section 8.4.1. Although tests are called for by an analyst, it is not the analyst's responsibility to define the test programme and equipment but it is important that the analyst and test engineer work together in drawing up the test requirement document – communicating by paper requests only opens the door to ambiguity and confusion! In essence, the analyst should be stating the outputs required from the test programme, namely:

- the quantified behaviour or performance of a specific component or parts of components;

- the experimental accuracy achieved during the test programme so that the values returned to the analyst have a known and bounded uncertainty values.

Outputs:

- Test requirements.
- Update QR.

Stage 10 Final reference model and analysis

This is the final analysis stage in the FEMEC process where all the information gathered in the previous stages is drawn together to allow the creation of a final reference model that can be used to predict the behaviour of the in-service structure. This analysis must be able to provide adequate bounds placed on the likely deviation of the in-service values of the qualification parameters from those predicted by the analysis using the final reference model.

Step 10.1 Final analysis plan

Inputs:

- Current QR document (note that this contains all the design and qualification information from Stage 1).
- Test results if required to select finite element models.
- Current AVP document.

The process

- All information on bounds from both analyses conducted using the methodologies in Chapter 8 and the test data now informs the analyst in developing the final finite element reference model that is to be used to create and supply the design information with an overall accuracy bound on the qualification parameters.
- If flags remain, these are reviewed and the final analysis reference model developed to allow these to be handled and appropriate bounds calculated. The results from this activity may require that the final model is adjusted to allow the methods of Chapter 8 to be employed.

- The guidelines of Chapter 6 and Chapter 7 are deployed to allow a high-quality finite element reference model to be developed.
- Special methods are considered and incorporated into the plan if required, e.g. the use of adaption methods, exploitation of symmetry, etc.
- Links are set up with other teams that are developing sub-structures which are to be incorporated into the current analysis or into which the current analysis is to be incorporated.

Outputs:

- CAD and finite element files created during this step are now available for exploitation together with a final analysis plan that describes how the analysis or analyses are to be undertaken. This information is directly linked to the QR through electronic data exchange or provided in a written form that adequately describes how the required analysis or analyses are to be undertaken.

Step 10.2 Final analysis or analyses

The final analysis plan is now executed and the results placed in the QR document using a direct link to the output finite element and CAD files and/or as a written contribution. These results should provide full information on the structural behaviour as required by the designer and be in conformity with qualification or certification rules. The output from this analysis must include bounds on the accuracy of these results.

If tests have been specified in order to quantify the impact of uncertainties on the qualification parameters that could not be found by computational methods, the results from such tests are to be included in the QR. In addition, certain assumptions made in the final reference analysis may have an impact on the operational service. These have to be interpreted in terms of maintenance requirements that ensure the structure does not enter an operational state that violates the analysis assumption.

The QR and AVP documents and all results are to be delivered to the design team and, in addition, placed in the company or team database of analyses successfully undertaken.

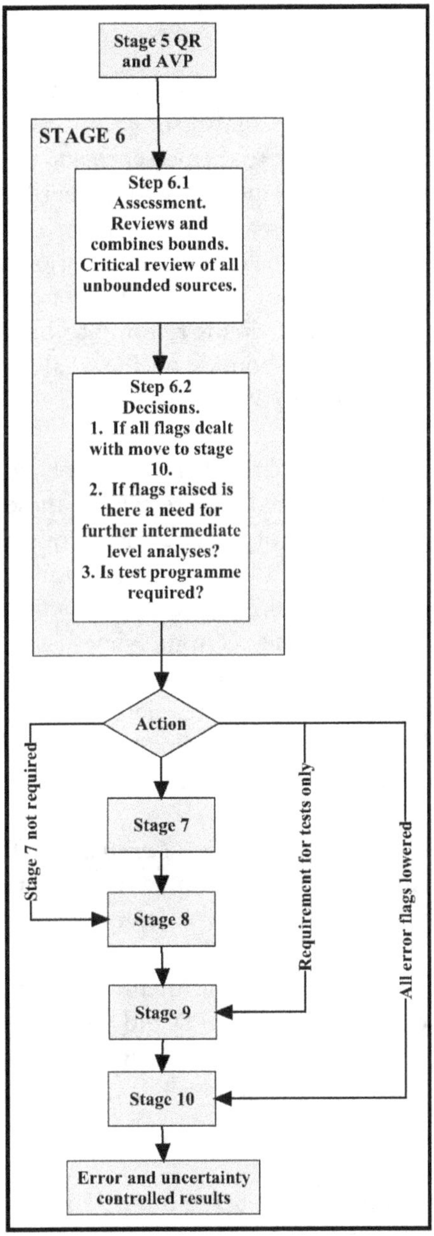

Figure 9.12 Outline of process from Stage 6 onwards.

Step 10.3 Error commentary

Although step 10.2 represents the end of the formal processes that make up the FEMEC procedure, the analyst must provide a synopsis of the conclusion reached with respect to the presence and impact of errors and uncertainties. This step can be merged with step 10.2 to form a single concluding step, but having a separate report on errors and uncertainties has the advantage of focusing the designer's attention on the key aspects resulting from the application of the FEMEC procedure. The commentary needs to cover three major items relating to both the internal and external error sources and the bounds on the qualification parameters. Thus, three sections are required:

1. *External error sources*: All the flags that have been raised are listed and the conclusions reached with respect to these flags commented on, indicating the nature of the flag and its raised or lowered state at the end of step 10.2. If a flag has been left in the raised state, the consequences of this condition require comment.
2. *Internal error sources*: This section reviews issues relating to mesh quality and the level of internal accuracy based on the internal error measures that most finite element systems now supply. Essentially this section is concerned with the points regarding the quality of the finite element analysis that are partially covered in Chapter 6.
3. *Error bounds on the qualification parameters*: This section draws together the analyst's conclusion with respect to bounding the impact on the qualification parameters of all the error sources, both internal and external. The section has two components: it first lists the bounds on all the qualification parameters due to uncertainty sources (from 1 above) and on potential internal errors (from 2 above). Second, it lists the estimates of the limit values for qualification parameters based on the values ascribed to the error measures.

An outline of the process from Stage 6 until the end of the method is displayed in Figure 9.12.

10

FEMEC Walkthrough Example

10.1 INTRODUCTION

This chapter illustrates the application of FEMEC to the solution of relatively small-scale analysis problems. The method is displayed through the steps in the Quality Report (QR) which, of necessity, is presented as the final and complete QR document. In a real application, the QR is built up as the various stages and steps in the FEMEC process are worked through. In addition, the FEMEC process is supported by a range of documents and information that the analyst may feel is important to have as a set of appendices. An example, in the case of a complex and large-scale structural design problem, would be the many drawings or CAD models associated with it that are already in existence. While Stage 1 of the FEMEC process does describe the design problem, it is only done in terms of those aspects of the design that the analyst thinks is relevant to the error controlling process. However, in a complex analysis it is likely that the team or person tasked with reviewing the QR would normally wish to have access to all the design information in order to be assured that the analyst in following the FEMEC path has not omitted important information that is directly related to the identification of error or uncertainty sources. Because the illustrative problems described in this chapter are small in scale, the designer's information is introduced in a 'design requirements' section below. In many commercial situations the complexity of the design often requires that the team or person performing the analysis would be included in the process of developing the design before FEMEC is entered. However, a reasonable modification in this situation would be to expand Stage 1 to include the development of the design requirement.

Because this chapter is a walkthrough of the FEMEC process, it does not provide a complete and comprehensive demonstration of the method and the reader will note that a number of the requirements detailed in Chapter 9 and some of the actions detailed in Chapter 8 are omitted. In addition, because the examples are walkthrough analysis problems and are intended as illustrations of how to apply the method, they do not cover all aspects of the analyses of these small-scale structural analysis problems. Where a particular discussion applies to more than one component in the structure, there is no benefit in repeating an already demonstrated process simply to have completeness.

10.2 FEMEC STATIC ANALYSIS ILLUSTRATIVE PROBLEM

The walkthrough uses the Z-stringer reinforced panel problem first introduced in Chapter 7. The methodology of Chapter 9 is demonstrated using a limited number of aspects of the full analysis problem but this subset does cover all of the error and uncertainty sources that this problem can encounter.

10.2.1 The Design Requirement

The design problem consists of creating a panel structure that is attached to a wall by an L-shaped bracket and is able to carry a uniformly distributed load of 30 Pa such that limits on the stresses within the entire structure and the displacements experienced by it do not exceed certain prescribed limits. There is a requirement that the structural layout be such that a relatively efficient form is employed. To this end the designer has elected to use a Z-string reinforced panel and the selected layout is shown in Figure 10.1 which repeats Figure 7.8. As can be seen, the dimensions of the structural components have already been selected and these cannot be changed without returning to the designer. The mean radius of the bracket corner, not shown in Figure 10.1, is 7.5 mm. The material used in the construction of the panel, the stringers and the bracket is aluminium alloy and that for the bolts and rivets is steel. There is little information on the supporting wall but it is known to be

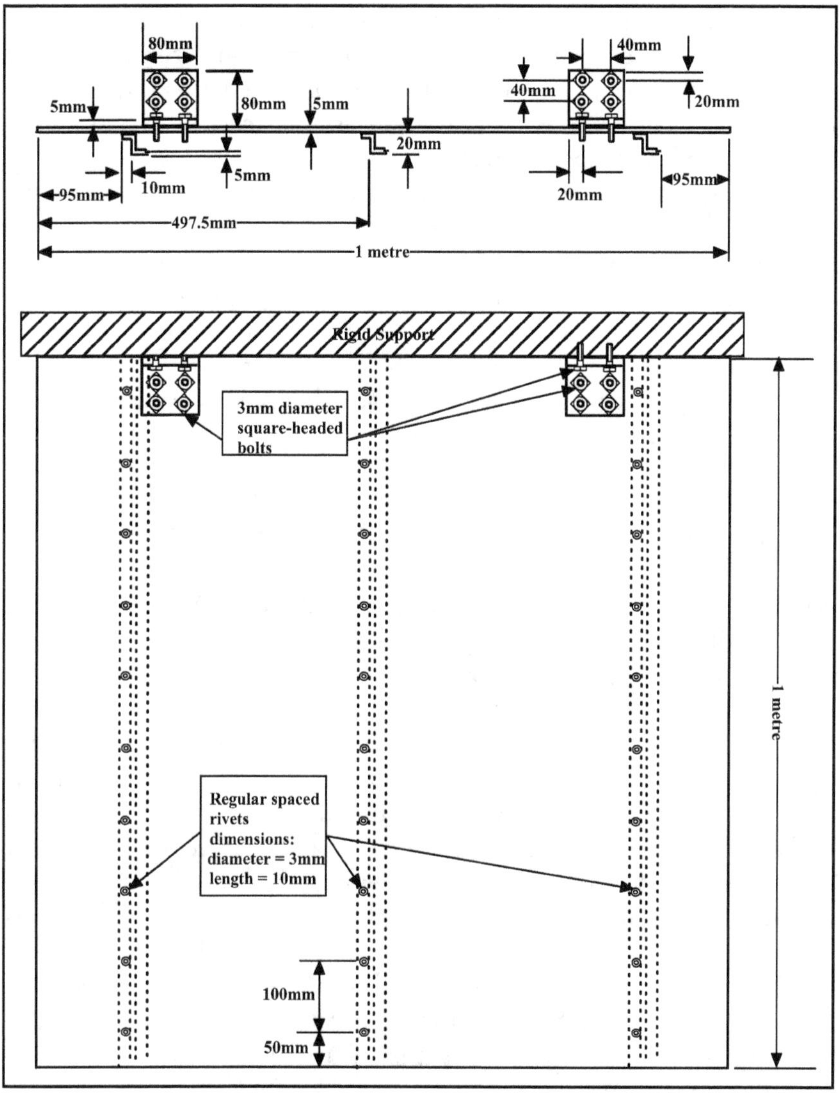

Figure 10.1 Reinforced panel layout.

very stiff so, throughout the discussion, the wall is considered as rigid and makes no contact with the panel or stringers, i.e. the Z-stringer reinforced panel is suspended below the attachment bracket. However, this assumption of rigidity will need to be taken into account at some point in the application of the FEMEC process to this design problem.

The design must be able to sustain the load without the maximum Von Mises stress exceeding 50 MPa in the aluminium components or 400 MPa in the steel components. In addition, the maximum displacement should not exceed 4 mm anywhere in the structure. These represent the qualification parameters.

In addition, there is a requirement that the accuracy of the bounds on these qualification parameters should be less than 5%; that is, they must be shown to lie between ±5% of the values that will be found when the structure is manufactured, placed in position and the loads applied. The 5% value is selected as this represents the normally acceptable level of experimental error and it would not be appropriate to expect that any final in-service measurements of stress or displacement would be more accurate than standard experimental error.

10.2.2 Application of FEMEC via the Quality Report

The sequel represents a part of the Quality Report (QR) for the finite element solution of the design problem described in Section 10.2.1. In the description of the FEMEC process given in Chapter 9, Section 9.4.2, each step has an input and an output task. The 'input' task describes what has to be available before a step can be initiated. In most cases the input information relates to information that must be taken from a QR that is in a state of being built up. Because the walkthrough is presented in the form of a completed QR, there is no need to highlight that information must be drawn into consideration at each step. The 'input' task is, therefore, not required and is only included when information is needed that is not included in the QR. Similarly, in the description of FEMEC in Chapter 9, the 'output' task normally refers to the necessity of updating the QR with information generated in a specific step and is also omitted below unless there is a particular need for its inclusion in the walkthrough.

The first task in the process is to establish a FEMEC implementation plan for the analysis being undertaken. Because the analysis is a demonstration of the FEMEC process it is implemented using the stages and steps displayed in Chapter 9, Section 9.4.2, without modification. If a modified process were required in order to fit in with company policies or because of the level of complexity of an analysis, then a modified form of the process given in Chapter 9 would be required. This modified process would then constitute a new implementation plan that would have to be written down and presented for review as a replacement for FEMEC.

Having elected to accept the FEMEC process in its standard form, this now represents the implementation plan for the present analysis. With this implementation plan ready to hand, the FEMEC process can be started and displayed in the form of a completed QR.

Quality Report Set-up Information

- QR is compiled by Professor Alan Morris.
- QR is reviewed by Dr Ahmed Rahman.
- Qualifications and experience of Alan Morris:
    - BSc, MSc, PhD (Cantab), FREng, FRAeS, CEng
    - Experience in the design of large-scale steam turbines; application of the Finite Element Method to aeronautical structures; research into the development of the Finite Element Method.
- Qualifications and experience of Ahmed Rahman:
    - BSc, PhD, CEng
    - Experience in the analysis of a range of products employing the Finite Element Method; employed by the Institute of Sound and Vibration at Southampton, Assessment Services Ltd, Siemens, Qinetiq.

Quality Report Stage Reports

Stage 1 Scoping the real-world problem

The aim of this stage is to link the real-world design problem with the finite element analysis; because the complexity is limited there is no requirement to involve design or test engineers at this stage.

Step 1.1 Analysis problem overview

Note is taken of the design requirement detailed in Section 10.2.1 and the qualification requirement contained therein.

This step quantifies the overall analysis requirements as they relate to the design requirements.

Design requirements

The design objective is to construct a Z-stringer reinforced panel structure shown in Figure 10.1. Note is taken of the fact that the support structure to which the panel is attached through two L-shaped brackets is assumed to be rigid and this may require examination as the QR is developed. There are no special design objectives but note is taken that the stringer reinforcement improves the efficiency of the structure.

Design constraints

The structure is not being designed in conformity with an externally imposed set of qualification or certification requirements. The qualification requirements imposed are a limitation on the maximum stress anywhere in the structure and on the maximum displacement. These equal a Class 2 qualification type as described in Chapter 5, Section 5.6.1.

The top surface of the panel is subjected to a static, uniformly distributed, downward pressure load of 30 Pa.

Design-imposed analysis limitations

The structure is subjected to qualification requirements that necessitate the calculation of stresses and displacements at all points in the domain of analysis. Although not specified in the design requirement in Section 10.2.1, the bolts and rivets shown in Figure 10.1 must not fail when the pressure load is applied. Calculations quantifying the bolt and rivet stresses must show that these components are not in danger of failing.

The overall analysis accuracy requirement is that the values of stresses and displacement measured in the operational structure should not deviate from those predicted in the results of the final analysis to within a defined error bound.

The fact that the support can be taken as rigid defines the domain of analysis as the panel, the Z-stringers, the brackets and all the bolts and rivets that make up the attachments and joints.

Initial uncertainty review

The domain of analysis is artificial and represents a potential uncertainty source which may require consideration later in the FEMEC process.

Although not formally an uncertainty source within the qualification criteria, the stress levels in the bolts holding the bracket to the panel and to the rigid wall together with the rivets attaching the panel to the Z-stringers require assessing for potential failure.

Step 1.2 Initial analysis assessment

This step draws on the basic design information to form an initial view of the type of analysis that will be required; however, no attempt is made to form an opinion on the details of the actual finite element analyses that will eventually be implemented.

Basic structural specification

The structure is built up from a panel, three Z-stringers that play the role of reinforcing beams and two L-shaped brackets. These are assembled to form the structure shown in Figure 10.1 through the use of bolts for the bracket attachments and rivets for the panel–stringer attachment.

It is assumed that the structural components are drawn from reputable suppliers and to be of such high quality that the given dimensions in Figure 10.1 do not contain any meaningful uncertainty associated with them and the assembly process does not cause any distortion that would introduce an assembly error source.

The material used for the panel, Z-stringers and bracket is an aluminium alloy with a Young's modulus of 60 GPa and yield stress of 50 MPa. The material for the bolts and rivets is steel with a Young's modulus of 200 GPa and yield stress 400 MPa.

Analysis type

At this stage a linear static analysis is required.

Uncertainty review

The following uncertainty sources are considered as applying to the problem associated with the linear static analysis of the reinforced panel structure:

- the artificial nature of the domain of analysis;
- level of abstraction for the panel, stringers and bracket;
- joint behaviour for the bracket attachments and the panel Z-stringer attachments;

- failure potential of bolts and rivets.

N.B. As explained above, because this is a walkthrough illustration of the method no further reference is made with respect to uncertainties or errors associated with the Z-stringers and rivets. The procedure required to handle the stringers and rivets can be derived from the consideration given to the bracket/panel behaviour that is fully covered below.

Step 1.3 Review of available information

This step draws together and reviews all the relevant information that might assist in forming an opinion on the value of any background information. It is noted that, for this analysis, there is no relevant company documentation relating to earlier analyses.

***Compilation of available information*:**

- No previous QRs are available for this analysis.
- No data sheets are to be employed.
- There is no existing test data for this analysis.

A number of reference sources are available for use in examining certain aspects of the structure including:

- *Roark's Formulas for Stress and Strain* (reference [1])
- F.R. Shanley's *Mechanics of Materials* (reference [2]).

A rule of thumb is available for defining structural behaviour for deciding if a panel falls within the category of thin plate theory. This states that a flat 'plate' structure behaves as a thin plate conforms to the classical Kirchhoff thin plate theory if the ratio of the plate thickness t to a typical edge length l falls within the range $8 \leq l/t \leq 80$.

Exploitation of available information

There is no codified experience database; however, the experience of the analysts, as defined in the QR set-up, will be employed.

Apply the 'flat plate' definition to the panel and the L-shaped bracket

In the case of the panel, the thickness is 5 mm and edge length is 1 metre, hence $l/t = 200$; this is above the normal upper limit for thin plate theory and if major in-plane loads had been applied, this structure would be categorised as a membrane. However, in the absence of in-plane loads and the presence of the Z-stringers, the level of abstraction is categorised as a thin plate with no uncertainty associated with this decision at this point.

In the case of the bracket the thickness is 5 mm and the length of each of the branches of the 'L' is 80 mm, giving an l/t ratio of 16 implying thin plate theory behaviour. However, if the presence of the bolts is taken into account and the length between bolts is employed, the ratio reduces to 8. It is, therefore, not possible to categorise, with certainty, the behaviour of the bracket with respect to its level of abstraction.

Step 1.4 Uncertainty list

This step draws together all the compiled information on potential sources of uncertainties and starts the process of either bounding them or identifying their existence for future consideration.

Flagged list of uncertainties identified in Stage 1:

- Flag 1 The use of a rigid wall as the support and outer limit of the domain of analysis.
- Flag 2 Level of abstraction for bracket (note: the level of abstraction for the main panel has been resolved in step 1.3).
- Flag 3 Joint behaviour for the bracket and attachments.
- Flag 4 Potential failure of the bolts used as bracket attachments.

Bounding of identified uncertainties:

At this stage the information available does not permit any action being taken with respect to flags 1–3. Flag 4 can be addressed using simple upper bounding analysis methods:

Apply the simple bolt calculation method from Shanley's Mechanics of Materials *as the reference model to check for the possibility of bolt failure at the bracket attachment points.*

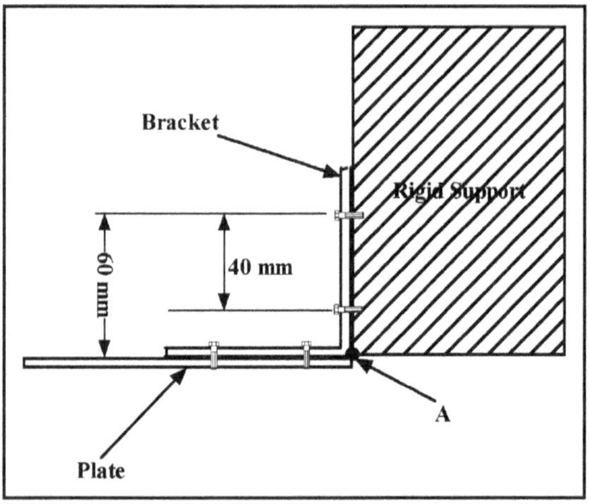

Figure 10.2 Bracket attachments.

In the case of the rigid-wall/bracket attachment, the lower line of bolts is neglected in calculating the tension forces in the bolts. Taking moments about point A in Figure 10.2 and noting that there are four bolts in the top line yields:

$$m_A = 60 \cdot 4 \cdot \pi \frac{d^2}{4} \cdot \sigma \tag{10.1}$$

where d is the diameter of the bolt, i.e. 3 mm, and σ is the bolt stress. The term m_A is the moment arm of the applied load of 30 Pa operating on the panel top surface and with this information equation 10.1 yields the reference value of the direct stress in the bolts, forming the qualification parameters, which can then be compared with the yield stress value give in step 1.2. This process indicates that there is a very large safety factor for the bolts that is hardly affected by the addition of the shear stress through which the set of eight bolts carries the vertical force of the applied load. *The gap between the calculated bolt stress values and the limit values required by the qualification criteria are so large that an uncertainty bounding process is not required.*

The same calculation process can now be applied to the bolts attaching the bracket to the main panel. Taking moments once again about point A, equation 10.1 provides the means for calculating the tension in the outer bolts (neglecting as before the inner bolts) due to the bending action of the applied load. In this case an additional load must be added that accounts for the fact that the direct action of the pressure load adds to the bolt direct tension. As in the case of the bolts attaching the bracket to the rigid wall, the safety factor for bolts made of a material with the properties given in step 1.2 is sufficiently large that bolt failure can be ignored. *Once again the gap between the calculated values and the limit values required by the qualification criteria is so large that an uncertainty bounding process is not required.*

On the basis of this information error flag 4 is lowered but error flags 1, 2 and 3 remain raised.

Step 1.5 Novelty, complexity and experience assessment

This step makes critical decisions on the future progress of the analysis based on the assessment of novelty and complexity.

- Novelty: Assessed as low.
- Complexity: Assessed as low.
- Experience base: The analyst has encountered a significant number of analysis problems of this type.
- Experience gap: There is no significant gap between the level of experience and the novelty and complexity of the analysis. In consequence, it is concluded that the gap is manageable and that no additional expert help is required.

Recommendation: No impediments exist that bar progress to Stage 2.

Step 1.6 Review

This is the point where decisions are made relating to the progress of the analysis taking into account all the information accumulated in this stage.

Go/no-go decision:

The recommendation from step 1.5 is accepted and the process can continue without any modification to the FEMEC process as defined in Chapter 9.

At this point in the FEMEC process there is a pause in order to construct the first issue of the Analysis Validation Plan (AVP). Because we are dealing with a relatively simple problem the AVP is also relatively simple but its creation does focus attention on the fact that the analysis has to be validated and that the real-world design does not yet exist to provide full experimental data. The AVP for the Z-stringer reinforced panel under consideration is now presented.

Analysis Validation Plan for Statically Loaded Reinforced Z-Stringer Panel

Issue 1

Dated 12/12/2006

1. Introduction

Responsible for assuring adequacy of the AVP: Dr A. Rahman.

2. Initial considerations

Qualification objectives

That the structure can sustain a static uniformly distributed load of 30 Pa without violating limits set on the values of defined qualification parameters.

Structural interpretation

Critical Components: The critical structural components are considered to be the panel, the Z-stringers and the brackets attaching the panel to the outside world support system.

Qualification Parameters

There are two overall qualification parameters, one being the maximum stress found anywhere in the structure, the other being the maximum displacement found anywhere in the structure. The required limits on these parameters are:

- maximum stress be less than 50 MPa;
- maximum displacement be less than 4 mm.

3. Validation criteria

Results accuracy

That the values obtained for the qualification parameters from the finite element analysis should not deviate by more than 5% from the operational values.

Internal process:

- Uncertainty sources to be located using the criteria and methods in Chapter 7 of this book.
- The influence of uncertainty on the qualification parameters to be evaluated using the methods in Chapter 8 of this book.
- FEMEC to be employed.

External process:

- Experience to be employed using handbook data and simple calculations.
- As no other team is involved there is no requirement to consider error injected into the analysis process from an external source.

4. Revision procedure

- The scope of the analyses may need to be extended if additional structural items or components require to be introduced into the critical component set.
- Revisions to be monitored by Dr A. Rahman.
- Initiation of request by Prof A. Morris.

Stage 2 Initial assessment

The aim of this Stage is to take the outputs from Stage 1 and develop the information so that assessments can be made of sources of uncertainties from a higher level viewpoint where bounds are sought without the direct application of supporting finite element reference models. It also sets up the analysis process so that an analysis plan can be constructed before moving on to a detailed assessment of uncertainty sources.

Step 2.1 Definition of domain of analysis

The analysis problem is sufficiently small in scope that the decisions already made in step 1.2 with respect to the location of the outer boundary of the domain of analysis (DoA) do not require further consideration. Although the supporting wall for the reinforced panel has been taken to be rigid, this needs to be reviewed and some assessment made of the impact that wall flexibility can have on the qualification parameters.

Thus no revision of the domain associated with the DoA is to be made but a revised definition of flag 1 is required.

Revised Flag 1

Flag 1: Influence of flexibility of the supporting wall on the maximum stress and displacement (qualification parameters) to be assessed.

Step 2.2 Loading actions

This step traces the relevant loading actions, experienced during operational service, through the various parts of the structure providing information to assist in defining and listing the features that play a central part in the uncertainty control procedure.

A single distributed load is applied to the upper surface of the panel and is transmitted to the supporting wall through the attachment brackets which are connected to the panel and supporting wall by a set of 3 mm diameter bolts. The load path involves the undefined contact surfaces between the panel and bracket together with the bolts. No self-weight is taken into account and uncertainties relating to the stringers are not considered in this walkthrough.

Step 2.3 Decomposition of structure

This step decomposes the structure into a set of features taking into account the loading actions.

Decomposition based on loading actions

Because uncertainties relating to the Z-stringers and rivets are to be ignored in this walkthrough, the loading actions illustrated in step 2.2 require that three features need to be considered and these are displayed in Figure 10.3. This simple decomposition raises a number of issues:

- Although the panel has been identified as being well within the range where thin plate theory can be applied, the behaviour pattern in the bracket/panel contact region is unclear.

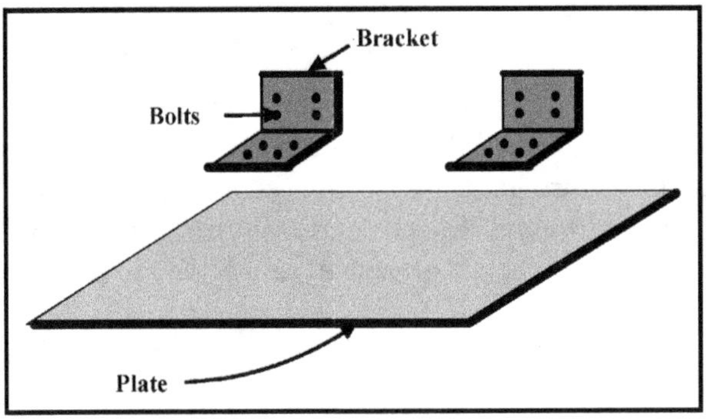

Figure 10.3 The three features explicitly accounted for in the walkthrough example.

- The same remark applies to the bracket.
- The panel/bracket contact zone is unknown.
- The interaction between bolts, bracket and panel is not known.

Revised or extended uncertainty sources

Flag 2: The level of abstraction for the bracket must allow for the fact that the error in the qualification parameters is a function of the uncertainty in accurately characterising the bracket behaviour and the extent of the contact region with the panel.

Flag 3: The mechanism by which the bolts make contact with the panel and the bracket bolt holes is an error source for the qualification parameters.

Flag 5: The thin plate model for describing the behaviour of the panel in the contact zone for assessing error impact on the qualification parameters may not be adequate.

Step 2.4 Initial uncertainty bounds

This step attempts to bound the impact that uncertainty sources have on the eventual error associated with the qualification parameters when the final analysis is undertaken.

Bounded sources

Flag 4 has been eliminated as the simple calculation process clearly indicates that bolt failure will not take place due to the application of the distributed pressure load.

All other flags remain raised and require action to bound the amount of error that their presence can have on the qualification parameters.

Stage 3 Analysis review and decisions

This stage reviews the analysis requirements in terms of models to be employed for bounding the impact of uncertainty on the qualification parameters.

Step 3.1 Review

This stage takes the information from the QR, suitably augmented by design requirements if necessary, together with requirements from the AVP and scopes the analyses required both to quantify the uncertainties flagged by earlier stages as unbounded and to design the structure. The dynamic nature of the analysis process means that the analyses envisaged at this stage may not represent the complete set that will eventually be required to complete the analysis process.

Global assessment using a reduced model

The first finite element reference model required by the FEMEC method for the Z-stringer reinforced analysis problem is one of limited size that:

- is able to assist in confirming that the design can, in principle, carry the applied loads without violating the qualification criteria;
- can assist in identifying any additional sources of error or uncertainty that have not been noticed or considered in the earlier steps.

Since thin plate theory has been taken as the characteristic behaviour, then the initial reduced global reference model should employ thin plate elements for all components.

Detailed assessments

In order to bound the terms identified in flags 1, 2, 3 and 5 a more detailed model using thin plate elements is required that can:

- allow the influence of the bolt interaction with the panel and bracket structures on the qualification parameters to be assessed;
- allow the impact of the extent of the contact zone between the panel and bracket on the qualification parameters to be assessed employing the sensitivity method described in Chapter 8 (however, this is not pursued in this chapter as the process of using sensitivities is adequately demonstrated in Chapter 8 and does not merit repetition);
- allow the level of abstraction for the bracket to be assessed and the impact of any uncertainty associated with the selected bracket model on the qualification parameters to be quantified.

It is noted that fully assessing the influence of the bolt actions and the level of abstraction for the bracket may require the development of more than one finite element reference model, i.e. resort may need to be made to the application of thick plate theory or 3-D solid elements.

It is possible to examine the area involving the bracket, bolts and the immediate region of the panel and Z-stringers by means of an explicit–implicit analysis by using the reduced global model as an implicit model to provide boundary conditions for a more detailed model that explicitly models the bracket, bolts and the panel and Z-stringers in the immediate region of the bracket. However, the current analysis problem is too small to follow this route but it would be an important consideration for larger scale problems in which there are a variety of detailed structural components.

Analysis requirement

Based on this review, two finite element reference models are required with a third envisaged if justified.

Reference Model 1:

- A reduced global model employing thin shell elements for the bracket, panel and Z-stringer fastener neglected.
- The bracket interfaces with the panel are modelled as surface-to-surface contact for the entire bracket lower surface areas, i.e. all

points on the interface surface are assumed to be in contact when the load is applied.
- This same assumption applies to the panel and Z-stringer contact area.
- The bracket corners are to be modelled as right angles with no curvature.

Reference Model 2:

- Model 1 extended to include explicit models for the bolts.
- The contact zone between the bracket and panel restricted to bolt attachment points only.

Reference Model 3:

- If models 1 and 2 cannot fully resolve uncertainties with respect to the bracket level of abstraction, an additional model is necessary.
- If models 1 and 2 cannot fully resolve uncertainties with respect to the behaviour of the structure in the contact zone between the panel and bracket, an additional model is necessary.

The level of novelty for these analyses is assessed as low while the level of complexity for model 2 is assessed as medium. These levels fall within the experience base of the analyst.

Step 3.2 Decisions

Based on the review and the analysis requirement it is concluded that:

- No loop-back to earlier stages is required.
- On the basis of the novelty/complexity/experience comparison, the FEMEC process may proceed to Stage 4.

Stage 4 Global assessment

The aim of this stage is to undertake an analysis of the complete structural analysis problem using a reduced finite element reference model. The targets for the analysis are to generate appropriate bounds for global-level uncertainties flagged as unbounded by earlier stages and to provide basic information on the behaviour of the structure as a steer to the analyst in generating more comprehensive finite element reference models.

Step 4.1 Initial calculations

This step attempts to generate an initial view of the likely behaviour of the structure to provide basic information to assist in assessing the quality of the outputs from the finite element reference models. Two initial calculations are undertaken based on simple reference models: one to estimate the maximum displacement of the structure, the other to estimate the maximum stress as these are the two qualification parameters.

Maximum stress estimate using the 'Shanley' reference model

The maximum stress in the structure (neglecting the stringers and stringer–panel fasteners) occurs at point A in Figure 10.2 where the bracket has to carry the bending forces and the direct forces from the panel. If b = the length of the bracket surfaces, d the bracket thickness, l the panel edge length with P the applied distributed load, then the maximum stress due to the moment arm created by the load σ_b, using Shanley's *Mechanics of Materials*, reference [2], is given by:

$$\sigma_b = 3 \cdot \left(\frac{P * l^3}{b * d^2} \right)$$

And the direct stress σ_d by:

$$\sigma_d = \frac{P * l^2}{2 * b * d}$$

The combined stress is given by adding these two terms and, for this problem, indicates that the maximum stress is seen at the corner of the bracket and has a value of 45 MPa.

Maximum displacement estimate using 'Roark's' reference model

Calculating the maximum displacement is a little more complicated so we have elected to calculate an upper bound on the estimated maximum displacement. This displacement occurs at the centre of the outermost edge of the panel measured from the supporting wall. The procedure adopted estimates the free edge displacement δ by taking the special case of an encastré beam with the same length as the panel edge length, of width equal to the width of the bracket and other relevant dimensions and terms as above, and then, using *Roark's*

Formulas for Stress and Strain, reference [1], the free end displacement is given by:

$$\delta = \frac{P * l^4 * b}{8 * E * I}$$

where E is Young's modulus and I the second moment of area. For the current problem this formula estimates the maximum displacement to be 6 mm.

Step 4.2 Initial global assessments

The aim of this step is to set up the reduced finite element analysis model for the structure lying within the DoA. The model is to comply with the reference model 1 definition in step 3.1 employing thin plate elements for modelling the panel, brackets and stringers. The attachment between the panel and stringers is taken to be continuous along the entire contact zone where these two components interface. Similarly, the contact between the panel and the brackets is assumed to be continuous within the interface surfaces. For both of these attachment regions the appropriate panel nodes are directly connected to the nodes of the attached components. A vertical uniformly distributed pressure is applied in a downward direction to the upper surface of the panel.

The finite element model constructed for the solution of this analysis problem is shown in Figure 10.4, where the bracket and panel have merged to become, effectively, double-thickness plate elements

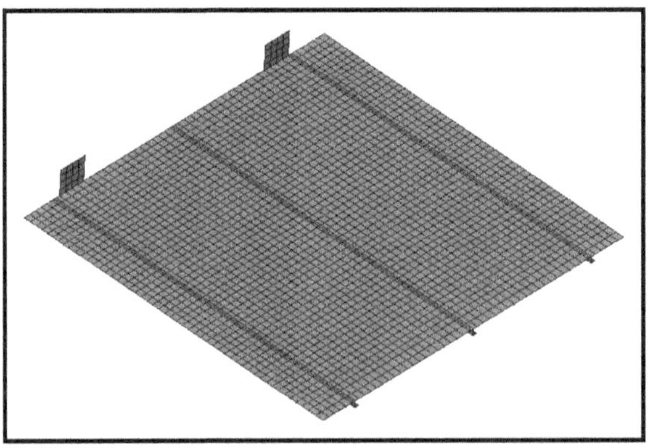

Figure 10.4 Initial reduced finite element model.

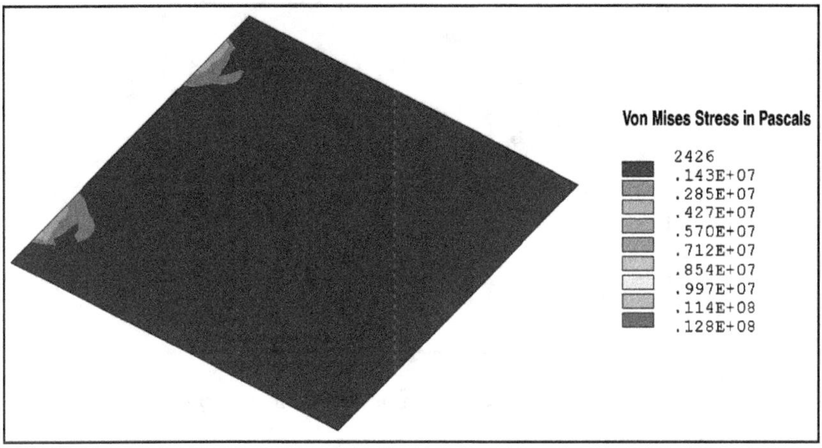

Figure 10.5 Stress contours for the panel. (See also Plate 4.)

where the bracket overlaps the panel. The bolts are not modelled as the entire interface surfaces are assumed to be in direct contact. Stress contours for the panel and the bracket are displayed in Figure 10.5 and Figure 10.6 respectively, from where it is seen that:

- The maximum predicted stress occurs in the bracket with a value of the Von Mises stress of 38.4 MPa while that in the panel is 12.8 MPa.
- The maximum displacement is located at the outer edge, i.e. at a point furthest from the location points where the finite element model predicts a maximum value of 2.515 mm.

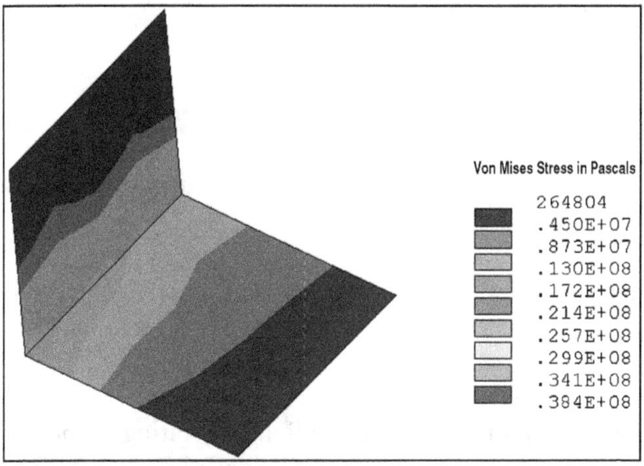

Figure 10.6 Stress contours for the bracket. (See also Plate 5.)

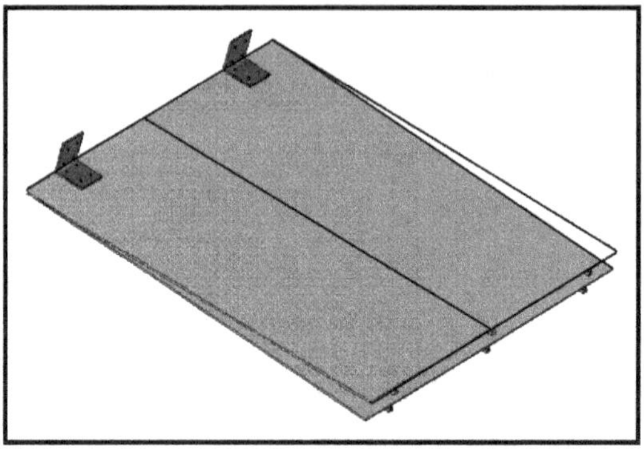

Figure 10.7 Illustration of displacement field.

As might be expected, the displacement field for this problem is very regular with the displacement sweeping to maximum values at the outer edge as illustrated in Figure 10.7.

Two main observations can be made from these initial results:

1. The finite element values of maximum stress and maximum displacement compare favourably with the maximum bracket outer fibre direct stress of 45 MPa and a maximum displacement of 6 mm given by the simple calculation in step 4.1.
2. The fact that the position of the maximum stress occurs at the bracket corner raises the question that uncertainty in the curvature of the corner may have an impact on the predicted values of the qualification parameters.

Step 4.3 Uncertainty assessments and bounding

Although the analysis has provided only limited information, a number of conclusions can be drawn:

1. The simple calculation and the current finite element values are in approximate agreement so that there is no clash between the two sets of predictions and no need, at least in this step, to require a return to earlier steps.
2. A new uncertainty has been highlighted relating to the curvature of the bracket at the corner which was modelled as a right angle in step 4.2, i.e. the level of abstraction is still an open issue.

3. Although the results appear reasonable as a result of point 2 above, they do not provide bounding information relating to the four flags currently raised.

Step 4.4 Review

The results from steps 4.2 and 4.3 have indicated the following:

1. The initial decisions used to define the DoA do not require modification.
2. A new feature is required that incorporates the curvature present at the bracket corner and that uncertainty associated with the corner radii requires assessing.
3. Although no information has been generated that allows any of the flags to be lowered, it is noted that flag 2 could be treated by developing the model used in step 4.2 to employ higher dimensionality elements.

The statement in point 2 identifies a need to raise a new flag, flag 6, that requires the impact of uncertainty with respect to the curvature of the bracket corner to be assessed. In addition, point 3 indicates that an additional finite element reference model is required that allows the level of abstraction for the bracket to be assessed. This can be handled by leaving this flag raised and returning to the matter at Stage 7. However, it is a relatively simple matter to change the elements of reference model 1 to 3-D elements and incorporate the required curvature immediately. (This move seems to imply that the process is jumping ahead as reference model 2, still employing plate elements, has not yet been deployed, but the reader will recall that such a change in procedure is noted in Chapter 9, Section 9.4.3.)

Decisions:

1. That the process loops back through step 4.2 employing a revised finite element model to provide the information required to lower flags 2 and 6. This to be undertaken before moving to Stage 5.
2. The revised finite element model is to use the same basic element layout as illustrated in Figure 10.4 but employing 8-noded solid elements with three layers of elements through the three main components of the structure as shown in Figure 10.8.

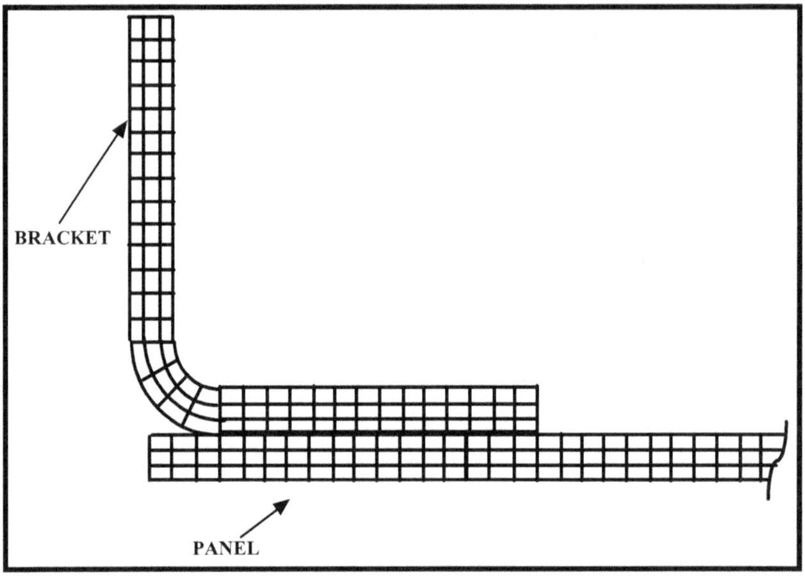

Figure 10.8 Three-dimensional solid elements.

Step 4.2 Initial global assessment (second pass-through)

The finite element model was constructed employing the Figure 10.4 layout using 8-noded elements with three elements through the component thicknesses. A transition set of solid elements is deployed to model the 7.5 mm mean radius corner as shown in Figure 10.8. Employing solid elements for the bracket requires that solid elements are used to model the panel and stringers. The 30 Pa load is applied to this model and gives the following key results:

- The maximum stress again occurs at the bracket corner with a value of the Von Mises stress of 36.8 MPa.
- The maximum displacement occurs at the same point as in the initial global model with a value of 2.544 mm.

Step 4.3 Uncertainty assessments and bounding (second pass-through)

The results from step 4.2 (second pass) indicate that changing the level of abstraction for the bracket and including the corner has not had a significant impact on the values predicted for the qualification parameters. Observing that both reference models predicted values for the stresses and displacements that satisfy the qualification criteria, the FEMEC process can continue.

Since the reference values to be inserted into equation 5.3 should represent 'best' estimates, these are taken, at this point, as the mean values for the two quantities making up the qualification criteria from the two finite element analyses. Thus, using q_1 to denote the maximum stress criteria and q_2 to denote the maximum displacement criteria then:

$$q_1(\text{ref}) = 32.6\,\text{MPa}$$

and:

$$q_2(\text{ref}) = 2.530\,\text{mm}$$

and the associated variations in these quantities from the mean values are given by:

$$\delta q_1 = 0.8\,\text{MPa and } \delta q_2 = 0.030\,\text{mm}$$

Inserting these into equation 5.3 for each qualification parameter yields the estimated bounds on the potential error as:

$$M_1 = |\delta q_1/q_1(\text{ref})| = \pm 0.02\,(2\%)$$

and:

$$M_2 = |\delta q_2/q_2(\text{ref})| = \pm 0.006\,(0.6\%)$$

These are within the required limits on the qualification parameter.

Step 4.4 Review (second pass-through)

The results from steps 4.3 and 4.3 (second pass-through) show that changing the level of abstraction for the structure, and in particular the bracket, from a thin plate model to a solid model has little effect on the values predicted for the qualification parameters. It is, therefore, possible to model the bracket as a thin plate structure. Although not considered as a separate issue, the limited impact on the qualification parameters of including the curvature implies that this feature is not significant.

As a result of the conclusion from step 4.3 (second pass-through) error flags 2 and 6 are lowered but all other error flags remain raised.

The FEMEC process can now move to Stage 5 with error flags 1, 3 and 5 still raised and requiring attention. The results achieved in this stage yielded predicted operational values for the qualification parameters based on the assumption that the contact zone between the bracket and the panel constitutes the entire interface surfaces where these two components overlap. The solutions obtained do not provide bounds on the qualification parameters with respect to the extent of the bracket–panel contact zone.

Stage 5 Detailed assessments of uncertainty sources at the feature level

Stage 4 has initially treated error sources that are due to the presence of certain uncertainties. The aim of this stage is to undertake a finite element analysis focused on the bracket as the main feature in view of the decision not to include the stringer and stringer fasteners within this control process. The analysis has now moved to considering the finite element model designated as reference model 2 in the analysis requirement section.

Step 5.1 Feature analysis set-up

The analysis to be undertaken at this stage has the following characteristics:

- Stage 4 has established that thin plate elements are appropriate for the analysis of this problem allowing these elements to be employed for the current analysis.
- The contact zone is restricted to the bolt locations; this represents the minimum zone in contrast to that deployed in the initial global analysis.
- In order to incorporate the bolts these are modelled by cylindrical elements.

The basic element layout for this is shown in Figure 10.9 where the mesh layout has increased the density in critical regions. Essentially this step is setting up an analysis that is focused on lowering flags 3 and 5.

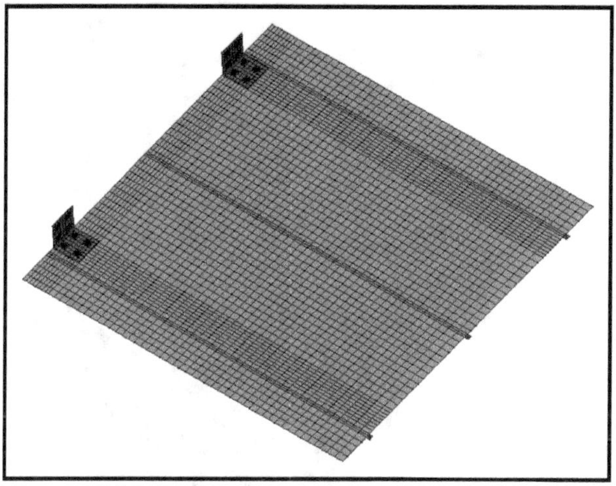

Figure 10.9 Panel element layout.

A detailed element layout configuration is required at the bolt locations in both the brackets and panel. The mesh and element layout for this region follows the advice given in Chapter 6 and is shown in Figure 10.10.

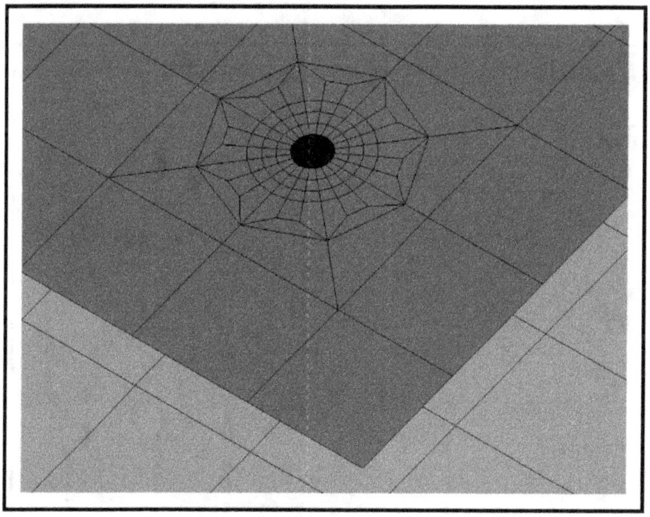

Figure 10.10 Detail of element configuration local to the bolts in the brackets and panel.

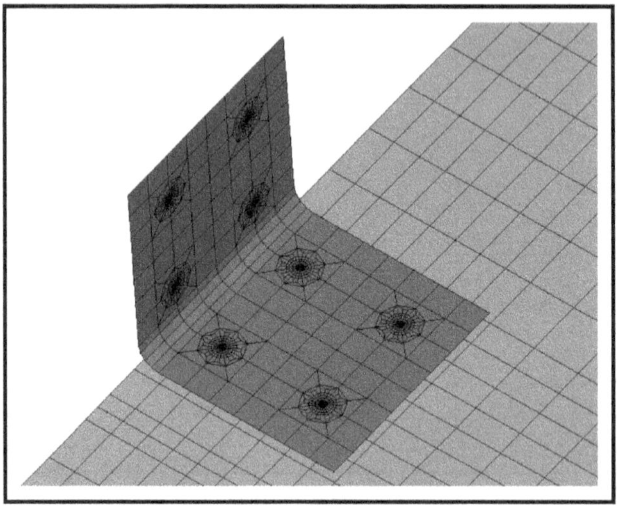

Figure 10.11 Bracket model.

A detail of the finite element model for the bracket is shown in Figure 10.11 and the details of the modelling for the brackets, panel and stringers are shown in Figure 10.12.

The distributed 30 Pa pressure is applied on the nodes of the upper surface of the panel and on the upper nodes of the horizontal part of the

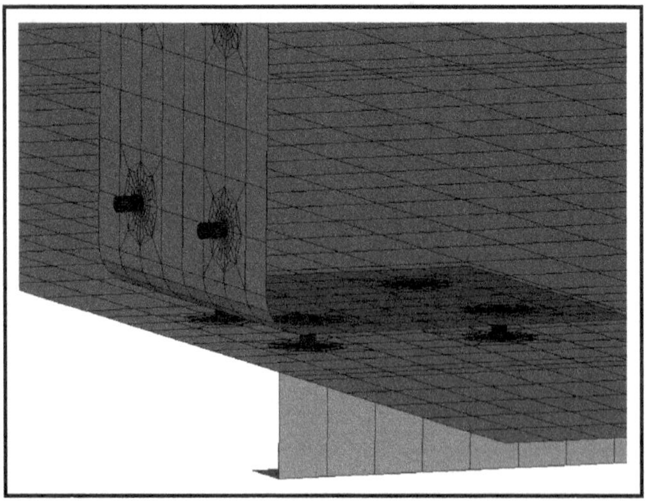

Figure 10.12 Detail of bracket, panel and Z-stringer finite element models.

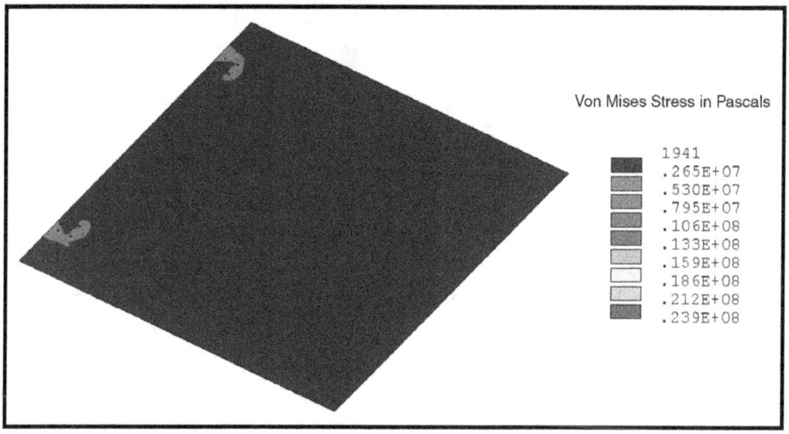

Figure 10.13 Stresses in the panel. (See also Plate 6.)

bracket. The model is now ready for the initiation of a static analysis solution run.

Step 5.2 Uncertainty assessment and bounding

As the walkthrough is only considering bounding the impact of errors on the qualification parameters related to uncertainties associated with two features, namely the panel and the bracket, consideration need only be given to the results shown in Figure 10.13 and Figure 10.14 respectively.

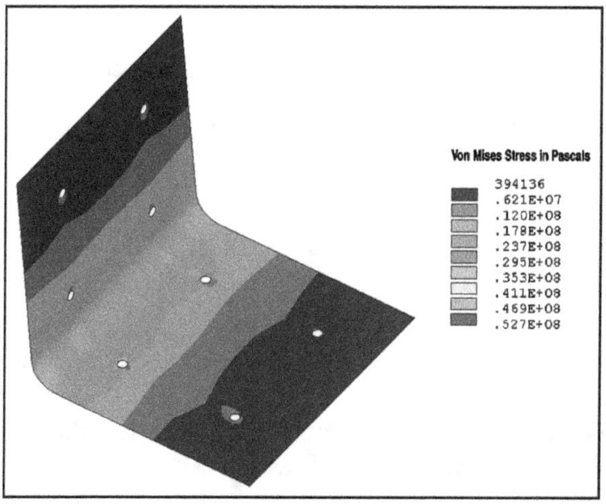

Figure 10.14 Stresses in the bracket. (See also Plate 7.)

It is noted that:

- The maximum stress is now located at the bolt attachment points and is predicted to be 52.7 MPa.
- Although the displacement field is not shown, the maximum value was found to be located at the same position as previously but having now increased to 3.654 mm.

Referring to the AVP, it is noted that the maximum stress from the analysis undertaken in this step exceeds the maximum acceptable value. In view of this, it is not possible to lower any of the outstanding flags.

Stage 6 Overall assessment

This stage represents the end of the first complete pass through the FEMEC process in that both global- and feature-level examinations have been undertaken with the objective of bounding the errors in the qualification parameters due to the presence of uncertainties. It notes the consequences resulting from these analyses and plans the next stages required to bring the analysis process to the point where a final analysis can be run and the results passed to the design team.

Step 6.1 Assessments

This step undertakes the direct assessments of the results allowing decisions to be made with respect to subsequent actions in step 6.2.

Assessments of inputs from earlier stages:

Stage 4 Analysis assessment:

- The global analyses undertaken at Stage 4 indicate that there is no risk to the structure from bolt failure.
- The finite element analyses indicate that the curvature of the bracket did not introduce significant stress raisers and that the brackets and panel could be modelled using thin shell elements. On this basis, flags 2 and 5 were lowered.
- The analyses at Stage 4 have not shown that the maximum displacement is close to the limits indicated in the AVP.

Stage 5 Analysis assessment:

- The analysis undertaken at Stage 5 has predicted that the value of the maximum stress has significantly increased from that obtained at Stage 4 and with a value of 52.7 MPa that exceeds the maximum acceptable value indicated in the AVP for this qualification parameter.
- The Stage 5 analysis, in addition, predicts that the position of the maximum stress moves from the bracket corner to the bolt positions.
- The analysis at Stage 5 confirms the conclusion from Stage 4 that the displacement qualification parameter is significantly below the value proposed as the maximum acceptable value in the AVP.

Flag assessment

On the basis of these results, flags 1, 3 and 5 cannot be lowered but flag 6 remains lowered and does not require further consideration. The lowering of flag 2 at Stage 4 requires reconsideration at step 6.2.

Test and maintenance requirements assessment

It is noted that not all levels of abstraction for modelling the panel, stringers and brackets have been explored. Until this point is reached there is neither a requirement for a supporting test programme nor any need to ask for the imposition of any maintenance conditions.

Step 6.2 Decisions and actions

On the basis of the outputs in step 6.1, the following decisions and actions are undertaken:

- Flag 2 is raised so that there are now four error flags in action, i.e. flags 1, 2, 3 and 5.
- A solid finite element model, i.e. reference model 3, is required in order to move the level of abstraction to the next available level because the maximum stress is located at the bolt locations; these regions must be given detailed consideration by reference model 3.
- The finite element model must have sufficient mesh density in the regions of high stress.

- Although the purpose of a subsequent analysis is to examine errors associated with uncertainties in the modelling of features, the analysis has to take place at the global level unless an implicit–explicit modelling strategy is followed. This latter strategy has not been employed as the models used are sufficiently small not to merit an implicit–explicit modelling process.
- On the basis of these decisions, Stage 7 is initiated.

Stage 7 Intermediate global assessments

The solid finite element model for this stage is displayed as an overall model in Figure 10.15 where 8-noded solid hexahedral elements are employed. Essentially the analysis is employing constant stress elements as the variation of stress in the bulk of the structure is not expected to be significant. Details of the models for the bracket, panel and Z-stringer are displayed in Figure 10.16. Where it is anticipated that high stress gradients are likely to be found, the mesh density is increased and the models for the bolts and the regions in both the panel and the bracket adjacent to the bolt holes, where past results have indicated high stresses, are displayed in Figure 10.17. In addition, to provide a worst case scenario, the connection between the brackets and the panel is assumed, in this analysis, to take place at the bolt locations, i.e. the panel hangs on the bracket through the bolts and there is no other surface contact between the brackets and the panel. In grading the mesh around the bolt location points in the panel and the bracket, due note is

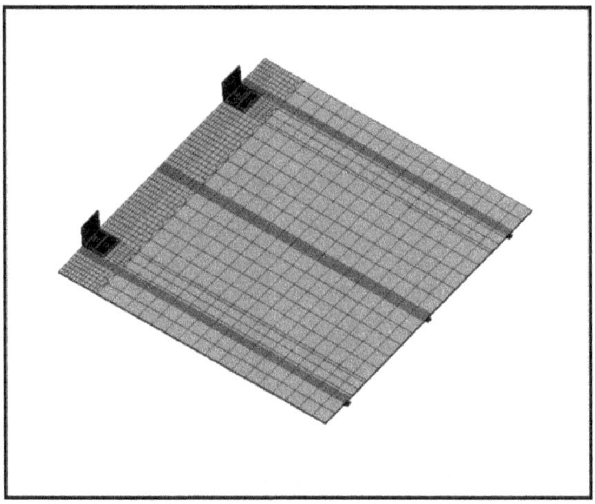

Figure 10.15 Overview of solid finite element model for reinforced panel.

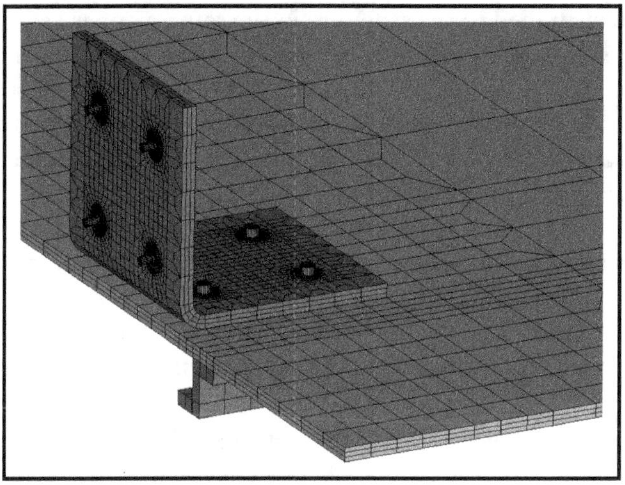

Figure 10.16 Detail showing the models for the bracket, Z-stringer and the panel adjacent to a bracket.

taken of the fact that the elements employed in this analysis are constant stress elements requiring sufficient attention to model the stress gradients in the region. This mesh is also graded following the advice given in Chapter 6 on grading meshes as the model approaches a surface or region of contact. The resulting model generated 43,677 elements. The

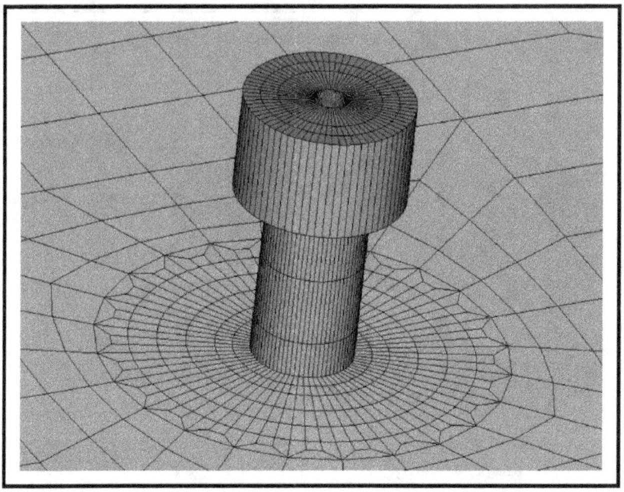

Figure 10.17 Detail showing the solid model for the bolts and the mesh around a bolt hole.

30 Pa load is applied to the structure in the same manner as that used in the earlier stages of this FEMEC process and the finite element analysis problem is then solved. It should be noted that due account was taken of the size of the problem and the anticipated run-time assessed using the run-times from the earlier models to ensure that the computing time required was acceptable.

The finite element reference solution model was then run using the Ansys code and the results relevant to the selected qualification parameters are displayed in Figure 10.18 and Figure 10.19 for the Von Mises stresses in the panel and the bracket. The stress levels in the panel are well below the maximum values stated in the AVP and can be discounted.

The results displayed in Figure 10.19 show that:

- the predicted stress levels in the bracket at 49.9 MPa have been reduced to a value that brings them within the maximum indicated in the AVP;
- although not displayed in the figures, the maximum displacement occurs at the same point as earlier analyses have indicated with the current set of results predicting a value of 3.73 mm.

The results from the first analysis in this step have indicated that the maximum predicted value for the stress in the bracket and for the

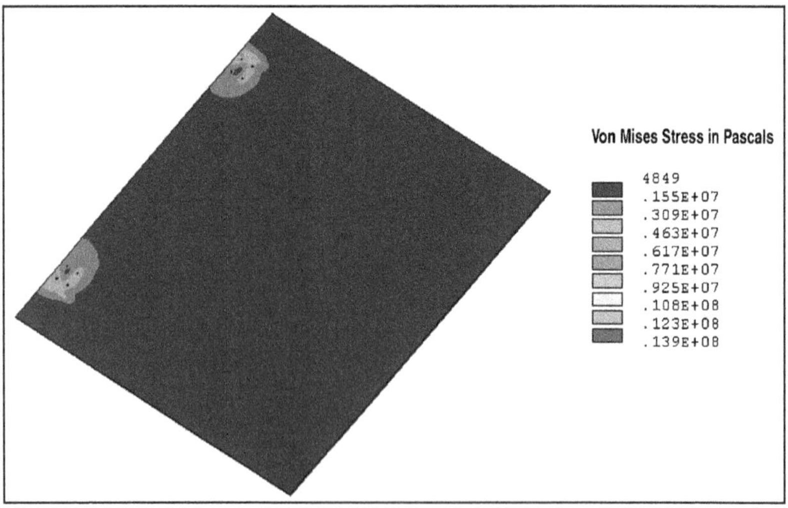

Figure 10.18 Stresses in the panel. (See also Plate 8.)

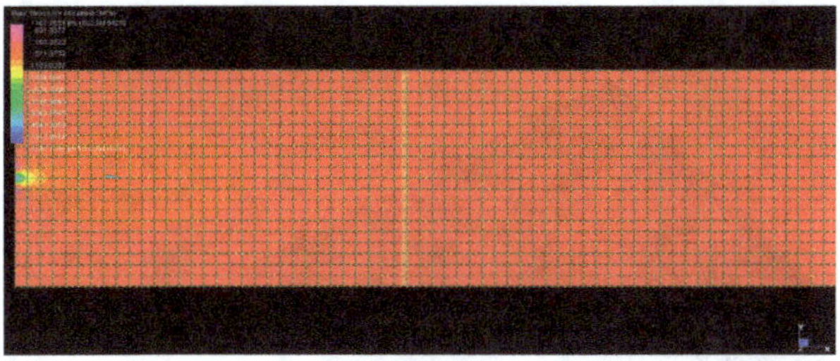

Plate 1 Plot of direct stresses in the x-direction for a model with 4000 elements. (See also Figure 6.26, p. 173.)

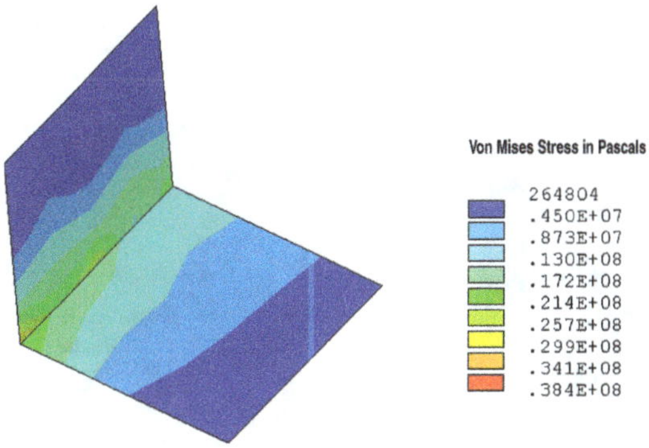

Plate 2 Bracket and panel assumed to be attached at all points on their interface surface. (See also Figure 7.15, p. 196.)

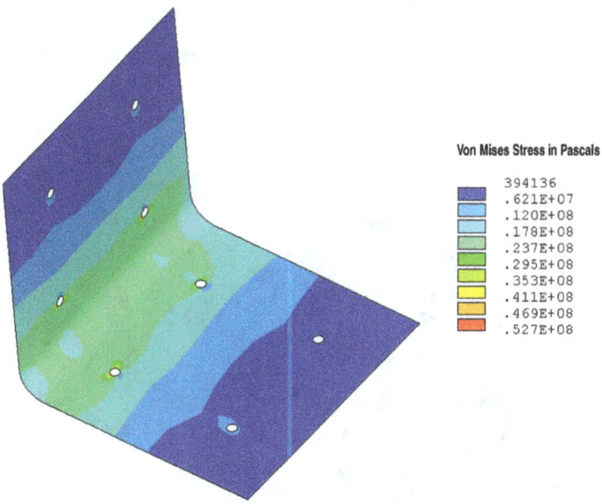

Plate 3 Bracket and panel assumed to be attached at bolt positions. (See also Figure 7.16, p. 197.)

Plate 4 Stress contours for the panel. (See also Figure 10.5, p. 323.)

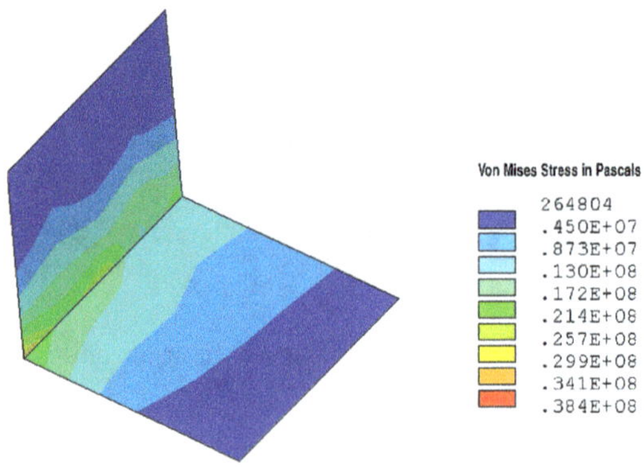

Plate 5 Stress contours for the bracket. (See also Figure 10.6, p. 323.)

Plate 6 Stresses in the panel. (See also Figure 10.13, p. 331.)

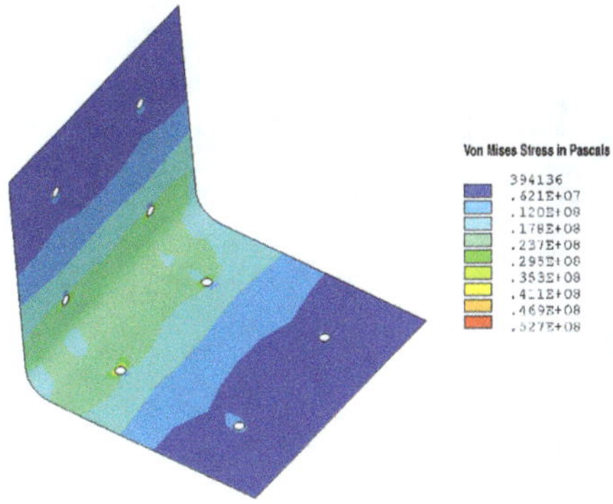

Plate 7 Stresses in the bracket. (See also Figure 10.14, p. 331.)

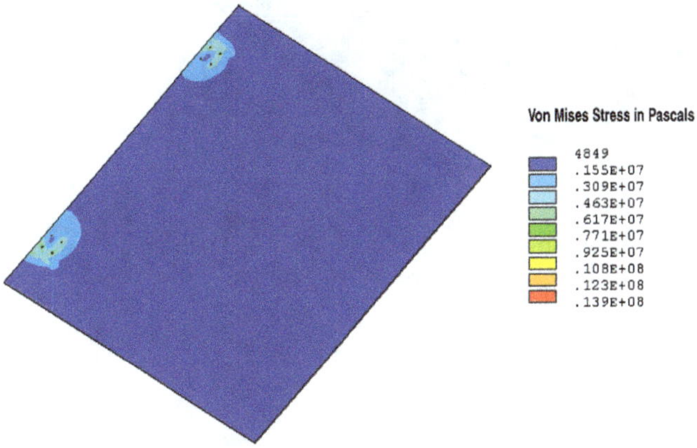

Plate 8 Stresses in the panel. (See also Figure 10.18, p. 336.)

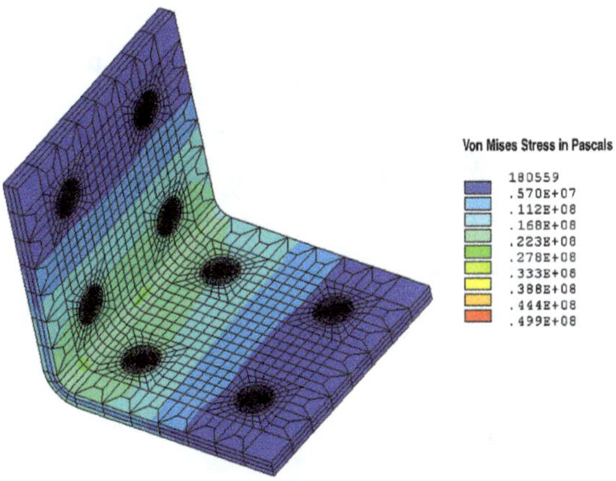

Plate 9 Stresses in the bracket. (See also Figure 10.19, p. 337.)

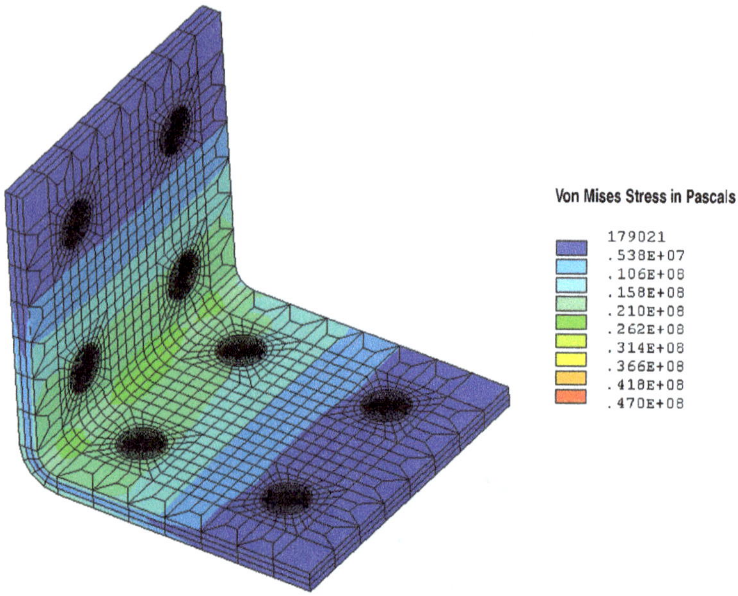

Plate 10 Von Mises stresses in the bracket with the contact zone covering the region between the bolt locations. (See also Figure 10.21, p. 340.)

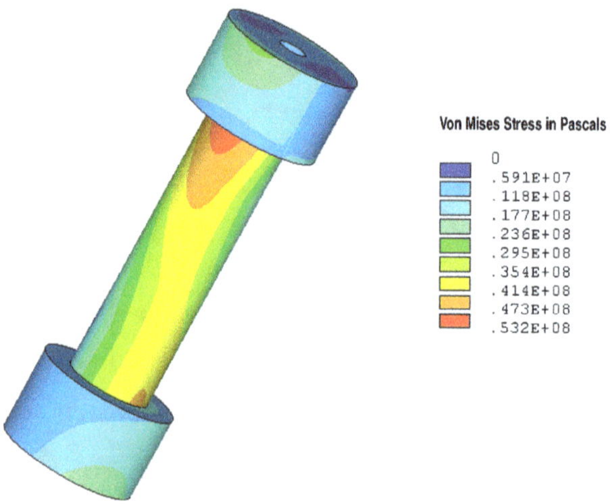

Plate 11 Bolt stresses. (See also Figure 10.22, p. 341.)

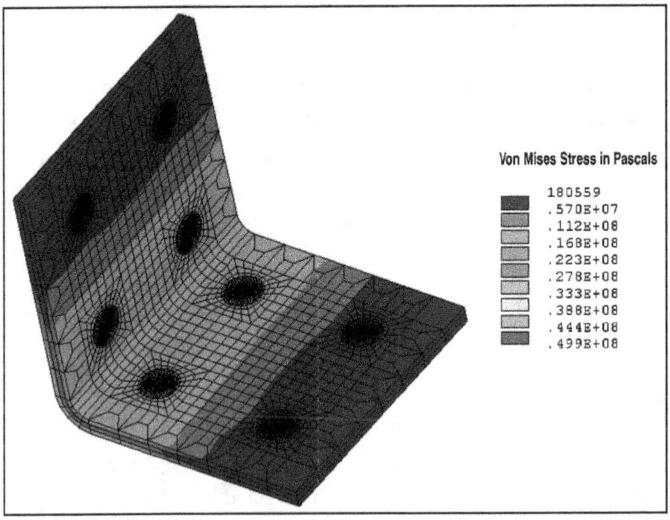

Figure 10.19 Stresses in the bracket. (See also Plate 9.)

maximum value for the displacement are now below the maximum values defined by the AVP as acceptable and the FEMEC process can continue. Since the worst case scenario is where the bolts represent the sole contact between the bracket and the panel, it is now clear that these maximum values represent the extreme values for the qualification parameters. On the basis of these results, the following conclusions can be drawn:

- Flag 1 still requires examination as the results do not shed any light on this parameter.
- Flag 2 can be provisionally lowered but has not been fully examined.
- Flag 3 cannot be lowered as the impact of the bracket/panel contact zone has not been examined.
- Flag 5 can be lowered as the solid element model is now considered as appropriate for the final model and the maximum values for the qualification parameters do not exceed the AVP values.

Because flag 1 relates to an assumption that has been imposed on the analysis, this aspect can be considered a separate aspect using a modification matrix linked to the final model. Thus there is no constraint imposed on the analysis process from flag 1 that restricts the setting up of a final analysis.

Stage 10 Final model and analysis

Because Stages 8 and 9 are not required, the final finite element model can now be constructed and run based on the results obtained from the earlier stages in the QR. The results obtained from this analysis must provide sufficient information to allow the impact of errors and uncertainties on the qualification parameters to be assessed.

Step 10.1 Final analysis plan

The output from Stage 7 has indicated that the maximum stress has reduced to a value that is within the required maximum value defined by the AVP. However, no estimate of the impact of the extent of the surface contact between the upper surface of the panel and the lower surface of the bracket on the qualification parameters has been obtained. Taking note of the analysis results from Stage 7 it is possible to argue that the contact zone between the bracket and the panel can be adequately assessed by connecting the nodes on the lower surface of the bracket and the upper surface of the panel in the region covered by the square defined by the four bolt positions and this represents the final finite element reference model.

The output from this step is a finite element model and loads set identical to that employed in Stage 7 with the addition that the nodes on the relevant surfaces of the panel and bracket in the region defined above are connected.

Step 10.2 Final analysis

The final model was run and because this analysis is the one upon which the final error assessment is made, it needs to be checked for convergence. This is done by following the procedure laid down in Chapter 6, Section 6.4.3, and applying the error expression in equation 6.4. In a real application a number of nodes at which a group of elements have common nodes would be selected. Because this is a walkthrough illustration, one point is employed at a position of relatively high stress for a node on the upper bracket surface as shown in Figure 10.20 where node number 8455 has been selected. It is surrounded by elements 5607, 5608, 5661 and 5662 and the nodal values for the stresses σ_{xx} and σ_{yy} generated are compared according to expression 6.4. The results from

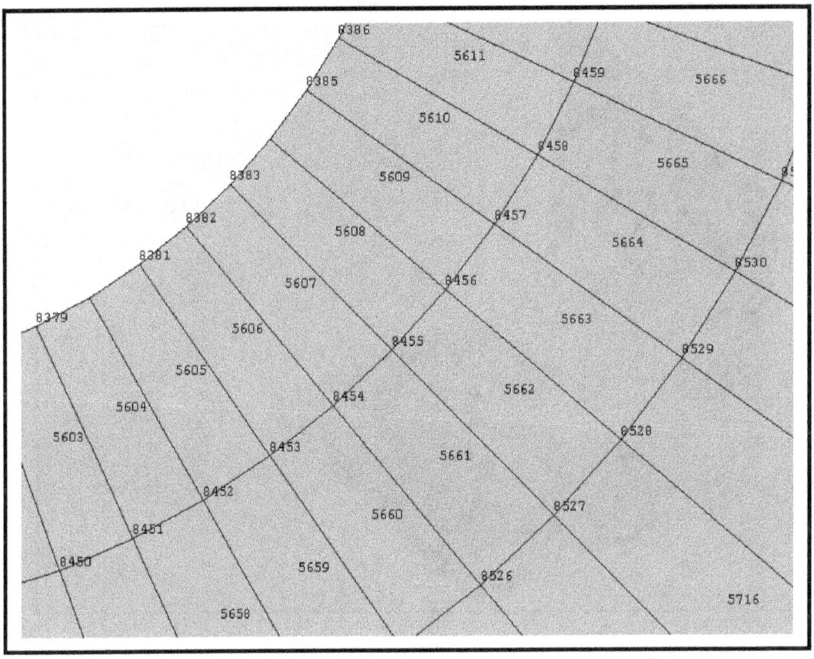

Figure 10.20 Element layout at assessment point.

this process are shown in Table 10.1 where it is concluded that a satisfactory level of internal accuracy has been achieved after taking into account that the elements employed are, essentially, constant stress elements.

NB. This is the point in the walkthrough sequence where adaptive methods might be employed to converge the finite element analysis to the highest level of internal accuracy; in this example a p-adaptive approach would be appropriate.

The results from this analysis are accepted and the critical results are shown in Figure 10.21 which gives the predicted values of the stresses in the bracket. The maximum value for the bracket stress has now reduced

Table 10.1 Convergence check.

The nodal value for σ_{xx} at node 8455	The nodal value for σ_{yy} at node 8455	Maximum error in σ_{xx} from Eq. 6.4	Maximum error in σ_{yy} from Eq. 6.4
6.5 MPa	7.9 MPa	0.1	0.1

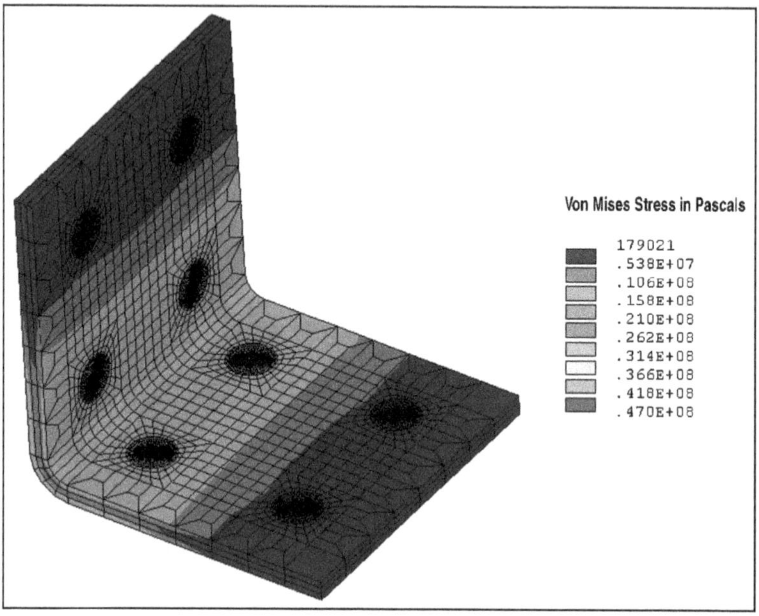

Figure 10.21 Von Mises stresses in the bracket with the contact zone covering the region between the bolt locations. (See also Plate 10.)

to 47 MPa and the maximum displacement has also reduced and was found to be 3.705 mm. These two values are the important numbers with respect to the qualification parameters and are the only factors being considered in this walkthrough example, but with a normal company or commercial analysis a complete set of stress and displacement results together with an appropriate commentary would be supplied to the design team.

The differences between the two values of stress and displacement given by this final analysis and that employed in Stage 7 can be taken as a measure of the impact that changing the contact zone has on the finite element estimates of maximum stress and displacement. It can, therefore, be taken as a measure of the uncertainty associated with the fact that the extent of the contact zone is not known but must lie between the two extremes presented by these analyses. Following the procedure in step 4.3 (second pass-through) the mean value of the stress and displacement values given above is accepted as the 'best' estimate. Hence, using the same nomenclature as before, bounds can be computed for this uncertainty source:

$$q_1(\text{ref}) = 48.45\,\text{MPa and } q_2(\text{ref}) = 3.717\,\text{mm}$$
$$\delta q_1 = \pm 1.45\,\text{MPa and } \delta q_2 = \pm 0.012\,\text{mm}$$

Inserting these into equation 5.3 gives:

$$M_{12} = |\delta q_1/q_1(\text{ref})| = \pm 0.03\,(3\%)$$

and:

$$M_{22} = |\delta q_2/q_2(\text{ref})| = \pm 0.003\,(0.3\%)$$

The subscripts 12 and 22 in the above equations indicate that the values M_{12} and M_{22} are additional values for the bounds on the uncertainties previously computed in step 4.3 (second pass-through) and identified as M_1 and M_2.

Because the analysis has been performed using solid elements the mechanism by which the bolts make contact with the bracket and panel can be examined. The stress field for the highest loaded bolts is shown in Figure 10.22 which indicates a smooth contact zone between the bracket and panel holes through which the bolts pass. It is noted that the highest stresses in the bolt are significantly below the qualification parameter maximum values for the steel used in the manufacture of this item. A key factor from the final analysis is, therefore, that flag 3 can now be lowered.

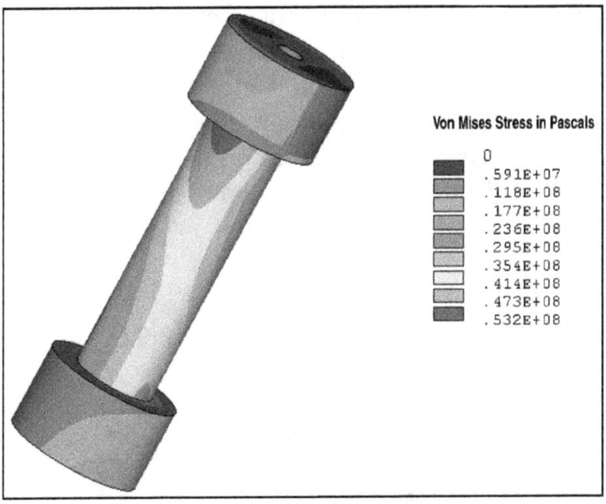

Figure 10.22 Bolt stresses. (See also Plate 11.)

Step 10.3 Error commentary

Although all the uncertainty flags have been lowered, indicating that the error impact of these uncertainties has been fully explored, the final bounds on the impact of these factors have not been provided. The report now considers each uncertainty flag and assesses its contribution to errors associated with the predicted values of the qualification parameters.

- Flag 1: Strictly speaking, this should not be a flag within this QR as the analysis has been defined with the reinforced panel attached to a rigid support. However, this is not a rational support condition and a comment by the analyst on the impact of support flexibility is required as part of the professional service and is provided as an addendum to this QR.
- Flag 2 and flag 5 (level of abstraction): In estimating the impact of the level of abstraction the results from Stages 5 and 7 provide the relevant information as these are models employing thin shell elements and solid elements respectively, but employing the same loading and contact assumption configurations. The results from these analyses show a reduction in maximum stress from 52.7 MPa to 49.9 MPa and an increase in maximum displacement from 3.684 mm to 3.729 mm. As discussed in Chapter 7, Section 7.4, these two models relate to two different levels of abstraction and can be considered as two separate measurements of the actual structural behaviour. The difference between them can, therefore, be considered as a measurement of the uncertainty in modelling this structure by these two different mathematical models. Following the procedure in step 4.3 (second pass-through) the mean value of the stress and displacement values given above is accepted as the 'best' estimate. Hence, using the same nomenclature as before:

$$q_1(\text{ref}) = 51.3\,\text{MPa and } q_2(\text{ref}) = 3.706\,\text{mm}$$
$$\delta q_1 = \pm 1.4\,\text{MPa and } \delta q_2 = \pm 0.022\,\text{mm}$$

Inserting these into equation 5.3 gives:

$$M_{13} = |\delta q_1 / q_1(\text{ref})| = \pm 0.027\ (2\%)$$

and:

$$M_{23} = |\delta q_2/q_2(\text{ref})| = \pm 0.006\,(0.6\%)$$

The subscripts 13 and 23 in the above equations as before indicate that the values M_{13} and M_{23} are additional bounds on the uncertainties to those previously derived.

- Flag 2 (contact zone): The solutions from the Stage 7 and Stage 10 models indicate that increasing the contact zone from the bolt location points to a region bounded by the four bolt positions results in a reduction in the maximum stress from 49.9 MPa to 47.0 MPa and a reduction in maximum displacement from 3.729 mm to 3.705 mm.
- Flag 3: The bolt contact zone is not considered as significant in view of the decision made at step 10.2.
- Flag 4: This flag has been lowered and has no impact on the qualification parameters.
- Flag 5: The impact of the curvature of the bracket corner was considered in step 4.3 (second pass-through) and was shown to have a maximum impact on the stress qualification parameter of ±0.8 MPa and of ±0.03 mm on the displacement qualification parameter. The introduction of a curved corner as opposed to a right angle reduces the maximum stress while increasing the maximum displacement.

Error bounds on the qualification parameters

Although all the flags have been lowered on the basis of results from the reference models, it is necessary to calculate the accumulated error term in conformity with the FEMEC requirements. Because the bounding terms '*M*' are fractional uncertainties, it is possible to adopt the simple-minded approach adopted in Chapter 8 and accumulate the bounds by adding the three components, thus:

$$M_1(\text{Final}) = M_1 + M_{12} + M_{13} = \pm 0.077$$
$$M_2(\text{Final}) = M_2 + M_{22} + M_{23} = \pm 0.015$$

Taking the mean values from the final reference analysis as the reference values for the qualification parameters and using M_1(Final) and M_2(Final) to generate the potential range of variation in the qualification parameters gives:

$$q_1 = 48.45 \pm 3.73\,\text{MPa}$$
$$q_2 = 3.717 \pm 0.056\,\text{mm}$$

Taking the upper bounds implies that the possible maximum values for the qualification parameters are:

$$q_1(\text{Maximum stress}) = 52.18\,\text{MPa}$$
$$q_2(\text{Maximum displacement}) = 3.772\,\text{mm}$$

The value for the maximum stress is not within the acceptable range defined by the AVP and it may be concluded that the design is not safe in its present form. A further violation of the requirement on this analysis is found if the maximum percentage error is computed which, for the stress component, is 7.7% and this lies outside the 5% values defined at the start of the FEMEC process.

It may be possible to argue that the M_1 and M_2 terms should not be accumulated into the final values as these are associated with modelling the curvature of the bracket and, at least for the stress assessment, this targets a different part of the structure from that associated with the other two error sources. If this is accepted, then the variation in the stress due to the accumulation of M_2 and M_3 only should be included in the qualification assessment. In this case the variation in stress is given by:

$$q_1 = 48.45 \pm 2.76\,\text{MPa}$$

and the estimated maximum value of this qualification parameter is now:

$$q_1(\text{Maximum stress}) = 51.21\,\text{MPa}$$

Again this violates both the required maximum value for this qualification parameter and, with a variation of 5.7%, exceeds the maximum error variation.

Addendum

The walkthrough has ignored two major sources of uncertainty that, in reality, would have to be included. One of these is the fact that the appropriate reference models are required to examine the impact of uncertainties in the modelling of the Z-stringers and the Z-stringer/panel attachments. As explained earlier, the approach adopted in handling this set of uncertainties would be similar to that presented above.

The other source is that the support for the structure has been assumed to be rigid and this could not be accepted in practice. There are two approaches possible, based on the methods presented in Chapters 8 and 9. One requires re-examining the domain of analysis and expanding it to allow part, at least, of the support structure to be included. The alternative is to use the artificial sensitivity method discussed in Chapter 8 and introduce multi-dimensional springs at selected points on the bracket–support interface surface to include the bolt locations.

10.3 A BRIEF LOOK AT THE DYNAMIC FEMEC WORLD

Although Section 10.2 provides a reasonably detailed walkthrough of the FEMEC procedure, it is useful to see it applied to a problem involving dynamic behaviour. However, there is no wish to test the reader's patience by presenting a second full FEMEC example so a very limited application is thought sufficient to illustrate the dynamic case. A number of options are open for making a limited presentation but the modelling of the attachment of the panel and Z-stringer is selected as this aspect of the analysis is ignored in Section 10.2. The problem has different design and qualification requirements given below and one step only of the FEMEC approach is illustrated which relates to step 4.

10.3.1 Modified Design and Qualification Parameters

The design problem consists of designing a panel structure shown in Figure 10.1 with the thickness of the Z-stringers now be equal to 0.1 cm; all other structural dimensions remain unchanged from Section 10.2. The attachment between stringers and panel is by rivets. There is no load applied to this structure and the design qualification parameters are the structure's free vibration natural frequencies. A somewhat artificial qualification requirement is imposed such that the structure has no natural frequencies in the 3.5–4.0 Hz and 28.5–29.5 Hz.

10.3.2 FEMEC Approach

It is assumed that steps 1, 2 and 3 have been completed and have established that using thin plate elements for this type of analysis is adequate for calculating the free vibration natural frequencies of the panel and Z-stringer configuration. Although a number of potential error or uncertainty sources are present in this analysis, following the task set for this dynamic section above, we shall now concentrate on dealing with the impact of uncertainties in the contact zone between the panel and the Z-stringers on the qualification parameters. That is, does the structure in its present design state satisfy the qualification requirements? To answer this and in view of the assumptions already made, it is possible to move immediately to step 4.2.

Step 4.2 Initial global assessment

As in Section 10.2, the aim of this step is to set up a reduced or number of reduced reference models for the structure that can assess the impact of potential uncertainties in the panel/Z-stringer joint attachment. The model is to comply with the reference model 1 definition set out in step 3.1 of Section 10.2 and employs thin plate elements for modelling the panel, brackets and stringers.

In Chapter 7, Section 7.4, it is argued that different levels of abstraction for a structural model or component can be considered as different 'measurements' of the structure or component behaviour. It is further argued that, in principle, it should be possible to define models or levels of abstraction sufficiently different that the 'in-service' behaviour of the structure is bracketed by their predictions. In the present case this argument is accepted and two reference models are advanced that should provide upper and lower bounds on the free vibration natural frequencies of this structure: one where the entire upper surface of each stringer is assumed to be in full contact with the panel lower surface and a second where attachment is enforced only at stringer and panel nodes coincident with the rivet positions. These have been selected on the basis that the fully attached joint represents the stiffest model for the joint and that when nodes only are used as the connection between panel and stringers, this represents the most flexible joint model. The same element layout is used for both models and is displayed in Figure 10.23 where the rivet attachments are also shown. The results obtained from these two free vibration analyses are displayed in Table 10.2.

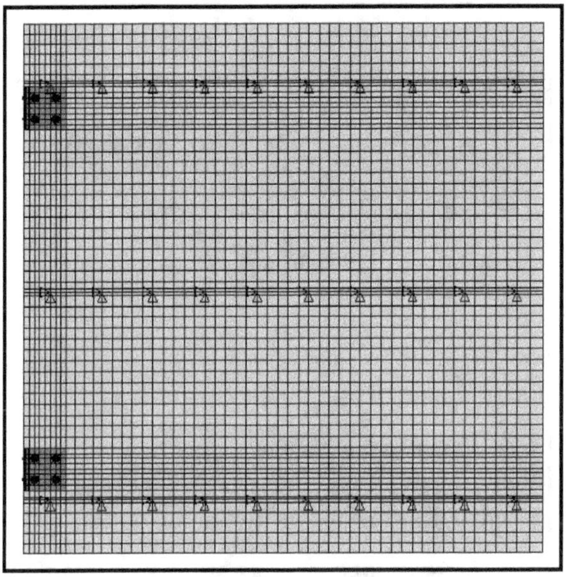

Figure 10.23 Element layout with panel and stringers attached at the rivet locations.

Table 10.2 Natural frequency results for both models.

Natural frequency no.	Frequency values for fully attached joint model (Hz)	Frequency values for joint attached at rivet locations (Hz)
1	4.13	3.81
2	9.67	9.42
3	29.67	29.56
4	31.98	31.54
5	43.49	42.93
6	54.58	54.26
7	75.20	74.60
8	75.46	75.14
9	101.52	99.14
10	105.57	103.50
11	110.57	106.15
12	135.93	131.98
13	143.73	143.60
14	146.48	146.26
15	154.36	147.15
16	173.92	174.08
17	185.27	184.40

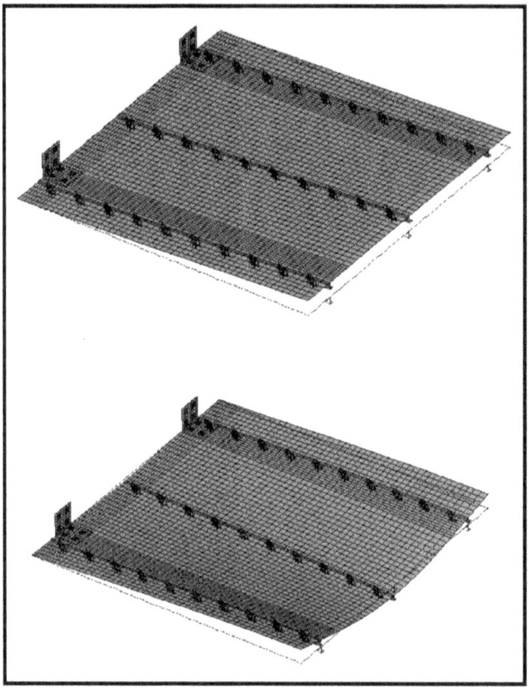

Figure 10.24 First and third mode shape for rivet position nodal attachment.

Although these results indicate very similar values for the first and third frequency for both models, because these are critical for the qualification parameters it is important that a check is made to ensure that the two sets relate to the same modes. The mode shapes for these two natural frequencies for the two joint models are shown in Figure 10.24 and Figure 10.25. It is clear that these two frequencies generate the same mode shapes for both models.

Although the gross mode shapes appear to be similar for both joint reference models, there are subtle but important differences that are revealed by detailed examination. If the behaviour of the stringer is displayed in detail at the end of the panel for the joint modelled by connections at the rivet points (Figures 10.26 and 10.27), it can be seen that the unconnected panel and stringer surfaces separate and then pass through each other during the vibration cycle. On the basis of this information it can be argued that the rivet attachment model represents a situation that cannot exist in the real world and must, therefore, represent an over-flexible model.

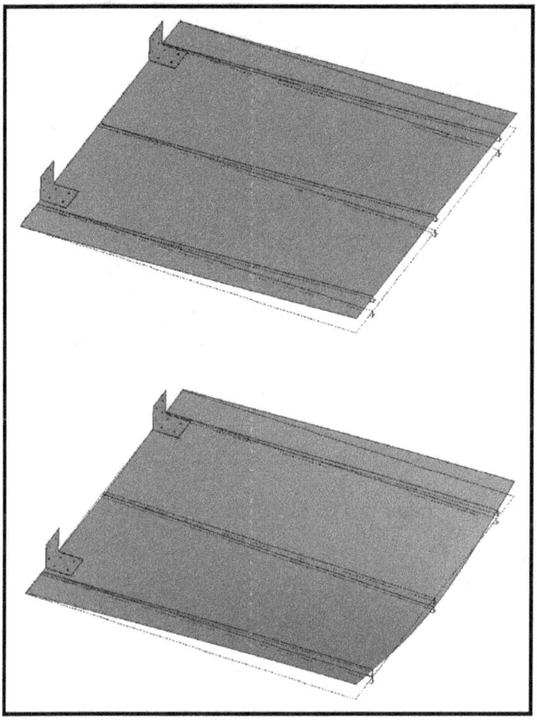

Figure 10.25 First and third mode shape for continuous joint attachment model.

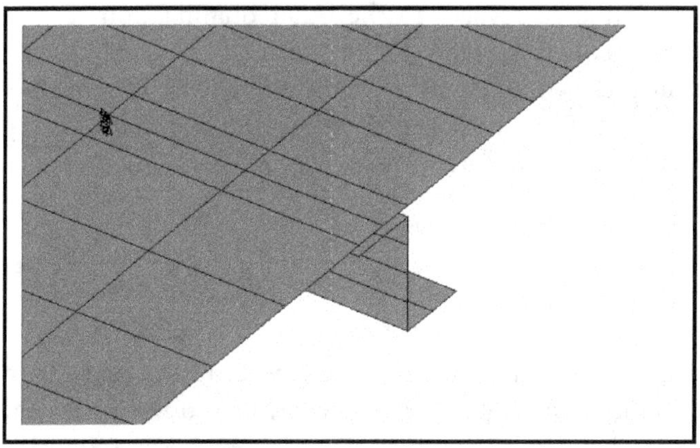

Figure 10.26 Surface separation.

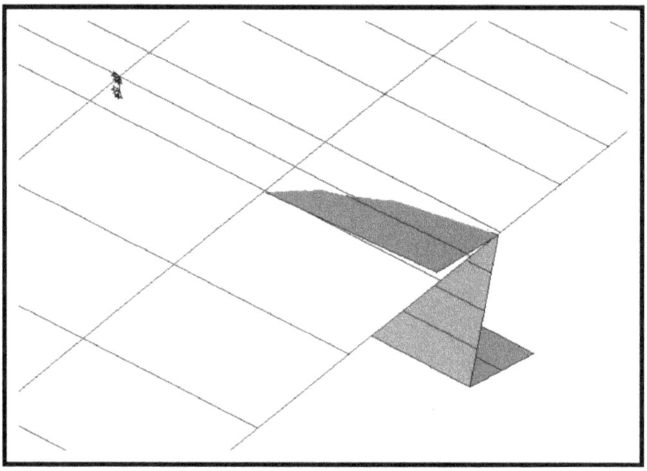

Figure 10.27 Surface crossing.

Step 4.3 Uncertainty assessments and bounding

Examining the qualification requirement that no frequency lies in the range 28.5–29.5 Hz, and comparing it with the finite element results when the rivet position joint attachment model is used, it can be argued that 29.56 given by this model is too low and that this value represents a genuine lower bound on this frequency. As none of the other frequencies come near to this critical range, it can be concluded that the structure, as designed, does not violate this qualification criterion.

The situation with respect to the other qualification requirement is unclear. The frequency predicted by the fully attached model predicts 4.13 Hz but the second estimate predicts 3.83 Hz. It is not possible to come to any firm conclusion with respect to the structure being able to satisfy the qualification requirement.

Step 4.4 Review

The claim that the qualification requirement stating that the in-service structure should not have any natural frequency in the range 28.5–29.5 Hz has been substantiated. However, the requirement that the structure should not experience any free vibration natural frequency in the range 3–4 Hz has not been substantiated. The two reference models used in the analysis indicated that the lowest natural frequency lies in the range 3.83–4.13 Hz. A more accurate value form of the model with nodal connections between the panel and the stringers could be achieved by employing a non-linear analysis that brought the surfaces into contact and released them at appropriate points in the vibration cycle. However, the closeness of the over-stiff model's predicted lowest frequency implies that the structure is, in fact, not fit for purpose and that a re-design is required.

Following the procedure adopted in the static analysis, the 'best' estimate for this frequency is taken as the mean values, hence:

$$q(\text{ref}) = 3.98\,\text{Hz and } \delta\text{q} = \pm 0.3\,\text{Hz}$$

Inserting these in equation 5.3 yields:

$$M = \pm 0.07\,(7\%)$$

This result clearly demonstrates that the predicted value for the frequency associated with the qualification parameter may be positioned so that it violates the qualification criteria.

REFERENCES

1. Warren, C.Y. and Budynas, R.G., *Roark's Formulas for Stress and Strain.* 2001: McGraw-Hill Professional, ISBN 007072542X.
2. Shanley, F.R., *Mechanics of Materials.* 1967: McGraw-Hill, ISBN 0070563934.

Index

3-noded membrane element, 148
4-noded element, 146
4-noded membrane element, 145, 165, 166, 169
6-noded membrane element, 148
8-noded hexahedral (brick) element, 149
8-noded membrane element, 152, 159, 160
20-noded hexahedral (brick) element, 152, 153

Acceptable magnitude of error or uncertainty, 132
Accumulation of errors and uncertainties, 241
Accuracy ladder, 164–173
Accuracy limits, 263
Accuracy margins, 263
Action plan, 258
Adams, V., 3
Adaptive meshing, 3, 173, 178, 214
Additional right-hand sides, 237
Adhesive bonded joint, 194
Ad-hoc method, 176
Aircraft attachments, 124
Aircraft moving surfaces, 124
Aircraft outer wing skin, 205
Aircraft wing, 47
Allowable values, 131, 132
Alternative level of abstraction, 186
Ambient stress levels, 285
Analysis and test, 198, 246
Analysis approach, 164, 165, 275
Analysis assessment, 264, 311, 335
Analysis-based estimates, 133
Analysis decisions, 180
Analysis features, 178
Analysis loop, 29–32, 37, 49, 59
Analysis novelty, 120, 122
Analysis problem overview, 262, 270, 309
Analysis qualification, 2, 125, 129, 256
Analysis review, 277, 281, 320
Analysis scope, 281
Analysis targets, 129, 231
Analysis type, 125, 136, 264, 311
Analysis validation plan (AVD), 135, 316
Analyst requested tests, 244
Anisotropic material, 122–124, 199, 211, 244, 245
ANSYS code, 338

A Practical Guide to Reliable Finite Element Modelling A. Morris

Anti-symmetric boundary conditions, 35
Anti-symmetric constraints, 35
Anti-symmetric displacements, 37
Anti-symmetric loads, 33, 34, 38
Anti-symmetry, 11, 12, 38, 40, 183, 184, 197, 198
A posteriori error, 166
Applicability of models, 199
Applied common sense, 138, 139
Applied loads, 1, 8, 24, 26, 30–32, 64, 68, 80, 120, 191, 200, 201, 212, 218, 237, 320
Approach, 2, 9, 10, 15, 16, 19, 20, 39, 52, 62, 64, 125, 178, 179, 204, 205, 213, 217, 263, 347
Approximate computational analysis process, 80
Artificial boundaries, 197
Artificial damping, 70
Artificial limitations, 266
Artificial sensitivities, 238, 240
Askenazi, A., 3
Assembled global stiffness matrix, 21, 41
Assembly details, 262
Assessment services, 309
Attachment options, 296
Automatic selection of slave degrees of freedom, 74
Automatically moving up the accuracy ladder, 165
Axially symmetric solids, 39, 71

Babushka I, 3
Back of envelope calculations, 261, 266
Banded matrix, 24, 54
Bar elements, 11, 13, 104
Barlow, J., 144
Barlow method, 145
Barlow points, 154
Barlow stress sets, 147–149, 151, 152, 164
Basic characteristic behaviour, 282
Basic design information, 264, 311
Bathe, K.-J., 3, 5, 19
Beam elements, 223
Beam theory, 129, 222
Belegundu, A., 3, 19
Belytchko, T., 10
Benchmarks, 253
Bending forces, 234, 321
Benign distortions, 161
Best estimates, 133, 134, 226, 232, 329
Best model, 117, 216
Best practice, 70, 250
Bill of materials, 264
Blevins, R., 209
Body forces, 84, 90, 92, 272
Bonded joints, 195, 295
Boundary conditions, 6, 7, 11, 15, 24, 25, 35, 38, 39, 86, 117, 138, 169, 192–194, 198, 208, 213, 218, 278, 286, 321
Bounded errors, 258
Bounded uncertainties, 246, 259, 299
Bounding parameter, 134
Breakout point, 225, 226, 232
Bridge analysis, 180
Bridge bracing beams, 124
Bridge deck, 124
Bridge girders, 124
Buckling mode, 40, 195, 234
Buckling modification matrix, 234
Budynas, R., 209, 312, 324
Buffeting, 119, 200
Build quality, 120, 136, 264, 266
Built-up domain, 210
Built-up structures, 210

CAD, 71, 158, 165, 264, 271–273, 283, 292, 300, 303
CAD displays, 257
CAD files, 271, 273, 300
Calibration tests, 255
Car analysis, 180

Carbon fibre reinforced composites, 201
Cataloguing process, 250
Central finite differences, 69
Certification parameters, 263, 300
Certification process, 130, 200
Certification requirements, 116, 125, 310
Challenge hypothesis, 184
Chandrupatia, T., 3, 19
Change in stiffness properties, 219, 221
Church–Turing theorem, 9, 10, 118
Civil airliner, 123, 205
Class 1 qualification codes, 130
Class 2 qualification codes, 308
Class 3 qualification codes, 131
Coherent sequence, 115
Commercial systems, 25, 74
Common sense, 10, 92, 127, 138, 139, 178
Compilation of available information, 265, 312
Complete strain compatibility, 156
Complete stress states, 146
Completeness in model selection, 213
Completeness of models, 119
Complex analysis, 38, 177, 215, 282, 284, 298, 305
Complex design problem, 33
Complexity, 2, 8, 14, 71, 116, 120, 123–125, 127, 128, 138, 184, 187, 279, 315, 322
Component level assessment, 130, 279
Component mode synthesis, 76
Computability, 10
Computational efficiency, 190
Computed measurements, 187
Computing power assessment, 280
Condensation, 12, 33, 40–42, 47, 72, 73, 77, 198, 255
Condensed stiffness matrix, 42
Confidence levels, 243
Connection quantities, 4, 5, 27, 29, 88–90, 212
Consistent loads, 30, 90, 92
Constant stress element, 104, 148, 149, 151, 164, 254, 255, 336, 337, 341
Constrained normal modes, 76
Construction details, 262
Contact zone, 319, 321, 322, 330, 340, 342, 343, 345, 348
Contour plots, 257
Convergent process, 13, 79
Coupon tests, 119, 137
Court of law, 213
Cranfield University, 116, 179, 185, 204
Critical damping, 68
Cyclically repeat symmetry, 39, 71

Damped forced responses, 66–70
Damping action, 53
Damping matrix, 53, 66
Data sheets, 137, 265, 312
Database information, 265
Database models, 126
Datum, 54, 126–128
  models, 126, 127
Decomposition based on loading, 318
Decomposition of structure, 273, 276, 318
Degree of complexity, 120, 123, 127, 184
Degrees of freedom, 27, 29, 33, 38, 43, 45, 48, 54, 55, 71–78, 152, 191, 226, 227
De-idealisation, 7, 115
Depth of analysis, 123, 192, 193, 195, 205
Design complexity, 125
Design constraints, 262, 310
Design environment, 116, 124, 136, 253, 263

Design imposed analysis limitations, 310
Design phase, 133
Design reduction, 116, 175–177
De-skilling, 113
Detailed assessment, 269, 277, 278, 289, 317, 321, 330
Dhatt, G., 20
Direct integration, 12, 62, 64–66, 69, 70
Direct mathematical solution, 253
Direct method, 20, 65, 231, 236–239
Direct sensitivity methods, 16, 240
  for buckling analysis, 234–235
  for dynamic analysis, 227, 231
  for static analysis, 218, 225
Discontinuities, 5, 116, 273
Discrepancies, 133, 187, 188
Discrete attachments, 194, 195, 295
Discrete points, 13, 21, 243
Discretisation, 14, 15, 117, 143
Displacement amplitudes, 54
Displacement boundary, 38, 83, 84
Displacement convergence, 106
Displacement elements, 3, 5, 104
Displacement field, 3–6, 30, 34, 35, 166, 193, 325, 334
Displacement sensitivities, 218
Distributed loads, 13
Distributed teams, 259, 266
D matrix, 20
Domain of analysis, 177, 180–184, 192, 195, 205, 238, 317
Domain reduction, 183, 184, 271
Dynamic analysis, 12, 16, 33, 51, 70, 71, 80, 111, 144, 184, 231
Dynamic displacement field, 64
Dynamic loads, 12, 71, 121, 183, 198, 231, 284
Dynamic modification matrix, 228
Dynamic strains, 59, 64
Dynamic stresses, 144, 228, 233
Eigenvalue, 54, 60, 77, 234
  problem, 12, 54, 58, 75, 78, 234
Eigenvector, 54, 61, 73, 109
Elastic behaviour, 9, 216
Elasticity modulus, 169, 221
Electrical networks, 21
Electromagnetic phenomena, 21
Element accuracy, 144
Element coordinate system, 27
Element delivery, 144
Element distortions group 1 (linear), 159
Element distortions group 2 (bi-linear), 160
Element distortions group 3 (quadratic), 161
Element distortions group 4 (cubic and higher order), 162
Element forces, 21, 23, 26, 30, 37
Element loads, 7
Element stiffness matrix, 11, 21, 24, 27, 30, 57, 72, 89, 96, 98, 158, 220
Element types, 3, 29, 140, 145, 152, 162, 177, 253
Elimination of errors, 258, 259
Elliptic membrane, 169
Empirical rules, 130
Encastré beam, 180, 222, 223, 324
Energy functional, 80, 88
Entry-into-service, 116, 203, 245
Equilibrium elements, 3, 5
Equilibrium equation, 23
Equivalent work, 34
Error bounding methods, 10
Error bounds on qualification parameters, 303, 345
Error commentary, 303, 344
Error control methodology, 243
Error control procedures, 8
Error controlled finite element analysis, 114

Error identification, 114, 117, 305
Error sources, 15, 17, 114, 133, 155, 176, 208, 241, 250, 260, 330
Error treatment, 9, 126, 127
Error type, 15, 175
Estimated change in element stiffness, 227, 233, 235
Estimating stress accuracy at common node, 154
Executable data files, 252, 253
Existing knowledge, 274
Expensive surprise, 277
Experience, 2, 14, 17, 66, 113, 114, 119, 121, 127, 130, 175, 209, 250, 271, 309, 312, 315, 318, 322, 353
  database, 126, 265, 266, 312
  gap, 315
Experimental accuracy, 299
Experimental data, 137, 243, 244, 246, 316
Experimental elements, 238
Expert resources, 267
Explicit reference model, 291
Extended domain of analysis, 183
Exterior world, 126
External error sources, 301
External medium, 53

Fabrication, 10
Fail-safe component, 132
Failure mechanism, 131
Failure modes, 136
Failure types, 136
Fatigue life assessment, 132
Feature analysis set-up, 289, 331
Feature level analysis, 278
Feature level assessment, 298
Feature level uncertainties, 287, 293
Feature level uncertainty sources, 287, 298
Feature properties, 179
Feature set, 206
Features
  single, 179, 205, 206, 213, 288
  five, 206
  three, 206
Feedback loops, 9, 115, 140
FEMEC, 17, 260–261, 264, 268, 272, 275, 277–278, 296, 299, 301, 305–310, 315, 329, 334, 338, 346, 348
Fidelity ladder, 8
Final analysis, 299, 300, 310, 319, 332, 337, 338, 340–341
  plan, 299, 300, 338
  reference model, 297–299
Final assessment, 298
Final reference model, 299
Final results model, 115
Finite differences, 62, 253
Finite element analysis, 1–3, 6–7, 11, 13, 19, 59, 66, 79, 113, 114, 118, 120, 125–127, 129, 130, 134, 139, 140, 174, 175, 178, 199, 200, 203, 204, 208, 241, 256
Finite element formulation, 4, 19, 21, 88, 122, 133, 151
Finite element library, 251
Finite element method, 1, 3, 13, 20, 21, 29, 52, 79–81, 84, 86, 88, 92–93, 113, 117, 184
Finite element world, 7, 133
Fitness for purpose, 16, 143
Flag assessment, 335
Flagged list of uncertainties, 311
Flags, 258, 266, 274, 288, 293, 296, 300, 313, 315, 320, 327, 330, 335, 344
Flat plate element, 30
Flatness of strain energy curve, 104
Fluid flow, 21
Fluid-structure coupling, 124
Force-balance equilibrium forced responses, 23
Fourier series, 40, 72
Fox, M., 129, 139
Fractional error, 134

Fractional uncertainty, 134, 240
Fraeijis de Veubeke, B., 5
Free surface, 155–157
Free undamped vibrations, 53
Free vibration, 12, 15, 53, 56, 59, 62, 72, 108, 227, 229, 348, 349
  analysis loop, 58
Frequencies, 53–54, 56, 59, 68, 75, 80, 108, 227–228, 232, 253, 348, 349, 352
Frequency range, 59, 263
Frequency spectrum, 77

Gas turbine engine, 123
Gaussian integration, 152–153, 161, 164
Gaussian integration points (Gauss points), 153
Generalised coordinates, 77–78
Generalised load vector, 90
General principle with respect to element distortions, 163
Generated error, 14, 118, 120, 140–141, 143
Geometric non-linearity, 211
Geometric shape, 179
Geometric stiffness matrix, 234–235
Geradin, M., 78
Global assessments, 278, 282, 297, 324, 336
Global coordinate system, 27
Global coordinates, 26
Global level analysis, 295
Global reduced reference finite element model, 282
Global stiffness matrix, 11, 21, 22, 24, 25, 41, 43, 98, 218, 219, 235
Global system, 24, 97, 218
Go/no-go decision, 268, 313
Godel's theorem, 10
Gradient information, 236
Graphical user interface, 113
Gravity force, 84
Gurdal, Z., 217

Haftka, R., 217
Hamilton's principle, 107
Handbooks, 265
Hand calculations, 16
Hard error bound, 133
Harmonic motion, 54, 67
Heat transfer, 21
Hidden information, 151
Hierarchical decomposition, 185, 201,
Hierarchical methodology, 16
Hierarchical modeling, 212, 216
Higham, N., 126
High-energy cases, 121
High-fidelity models, 8
Higher vibration modes, 67
Hilburger, M., 11
Homogeneous material, 176, 213
Hooke's law, 20, 85
Horses for courses, 151, 189
h–p adaptive codes, 173
h-type adaptive codes, 165
Hybrid elements, 3

I-beams, 124
Idealisation process, 7, 15, 16, 92, 125, 177, 249, 297
Idealised model, 115–117, 153, 177
Idealised world, 7, 203
Illustrative connecting rod, 284
Impact of errors, 140, 175, 184, 209, 210, 240, 249, 301, 340
Impact of uncertainties, 176, 178, 271, 277, 283, 291, 301, 347, 348
Implementation plan, 260, 261, 307, 309
Implicit–explicit analysis, 279
Imposing maintenance standards, 293
Inappropriate element shapes, 191
Incomplete knowledge, 119

Indirect evaluation of sensitivities, 236, 237
Indirect sensitivity methods, 16, 240
Inertia characteristics, 8
Inertia loaded spring, 52
Inertia loads, 12, 272, 284
Inexperience, 2, 66, 120, 127, 199
Information accumulation, 268, 315
Initial analysis assessment, 263, 309
Initial assumptions, 140, 177
Initial calculations, 281, 323
Initial considerations, 136, 209, 316
Initial global assessment, 282, 324, 328, 348
Initial review, 205, 266, 277
Initial set of bounds, 135
Initial sizing, 8, 285
Initial steps, 129, 131, 133
Initial uncertainty bounds, 273, 319
In-service failures, 250
In-service operation, 6
In-service structural behaviour, 135
In-service structure, 7, 8, 118, 133, 178, 211, 260, 299, 350
Inter-element compatibility, 5
Interface behaviour, 194, 195
Interface boundaries, 5, 77, 195
Interface boundary nodes, 76
Interior inertia forces, 21
Interior world, 7
Intermediate analysis, 8
Intermediate level global assessment, 297
Intermediate level global reference model, 215
Internal boundaries, 176, 179, 192, 193, 195
Internal company code, 262
Internal damping, 53
Internal error sources, 302
Internal friction, 53
Internal loading actions, 278
Interpolation, 105, 106
Isotropic material, 124, 199, 244

Joints, 5, 124, 136, 166, 238, 245, 273, 284, 310

Kamat, M., 217
Kardestuncer, H., 19, 51
Kinetic energy, 13, 56, 80, 81, 107–111
Knight, N., 11

Lack of experience, 127, 280
Lagrange multipliers, 6
Lagrangian function, 13, 107
Large safety factor, 200, 314
Lateral vision, 17
Legal defence, 129, 256
Level of abstraction, 277, 279, 290, 311, 319, 327, 329, 344
Level of commitment, 277
Levels of fidelity, 8
Limit state, 131, 132
Limiting physical state, 131
Line distribution of displacements, 5
Linear dynamic analysis, 124
Linear interpolation, 105
Linear static analysis, 13, 70, 71, 176, 183, 212, 309
Linear stress variation, 147, 151, 152
Load path, 5, 8, 177, 195, 235, 283, 291
  movement, 271
Load-reaction consistency check, 30
Load transfer, 271
Loaded spring combination, 87
Loading, 7, 9, 10, 12, 33, 35, 64, 116, 119, 122, 131, 176, 185, 187, 193, 200, 237, 273, 278, 316, 342
  actions, 116, 271–273, 278, 282, 283, 289, 316
Local axis facility, 39

Logical sequence, 9, 273
Loss of applied load, 93
Low-fidelity models, 8
Low-order elements, 8
Lumped loads, 201

MacNeal, R., 144, 155
Maintenance requirement assessment, 335
Manufacturing details, 264
Mass matrix, 51, 52, 108
Master degrees of freedom, 33, 43, 72, 73, 75
Master model, 47
Matching curvature and node positions, 163
Material properties, 119, 198–200
  linear dynamics problems, 199
  linear static problems, 199
  non-linear problems, 199
Material property uncertainty, 198
Mathematical model, 15, 177, 185, 192, 211
Matrix analysis, 19
Maximum deflection of beam, 222
Maximum displacement, 196, 306, 314, 323, 329
Maximum kinetic energy, 56
Maximum natural frequency, 61
Maximum strain energy, 56
Maximum stress, 188, 196, 208, 316, 323
Mean values for stresses, 104
Measurable error, 9
Membrane element, 30, 172, 277
Membrane forces, 234
Memory requirement, 254
Mesh distortion, 155–164
Mesh grading, 155–164
Mesh ladder, 165
Meshed models, 257
Meshed world, 7
Meshing, 117, 143
  errors, 155
Minimum natural frequency, 61
Minimum potential energy, 84–86, 88, 98
Minimum weight structure, 262
Mish, K., 10
Missing frequencies, 60
Misunderstandings, 140, 246
Modal analysis, 62–64
Modal analysis with damping, 66–68
Modal damping constant, 67
Modal damping ratio, 68–69
Modal loading function, 64
Modal methods, 64
Modal participation, 64
Modal vector, 54, 57, 63
Mode shape sensitivities, 228, 229
Model assessment, 216
Model build-up, 140
Modelling assumptions, 197, 204, 233
Modification matrix, 220, 222, 224, 225
Modified load vector, 25
Morris, A., 9, 179, 185, 217
Multi-company design, 263
Multi-component structure, 176, 179
Multi-disciplinary analysis, 124
Multi-element arrangement, 22
Multi-stage process, 261
Multiple analysis teams, 12
Multiple symmetry, 39, 71

NAFEMS, 139, 169
NASA, 11
Natural frequency, 54, 57, 61, 227, 228
Negative damping, 70
Nemeth, M., 11
Newmark's method, 69
Nodal accelerations, 52
Nodal connection quantities, 4, 29, 88, 89
Nodal displacement vector, 52, 62
Nodal displacements, 4, 21, 25, 48, 200

Nodal forces, 21, 27, 28, 55
Nodal rotations, 4, 30
Nodal values, 4, 168, 338
Nodal velocities, 52, 65
Non-homogeneous material, 211
Non-linear analysis, 121, 183, 187
Normalisation process, 56
Normalised mode shapes, 57, 62, 76, 230
Normalised modes, 56, 63, 67, 111
Novelty high level, 116, 121–123, 209
Novelty low level, 116, 121–122
Novelty medium level, 122
Nuclear electric, 130, 204
Nuclear power station, 130
Numerical analysis, 117
Numerical errors, 117
Numerical instabilities, 66

Oden, J., 9
One-dimensional elements, 26
Operational environment, 263
Optimal stress sample points, 152, 153, 164
Orthogonal vectors, 57
Orthotropic plate, 179, 209
Overall accuracy bound, 300
Overall assessment, 293, 334
Overdamping, 68

Partitioned matrix, 41, 77
Partitioned stiffness, matrix, 41, 43, 77
Peer review, 257
Permissible states, 131
Petyt, M., 51, 68, 74
Phase angle, 68
Physically symmetric structure, 33
Plate analysis, 6, 119, 176, 205
Plate bending theory, 119
Plate elements, 320, 321, 324
Plate/shell model, 211
Point-wise convergence, 80, 93
Polynomial approximations, 3
Positive damping, 70
Post-failure behaviour, 131, 132
Post-mesh-generation examination, 173
Potential energy, 13, 80, 82–83, 90, 103
Pre-defined analysis, 292
Pre-defined reference analysis, 292
Preliminary error assessment, 17
Preliminary review, 205
Primitive, 179
Professional test organization, 243
Progressive non-linearity, 211
Propagation bands, 71
Propagation of errors, 10
p-type adaptive codes, 3
Pzemieniecki, J., 20

Quadratic stress variation, 152, 153
Qualification criteria, 2, 130, 133, 329
Qualification information, 262, 299
Qualification objectives, 316
Qualification parameters, 133, 134, 203, 220, 242, 274, 316, 346
Qualification process, 2, 129–132, 205
Qualification requirements, 7, 310
Qualification rules, 130
Quality of results, 139, 168, 169, 250
Quality report, 16, 250, 256–259, 308
Quality system, 129

Rahman, A., 309, 316, 317
Random variability, 119
Rao, S., 64
Rapid re-analysis, 238
Rate of change of buckling load, 217
Rate of change of displacements, 217, 219, 220

Rate of change of frequencies, 217
Rate of change of natural frequency, 228, 230
Rate of change of stresses, 217
Rayleigh quotient, 59, 61–62
Reactions, 26, 30, 46
Real-world, 6, 7, 93, 116, 129, 131, 174, 204, 212, 213
  structure, 6, 9, 108, 116, 212
Reduced cost, 250
Reduced finite element reference model, 280, 282, 286, 322
Reduced matrix, 25
Reduced qualification criterion, 268
Reduced real-world, 7
Reference books, 137, 209, 265
Reference model, 134, 212, 218, 289, 290, 321–324
Reference value, 133, 134, 240, 346
Regulated analyses, 250
Regulatory authority, 256
Reissner–Mindlin model, 214
Relative values, 54
Reliable standard, 1
Removal of ambiguities, 243, 244
Repeated symmetry, 39, 71
Required level of accuracy, 15, 17, 135
Response to external loads, 179
Response type, 117, 124, 126, 185
Results accuracy, 316
Results conflict, 283
Results model, 114, 291, 343
Review of available information, 265, 312
Revised design requirements, 268
Revision procedure, 138, 317
Rigid body motion, 21, 56
Rigid body movement, 24, 54, 146, 239
Rigid boundary, 180
Rigid format, 260
Rivet locations, 349
Rivets, 7, 195, 306, 310–312
Roark reference model, 321
Roark's formulae, 209, 265, 285, 291
Role of testing, 128–129, 242–244
Rolls Royce, 143
Rossettos, J., 5, 81
Rules of thumb, 210, 211
Run-time assessment, 338

SAFESA, 11, 139, 179
Safety critical environment, 125, 130, 245, 250
Safety factor, 7, 132, 200, 314, 315
Sander, G., 5
Scoping information, 282
Scoping the real-world problem, 261–269, 309–312
Sensitivity methods, 16, 216, 217, 244–245
Serviceability limit state, 131
Set-up information, 257, 260, 309
Shanley, F., 312, 313, 323
Shanley reference model, 322
Shape function, 30, 33, 88, 90, 91
Shell analysis, 6
Shell elements, 29, 151, 188
Siemens, 307
Significant thickness, 211
Similarity parameters, 206, 329, 333, 344
Similarity rules, 126, 127
Simple analysis programs, 282
Simple engineering formulae, 134, 209, 210
Single level of decomposition, 193, 206
Single span bridge, 124
Singular matrix, 21, 24, 53
Singularities, 191
Size limitations, 116, 131, 266
Size of modal, 67
Slave degrees of freedom, 72–74
Slender bars, 234
Small thickness, 211
Solid model, 211, 330

Solution process, 49, 63, 117, 251
Solution world, 7
Sophisticated models, 297
Sources of error, 140, 198, 205
Sources of uncertainty, 177, 216, 273, 292
Specified functionality, 262
Spring element, 12, 27, 51, 55
Spurious modes, 191
Stage reports review meetings, 258
Stage reviews, 293, 320
Static analysis illustrative problem, 306
Static condensation process, 77
Statically determinate structure, 222, 242
Step inputs, 269
Step outputs, 269
Step review, 265, 267, 278, 288, 312, 315, 318, 327, 329, 330, 353
Stereotype loads, 7
Stiffened plate, 124
Stiffness matrix, 11, 21, 27, 29, 43, 57, 78, 90
Strain energy, 56, 81–82, 89, 95
Strand 7 code, 172
Stress at a common node, 169, 341
Stress concentration factor, 209
Stress contours, 325
Stress field, 21, 159, 160, 343
Stress gradients, 165, 336, 337
Stress improvement, 104–106
Stress in extreme fibres, 211
Stress intensity factor, 132, 290, 291
Stress jumps, 104, 105, 169
Stress ladder, 164
Strouboulis, T., 3, 167
Structural behaviour, 6, 118, 129, 236, 238, 312
Structural components, 8, 47, 238, 306, 311
Structural design, 6, 8, 125, 135, 243
Structural dimensions, 172, 236, 262
Structural failure, 125, 129, 130, 136
Structural interpretation, 314
Structural optimization, 16, 217, 236
Structural parameters, 8, 218, 236
Structural performance, 15, 130, 177, 283
  criteria, 130
Structural properties, 116, 120, 208, 218, 221
Structural theory, 282
Sturm sequence check, 59
Sturm sequence property, 59
Stylised loads, 116
Sub-matrices, 43, 77
Sub-processes, 114, 117, 118, 176
Substructure reference model, 47, 289
Substructures, 12, 76, 217
Superelements, 33, 40, 217, 278
Supporting test specification, 298–299
Supports, 238, 239
Surface crossing, 352
Surface distribution of displacements, 5
Surface separation, 352
Symmetric boundary conditions, 35
Symmetric constraints, 35
Symmetric displacements, 37
Symmetric loads, 33, 34, 38
Symmetric matrix, 21, 24
Symmetry, 33–40
  line, 35
System functionality, 252
Systematic errors, 120
Szabó, B., 3, 19, 81, 167, 216

Taig, I., 211
Tangential strains, 156, 157
Target error limits, 226, 233
Taylor, R., 3, 19, 81
Template, 136, 250, 260
Temptations, 46
Termination decisions, 258
Test data, 2, 114, 131, 243, 245, 300
Test engineer, 200, 243, 260, 299

Test errors, 128
Test requirements assessment, 299
Testing the unknown, 245
Theoretical manual, 252
Theory of experimental uncertainty, 246
Thermal analysis, 21
Thermal stress field, 21
Thick beam model, 129
Thin-walled structures, 234
Third party evaluation, 256
Thompson, W., 64
Three-dimensional beam bending element, 30
Three-dimensional elements, 148–151, 192
Through-thickness stress, 188, 211, 214
Time requirement, 254, 255
Tolerances, 135, 203
Tong, P., 5, 81
Torsion bar element, 30
Total analysis process, 14, 275
Touzot, G., 20
Traction boundary, 83, 84
Traction forces, 84
Traction surface, 83, 84
Transformation matrix, 28, 29, 44, 73, 77
Transformed stiffness matrix, 29
Trial functions, 63
Turbulence, 119
Tutorial, 252
Two-dimensional axis system, 27
Two-dimensional elements, 145–148

Uncertainty, 10, 114, 117–120, 132, 140, 176, 215, 233, 241–242, 259
  assessment and bounding, 333
  associated with modeling considerations, 192
  captured, 225, 258, 297
  control methods, 246, 252
  identification, 114
  list, 266, 313
  review, 179, 263, 310, 311
  unbounded, 266, 271, 295
Unconditional stability, 69
Undamped system, 52
Underdamping, 67, 68
Underpinning mathematical theory, 251
Unit load, 219, 220, 223, 226
User manual, 252

Validated finite element analysis model, 114
Validation, 135, 253
  criterion, 137, 314
  process, 137, 243, 274
  test, 243, 245–246, 251
Variability, 10, 245
Vector of nodal displacements, 21, 220
Verification, 251, 253
Vignjvec, R., 9, 116, 179, 185, 204
Violation of qualification criteria, 351
Von Mises criterion, 308, 323, 329, 338

Walkthrough, 17, 304, 306, 318, 333, 347
Warren, C., 209, 310, 322
Washizu, K., 81
Weighting functions, 63
Welds, 7, 116
Wrong problem, 111

Young's modulus, 82, 85, 93, 188, 311, 324

Z-stringer reinforced panel, 188, 304, 307, 310, 316

www.ingramcontent.com/pod-product-compliance
Lightning Source LLC
LaVergne TN
LVHW011517200926
848484LV00001B/58